BREEDING FIELD CROPS

The Author

Dr. R. A. Gami is currently serving as Associate Professor (GPB) and Unit Head, in addition to overseeing the All India Coordinated Research Project (AICRP) on sorghum. He has over a decade of teaching and research experience, having fulfilled various roles throughout his career. Throughout his academic tenure, he has taught more than twelve courses at both undergraduate and postgraduate levels. He has successfully mentored over twelve doctoral and master's students. His scholarly contributions include publishing two important academic books: **'Fundamental Genetics (2nd Ed),'** and **'Plant Breeding,'** along with two reference books focused on millets. He has authored more than eighty research articles published in esteemed scientific journals, over twenty abstracts, two booklets, three conference books and six practical manuals. He has played a pivotal role in developing and releasing ten varieties and two hybrids within the Gujarat State Agricultural Universities (SAUs). He has been honored with the ICAR-sponsored **'Best Researcher Award'** for the academic year 2021-22 by Sardarkrushinagar Dantiwada Agricultural University (SDAU), as well as two *Sadvichar Parivar* Awards in 2019 and 2024 conferred by the Gujarat Association of Agricultural Sciences (GAAS), in addition to receiving five accolades for best poster and presentation at various national and international conferences, symposiums, and workshops.

BREEDING FIELD CROPS

— Author —

Dr. R.A. Gami

Associate Professor and Unit Head
Centre for Millets Research
Sardarkrushinagar Dantiwada Agricultural University
Gujarat

2026

Daya Publishing House®
A Division of
Astral International Pvt. Ltd.
New Delhi – 110 002

ISBN 978-93-5919-225-3

Published by : **Daya Publishing House®**
A Division of
Astral International Pvt. Ltd.
– ISO 9001:2015 Certified Company –
4736/23, Ground Floor, Ansari Road
Darya Ganj, New Delhi-110 002
Ph. 011-43549197, 8130496929
e-mail: info@astralint.com
Website: www.astralint.com

Printed at : **Replika Press Pvt Ltd**

Foreword

Crop breeding is a mutual act of human art, knowledge of science and technology. Crop breeding aims to improve yield performance, i.e., food, fodder, fibre, and industrial components, enrich the quality and biodiversity of economically essential crop plants, and enhance resilience against biotic and abiotic stresses. The botanical features, reproductive systems, genetics, and essential breeding techniques are necessary for any crop improvement programme. The book '**Breeding Field Crops**' is prepared according to the syllabus outlined by ICAR on the recommendation of BSMA, Plant Sciences.

The book covers four sections: Introductory Crop Breeding, Breeding cereals, Breeding pulses, Breeding Oilseeds, Fibre & Sugar crops and Breeding forages. Encompasses each crop breeding chapter with basic to advanced breeding objectives, conventional and molecular breeding approaches and achievements. I congratulate the Author, Dr. R.A. Gami, for writing the book and hope that the book will be elementary material for students, research scholars and teachers/scientists associated with breeding field crops.

R.M. Chauhan

Vice-Chancellor

S. D. Agricultural University,
Sardarkrushinagar – 385506,
Dist. Banaskantha (Gujarat)

Preface

Over the six decades, crop breeding principles have been enormously used in agricultural sciences to increase the yield of many important crop plants. The efforts continue to develop phenotypically and genotypically superior crop plants from available genetic resources or variations better adapted to human requirements. The crop breeding objectives and methods have been enormously used in field crop improvement to enhance crop yield, quality and resilience traits against biotic and abiotic stresses. Breeding field crops have been taught at undergraduate and post-graduate levels in all SAUs. This book is written to help students, research scholars, and teachers/scientists gain essential perspectives on field crop breeding theory and practice.

The book **'Breeding Field Crops'** furnished in 25 compact chapters covering fundamental topics of plant breeding, contains historical perspectives of crop breeding, major objectives and consequences, pollination systems and associated mechanisms, classification of various breeding approaches, each crop breeding chapters includes introduction, background information of crop plant, floral biology, selfing and crossing techniques, basic to advance breeding objectives, conventional and molecular breeding approaches and achievements. Efforts have been made to assemble information in appendages of essential key terms related to crop breeding, Nobel laureates with their essential contributions and a list of crop plant genome sequenced. The information in these books has been compiled from reliable published sources, apparent reference books, research articles and online literature libraries. At the end of the book, the compiled original reference lists are cited, which will also help readers access original works. I hope the book's content will benefit students, research scholars and teachers/scientists.

In the process of drafting the manuscript, I was afforded invaluable assistance & suggestion from my P.G scholars, Sandeep Kumar, Ayushi Acharya, Ritika Dadhich and Rajkumar Patel. I would like to express my profound gratitude to Dr. R. N. Patel, Research Scientist (Seed Technology), and Dr. Kiran Kugashiya, Assistant Professor (Millets), as well as Dr. Hiren S. Patel, Assistant Professor, for their unwavering support and constructive collaboration in proofreading and providing insightful suggestions. I want to prompt my sincere thanks to Dr. S. D. Solanki, Principal & Dean (Agri) and Dr. P. T. Patel, Registrar, SDAU to always encourage and help me to write the book.

I express my sincere gratitude to the Hon'ble Vice-Chancellor, Dr. R. M. Chauhan, S.D. Agricultural university, Gujarat for granting the necessary permission to publish this academic book. I extend my indebtedness to **Astral International Pvt. Ltd, New Delhi (India)**, for their meticulous attention and commitment in bringing this publication to fruition. Although I have exerted considerable effort to ensure accuracy and precision in the composition of this work, it is possible that certain technical or typographical errors may persist. Consequently, I would greatly appreciate receiving feedback and constructive suggestions from readers to facilitate further enhancements of the book.

R.A. Gami

Contents

Foreword *v*

Preface *vii*

Section–I
INTRODUCTORY CROP BREEDING

1. Crop Breeding: An Introduction 3
2. Pollination Nature and Associated Mechanisms 5
3. Hybridization and Selfing Techniques in Crop Breeding 13
4. Crop Breeding Methods: An Overview 19

Section–II
BREEDING CEREALS
(WHEAT, RICE, MAIZE, PEARL MILLET, SORGHUM AND SMALL MILLETS)

5. Breeding Wheat (*Triticum* spp.) 33
6. Breeding Rice (*Oryza sativa* L.) 45
7. Breeding of Maize (*Zea mays* L.) 63

8. **Breeding Pearl Millet (*Pennisetum glaucum* L.)** **77**

9. **Breeding Sorghum [*Sorghum bicolor* (L.) Moench]** **91**

10. **Breeding Small Millets** **107**

Section–III
BREEDING PULSES (CHICKPEA, PIGEONPEA, GREEN GRAM, BLACK GRAM, COWPEA AND GUAR)

11. **Breeding Chickpea (*Cicer arietinum* L.)** **125**

12. **Breeding Pigeonpea (*Cajanus cajan* L.)** **135**

13. **Breeding Green Gram (*Vigna radiata* L.)** **147**

14. **Breeding Black Gram (*Vigna mungo* L.)** **155**

15. **Breeding Cowpea (*Vigna unguiculata* L.)** **161**

16. **Breeding Guar [*Cyamopsis tetragonoloba* (L.) Taub]** **167**

Section–IV
BREEDING OILSEEDS, FIBRE AND SUGAR CROPS (GROUNDNUT, MUSTARD, SOYBEAN, SUNFLOWER, SESAME, CASTOR, COTTON AND SUGARCANE)

17. **Breeding Groundnut (*Arachis hypogaea* L.)** **175**

18. **Breeding Mustard (*Brassica juncea* L.)** **189**

19. **Breeding Soybean (*Glycine max* L. Merrill)** **199**

20. **Breeding Sunflower (*Helianthus annuus* L.)** **205**

21. **Breeding Sesame (*Sesamum indicum* L.)** **215**

22. **Breeding Castor (*Ricinus communis* L.)** **223**

23. **Breeding Cotton (*Gossypium* spp.)** **237**

24. **Breeding Sugarcane (*Saccharum officinarum* L.)** **253**

Section–V
BREEDING FORAGES (LUCERNE, BERSEEM AND OAT)

25. Breeding Forage Crops **269**

Appendices **283**

References **319**

Index **349**

Section-I

Introductory Crop Breeding

Chapter 1

Crop Breeding: An Introduction

The crop breeding principles have been enormously used in agricultural sciences to increase the yield of many important crop plants over the seven decades. The efforts continue to develop phenotypically and genotypically superior crop plants from available genetic resources or variations better adapted to human requirements.

What is Crop Breeding?

An application of genetic principles improves the desired hereditary characteristics of a crop to produce crop varieties for the benefit of humankind. It is a combined action of human art, knowledge of science and technology. The various scientists and authors gave different terms concerning plant breeding; as per N. I. Vavilov (1935), an evolution directed by human determination. D. C. Smith (1966), plant breeding is an art and science. Crop breeding is a technology of producing improved varieties of crop plants according to human needs (R. Riley 1978). Among all definitions, the well-accepted definition given by John. M. Poehlman and David. A. Sleper. According to them, plant breeding is *'an art and science of improving the heredity of crop plants for the economic benefits of humankind'*.

Objectives of Crop Breeding

The fundamental aim of crop breeding is to improve yield performance, *i.e.*, food, fodder, fibre, and industrial components and enrich the quality and biodiversity of economically essential crop plants as well as improve resilience against biotic and abiotic stresses.

The primary breeding objectives of recent crop breeding are briefly discussed below:

1. To improve the economic yield at the maximum level
2. To reduce the crop duration (earliness)
3. To improve yield attributing traits with an agronomic advantage
4. To improve quality traits
5. To breed biotic and abiotic stress resilience crops
6. To breed crop variety with uniform maturity
7. To breed varieties for thermo and photo insensitivity
8. To breed shattering/bursting resistance variety
9. To improve physiological seed dormancy
10. To breed varieties with zero or very low toxicants content
11. To breed ideal crop plants
12. To breed crop varieties that fit for mechanical harvesting and fit in new agronomical practices

Undesirable Effects Allied with Crop Breeding

Crop breeding has been practiced to improve crop plants for the economic benfit of humankind. However, continuous breeding work may also invite some unwanted consequences.

- ☆ Reduction in crop biodiversity
- ☆ Narrowing of a genetic base, which is a danger for uniformity,
- ☆ Increasing the problem of minor disease pests.
- ☆ Raise undesirable plant combinations.
- ☆ The linkage drags among undesirable traits with desirable ones.

Future Challenges with Crop Breeding

- ☆ Growing human population Vs Crop productivity
- ☆ Impact of global warming
- ☆ Gene for new races of biotic and abiotic stresses
- ☆ Varieties for indoor- vertical farming
- ☆ Organic breeding and agriculture sustainability
- ☆ Globalization may lead to seed conflict
- ☆ Breeding for biofuel/energy crops
- ☆ Justifiable use of biotechnological tools
- ☆ Molecular farming, *etc.*

Chapter 2

Pollination Nature and Associated Mechanisms

A bi-nucleate stage pollen grains (male gametophyte) are released from anther and transferred to stigmas. The pollen belongs to the same flower or a different flower born on the same plant or a different flower born on another plant.

Mode of Pollination in Crop Plants

According to pollen belonging, there are two modes of pollination mechanisms found in crop plants. Each is briefly described below:

1. Autogamy (Self-pollination)

Sexual fertilization followed by seed development through pollen belongs to the same flower or the different flowers born on the same plant. Therefore, the crop plant with such a pollination mechanism is commonly termed self-pollinated or autogamous crops.

An essential features of autogamy:

- ☆ It creates homozygosity.
- ☆ Autogamy occurs in the flowering plant due to own male gametes (pollen); hence it is also considered inbreeding.

- ☆ Primarily autogamous crops do not exhibit inbreeding depression but may show a considerable amount of heterosis.
- ☆ The crop species are highly homozygous.

2. Allogamy (Cross-pollination)

Sexual fertilization followed by seed development through pollen belongs to the different flowers born on a different plant. Therefore, the crop plants with such pollination mechanism is commonly called cross-pollinated or allogamous crops.

An essential features of allogamy:

- ☆ It creates heterozygosity.
- ☆ Allogamy occurs in a flowering plant by genetically different male gametes (pollen). Hence it is also considered outbreeding.
- ☆ Primarily allogamous crops exhibit moderate to high inbreeding depression due to self-pollination.
- ☆ The crop species are highly heterozygous.
- ☆ It displays a significant amount of heterosis and hybrid vigour.

Mechanisms Associated with Pollination Nature

The essential factors (mechanisms) primarily associated with self and cross-pollination are briefly described below:

A. Mechanisms Associated with Autogamy (Self-pollination)

The list of different mechanisms associated with autogamy is mentioned below;

1. Homogamy

A situation in plants where both male and female organs matured simultaneously. In other words, anthesis and stigma-receptivity occur at the same time.

2. Bisexuality

The flower is said to be perfect as it has both sex organs, *i.e.*, male and female organs. *e.g.*, *Solanaceae*, cereal, pulse crops, *etc.*

3. Cleistogamy

In this situation, the bisexual flowers remain closed. **M. Kuhn,** 1867 first time described the mechanism of cleistogamy in a plant. Eventually self-pollination takes place in closed flowers. Cleistogamy mechanism is found in various crops like groundnut, mungbean, wheat, barley, oats, viola, oxalis species plants, lettuce, periwinkle, Benghal dayflower (Commelina) and various grass species.

4. Chasmogamy

Chasmogamy is a slightly different situation, where flowers never remain closed. The flower opens in a majority of cases only after pollination is over. Some degree of cross-pollination may take place after the opening of a flower. This condition occurs in various crops such as wheat, rice, cluster bean, sunflower, cotton, oats, and rose.

5. The Male and Female Organs Enclosed by Adjunct Whorls

In pulses like pea, cowpea, pigeonpea, chickpea, black gram, green gram, soybean, *etc.* (family *Papilionaceae*), stamens and sigma are covered by adjuncts whorls, *i.e.*, petals. In this condition, stamens and stigma are fully covered by two petals forming a unique structure like a keel of a boat, commonly referred to as a keel. In some species, flowers open, but stamens and stigma are hidden by other floral organs.

6. Position of Male and Female Organs inside a Bisexual Flower

In Solanaceae crop species like brinjal, tomato, chilli, *etc.*, female parts (*i.e.*, stigmas) remain strictly covered by anthers. This situation efficiently confirms the autogamy even pollination takes place in open flowers.

B. Mechanisms which Favour Allogamy (Cross-pollination)

Various flower mechanisms favour cross-pollination. These mechanisms broadly divided into two categories

1. Mechanisms which enhance the cross-pollination, and
2. Mechanisms which compulsory favour the cross-pollination.

Each category is briefly described below:

1. Mechanisms Enhance the Cross-Pollination

Several mechanisms enhance allogamy (cross-pollination) but do not guarantee to prevent self-pollination. In addition, in some cases, a certain degree of autogamy may occur. Each kind is briefly mentioned below.

1.1. Dichogamy

In a hermaphrodite flower, the reproductive organs, *i.e.*, stamens and pistils, mature at different times. There are two conditions relating to dichogamy

1.1a) Protogyny: When a female organ (*i.e.*, stigma) become receptive earlier, in comparison to a male organ (*i.e.*, stamens: dehiscence of pollen grains). This condition is found in the pearl millet (bajra) crop, also reported in crops like amaranth, sugarcane, fig, banana, strawberry, plum, pomegranate, arum lily (*Zantedeschia aethiopica*), Cherimoya (*Annona cherimola*), sapota, *etc.*

1.1b) Protandry: When a male organ (*i.e.*, stamens) releases the pollen grain before a female organ (*i.e.*, stigma) becomes in a receptive state. This condition is found in crops like maize, cotton, sunflower, coriander, carrot, rose, marigold, sugar beets, walnuts, coconut, ber, aonla, *etc.*

1.2. Unisexual Flower (Dicliny or Unisexuality)

Unisexuality facilitates allogamy in many crop species. The flower is referred to as an incomplete. The male and female sexes are present in a different flower. The flower that bears the male sex is called staminate or male flower. While a flower that bears the female sex is called pistillate or female flower. In a condition, staminate and pistillate flowers are either found on the same plant or on a different plants. These two conditions are briefly described below:

1.2a) Monoecism (Monoecy): The staminate and pistillate flowers are borne on the same plant. However, each flower occurs in a different location as an individual flower or a single inflorescence, *e.g.,* maize, cucumber, bottle gourds, *etc.* In some crop species, both flowers are borne on the same inflorescence, *e.g.,* castor, sunflower, coconut, banana, *etc.*

1.2b) Dioecism (Dioecy): The staminate and pistillate flowers are borne on different plants. When a plant bears only staminate (male) flowers is called a male plant. While, plants that bear only pistillate (female) flowers are commonly known as female plants. The best-known examples of dioecism are date palm, papaya, spine gourd (*Momordica dioica*), hemp, spinach, kiwifruit, yam, asparagus, *etc.*

1.3. Herkogamy: (Herkos = barrier)

The condition refers to any kind of physical barrier found between male (stamens) and female (pistil) organs in bisexual flowers. The physical barrier (transparent membrane) remains unbreakable until and unless any external forces are not applied. The best-known example is leguminous forage crop lucerne (alfalfa), where stamens and stigma are fully covered by two petals forming a unique structure like the keel of a boat, commonly referred to as a keel. Sometimes stigmas remain unreceptive due to the presence of a waxy film. Such physical barriers are removed or broken by external forces like wind, insects (honey bees), birds, *etc.*, which mechanically facilitate allogamy.

1.4. Heterostyly

A genetic condition is found in a species where hermaphrodite flowers have variations in style and length of filaments–commonly known as a pin (long style and short filaments) and thrum (short style and long filaments) flower. Such condition hampers self-pollination and enables cross-pollination, *e.g.*, Linseed, primrose, buckwheat, brinjal, chilli, *etc.*

2. Mechanisms Compulsory Favour the Cross-Pollination

Some genetically/physiological control mechanisms are found in many crop species that enforced allogamy (cross-pollination) and completely prevent self-pollination. Each kind is briefly mentioned below.

2.1. Self-Incompatibility

The self-fertilization is prevent may be due to rejection of self-pollen by stigma or non-receptive state of the stigma of the same flower (hermaphrodite) or another flower of the same plant (monoecism). The incompatibility is found in both heteromorphic and homomorphic kinds of flowers. In the case of heteromorphic, the incompatibility is caused due to a heterostyly situation where variation is found in style and filaments length–commonly known as a pin (long style and short filaments) and thrum (short style and long filaments) flowers. The incompatibility occurs between similar types of flowers (*i.e.*, Pin × Pin OR Thrum × Thrum).

While in the case of a homomorphic system, the incompatible reaction is governed genetically by a single gene 'S', but the gene 'S' has polymorphic. It has a series of multiple alleles designated by S_1, S_2, S_3, S_4, *etc.* The specific position of the S gene in a chromosome is called S-locus. Based on pollen- pistil (stigma) alleles interaction-reaction, it can be classified into two types

2.1.1 Gametophytic self-incompatibility (GSI)

Its haploid genetic constituent determines (genotype of pollen itself) pollen incompatibility. The alleles of the incompatibility gene act independently in style. They show no dominance.

Description

i. **Fully incompatible** → The Pollen grain (*i.e.*, S_1S_2) and stigma (pistil) alleles (*i.e.*, S_1S_2) are fully matched. Pollen grain failed to germinate; hence no fertilization (seed set) occurs by self-pollen.

ii. **Partial incompatible**→ The Pollen grain (*i.e.*, S_1S_2) and stigma (pistil) alleles (*i.e.*, S_1S_3) are partially matched. The partial germination of the pollen grain is possible. Hence partial fertilization (seed set) occurs by self-pollens.

iii. **Fully compatible**→ The Pollen grain (*i.e.*, S_1S_2) and stigma (pistil) alleles (*i.e.*, S_3S_4) are completely differed. The self-pollen grain is capable of germinating and fertilizing the ovule. Hence regular fertilization (seed set) occurs.

2.1.2 Sporophytic Self-Incompatibility (SSI)

The pollen incompatibility is determined by the diploid genome of pollen-producing (parent) plant (sporophytic generation) and shows a dominant relation. The allelic action is complex and is independent in both style and pollen grain.

Description

i. **Fully incompatible** → The Pollen grain (*i.e.*, S_1S_2) is unable to germinate on stigma contain either of like alleles (*i.e.*, S_1S_2 or S_1S_3 or S_1S_4). The allele exhibits a dominance relationship, *i.e.*, $S_1>S_2>S_3>S_4$, *etc.* So S_1 pollen produced by S_1S_2 or S_1S_3 parent genotype would act as S_1.

ii. **Fully compatible→** The Pollen grain (*i.e.,* S_1S_2) and stigma (pistil) alleles (*i.e.,* S_3S_4) are completely differed. The Pollen grain (*i.e.,* S_1S_2) can germinate on stigma containing unmatched alleles (*i.e.,* S_3S_4).

2.2. Male Sterility

Male sterility in plants refers to a non-functional pollen grain. Male sterility in a plant is considered a typical example of maternal inheritance. In-plant breeding, the male sterility system is more commercially viable in many crop plants. It is an essential out-crossing device and genetic emasculation, essential for large-scale, low-cost hybrid seed production. Male sterility may be further divided into three types based on its inheritance pattern and governing by hereditary unit.

2.2.1a. Genic (Nuclear) Male Sterility (GMS)

In most cases, this type of male sterility governs by a single recessive gene; however, in some cases, more than two recessive or dominant genes are also reported. This sterility can be maintained by hybridizing it with heterozygous (MSms) male fertile line, resulting in a maximum of 50 per cent sterile and 50 per cent male fertile plant population.

2.2.1b. Environmental Sensitive Genic (Nuclear) Male Sterility (EGMS)

The expression of a genetic male sterility system is also altered due to environmental changes. For instance, change in temperature or photoperiods directly affect male sterility.

i. **Thermo-Sensitive Genetic Male Sterility (TSGMS):** When the range of temperature markedly influences male sterility is called thermo-sensitive genetic male sterility (TSGMS). In rice, it was first discovered during 1997 in a japonica rice species as a spontaneous mutant [AnnongS-1 (Ans-1)].

ii. **Photoperiod-Sensitive Genetic Male Sterility (PSGMS):** When male sterility is regulated by photoperiod (day length) is called photoperiod-sensitive genetic male sterility (PSGMS). In rice species, it was first reported during 1973 in a japonica rice as a spontaneous mutant [Nongken 58S (NK58S)].

2.2.2. Cytoplasmic Male Sterility (CMS)

In crops like maize, a cytoplasmic control of male sterility is known. The name itself indicates that cytoplasmic genes control this type of male sterility. **Marcus M. Rhoades** (1931) reported the first case of cytoplasmic male sterility in maize crops. Based on the origin of cytoplasmic male sterility, it is of two types (1) Autoplasmic CMS, and (2) Alloplasmic CMS.

i. **Autoplasmic CMS**: When cytoplasmic male sterility arises due to spontaneous mutation in the mitochondrial genome of a species called autoplasmic CMS.

ii. **Alloplasmic CMS**: When cytoplasmic male sterility arises due to the fusion of diverse cytoplasm through intraspecific or interspecific or intergeneric hybridization. The cytoplasm is produced sterility due to the nuclear genome and cytoplasmic genome of two diverse species. Occasionally protoplast fusion techniques being used to get alloplasmic CMS lines in different crop plants.

2.2.3. Cytoplasmic Genetic Male Sterility (CGMS)

In the CGMS system, male sterility is genuinely controlled by cytoplasmic genes. Besides the cytoplasmic male sterility, the nuclear gene, often termed the restorer gene is available to restore fertility.

The fertility restoration occurs when male sterile (A-line) cross with a fertility restorer gene (often called R-line). The resulting hybrid would be fully fertile–this type of cytoplasmic male sterility results from a combined action of both nuclear and cytoplasmic genes.

Chapter 3

Hybridization and Selfing Techniques in Crop Breeding

Introduction

The crossing in plants is done either naturally or artificially. The process is usually termed hybridization, and the resulting progeny is called a hybrid. Hybridization facilitates combining genes or transferring a gene into a new individual from genetically dissimilar parents and has evolutionary significance. Thomas Fairchild, in 1717, produced an artificial hybrid by attempting hybridization between a Carnation (*Dianthus caryophyllus*) and Sweet William (*D. barbatus*); it was a sterile so-called *'Fairchild's mule'*. J. G. Kolreuter, in 1761-62, produced hybrid and noted hybrid vigour in various species, *viz. Nicotina, Mathiola, Dianthus, etc.,* concluded the equal contribution of both male and female parents in the inheritance. He was also described the importance of cross-fertilization against the self-fertilization in a plant.

Applications of Hybridization in Crop Breeding

1. Transfer-introgression of a desirable gene
2. Create and maintain genetic variation
3. Gain and utilize heterosis/hybrid vigour
4. Overcome inbreeding depression
5. Production of hybrid seed

6. Maintain and develop parental lines
7. Evaluation of combining ability of parental lines
8. Various genetic analyses
9. Breaking undesirable linkage
10. Hybrid speciation and polyploidy

Types of Hybridization (Crossing)

A. Normal Hybridization

The hybridization (crossing) is grouped into different categories based on the nature of the plant species/varieties involved, each briefly described below:

1. **Intra-varietal Hybridization:** The crossing is performed between two plants which belong to the same variety. This type of hybridization has limited use. However, it may be performed in the case of some land varieties by breeder own concern and use.
2. **Inter-varietal Hybridization:** It is a most basic hybridization type, which is frequently used in plant breeding programs. The crossing is performed between two varieties which belong to the same species. It is also known as intra-specific hybridization as the hybridization performed within a species.

B. Wide/Distant Hybridization

Hybridization performed among different species of the same genus or two different genera is called wide hybridization. Since hybridization is also termed distant hybridization between distantly related species, the resultant hybrids (crosses) are known as wide crosses or distant crosses.

Applications of Wide/Distant Hybridization

A. To produced novel genotypes/species: The new unique genotypes or species are produced by distant crossings, such as *Raphano brassica* and *triticale*.

B. Exploitation of heterosis. In vegetatively propagated/ornamental crops, the distant hybrid exhibits superiority in vegetative growth and vigour but declines reproductive ability (fertility). It also termed as pseudo-heterosis.

C. Haploid production: The haploid produced by chromosome elimination in wide crossing, *e.g.*, Cross between *Hordeum vulgare* × *H. bulbosum*; *Avena sativa* × *Zea mays*

D. To transmit desirable genes/unique traits from wild relatives that do not exist in cultivated crop species. Such desirable genes/unique traits may be for stress resistance, quality-related or other agronomic related traits, sterility related traits, *e.g.*, sterile cytoplasm.

E. Broadening the gene pool and increasing genetic variation: It can be possible by transferring a useful gene from wild relatives into cultivated species.

F. Produced complex useful hybrids, *e.g.*, Sugarcane hybrids involving *S. officinarum and S. spontanium*.

G. Produced alien addition lines or alien substitution lines in crop plants.

Types of Wide/Distant Hybridization

1. Inter-Specific (Intra-generic) Hybridization

The hybrid or cross resulting from hybridization between two species belongs to the same genus. The hybrid is more often sterile; Such hybridization is performed to obtain

a. **Novel hybrids,** *e.g.*, tetraploid fertile cotton hybrid, three allotetraploids (Amphidiploids) species of *Brassica viz., B. napus* (2n= 4x= 38, AACC), *B. juncea* (2n= 4x= 36, AABB) and *B. carinata* (2n= 4x= 34, BBCC).

b. **Haploid production,** *e.g.*, Cross between *Hordeum vulgare* × *H. bulbosum*.

2. Inter-Generic Hybridization

The hybrid or cross results from hybridization between two different genera of the same family. The hybrid mostly sterile. Hence such cross is performed in exceptional cases like transferring biotic or abiotic stress resistance genes into cultivated varieties. Examples of synthetic allopolyploids, *e.g.*, *Raphanobrassica*, hexaploid triticale (2n=6x=42).

Problem Associated with Wide/Distant Hybridization

1. Failure of zygote development
   a. Cross incompatibility (pre-fertilization barrier)
   b. Hybrid inviability (post-fertilization barrier)
2. Failure of F_1 seedling development
3. Hybrid sterility, and Hybrid breakdown
4. Absence of Homoeology among chromosomes of a parental species
5. Associated with undesirable agronomic traits like the absence of flowering, delayed maturity, seed dormancy, shrunken seed, *etc.*,
6. Sometimes produced useless combinations
7. Undesirable linkages, and
8. Problems in the transfer of traits governed by recessive genes.

C. Introgressive Hybridization

Modifying the genetic architecture of one species by long term rigorous hybridization, *i.e.*, repeated backcrossing with another species. The primary aim of transfer a gene from one species to another. The line is derived through introgressive hybridization, commonly termed an introgressive line. For example, transfer of cytoplasmic male sterility and restorer gene in crop plants. The introgressive line differed from common hybridization in terms of complexity and mixture of genes/ genome parts or sometimes a few segments of a genome from donor species.

Emasculation Techniques

Emasculation can be achieved through different techniques. Each describes below:

A. Hand Emasculation (Direct Method)

This technique is most common and practicable among other techniques of emasculation. It is most suitable for large and hand-operated flowers of crop species. In this method, an emasculation is performed with the help of forceps and scissor. Hence often referred to as a forceps and scissor method.

B. Other Methods (Indirect method)

In addition to hand emasculation, many techniques that have been used for emasculation are described below:

i. Genetic Emasculation

Certain crop species having self-pollination control mechanisms, *viz.*, male sterility, self-incompatibility, monoecism, dioecism or protogyny conditions, *etc.*, facilitates the direct hybridization without removing the male part (*i.e.*, anther). The male sterility system in crop species like pearl millet, maize, sorghum, cotton, rice, onion, *etc.* is rewarding in hybrid seed production.

ii. Cold/Hot Water Treatment

This method is also suitable for crop species having small flowers in which hand emasculation is difficult and tedious. The androecium part is sensitive to cold or hot water compared to gynoecium. The cold/hot water temperature and duration are standardized based on male part sensitivity.

iii. Suction Pressure Method

This method is suitable for crop species having tiny flowers. The pressure of pollen suction standardizes so that only anthers are removed, and female parts remain undamaged.

iv. Gametocides

Some commercial chemicals designate to kill the anther without damaging female reproductive organs, commonly known as chemical hybridizing agents (CHA), *e.g.*, Sodium methyl arsenate, Maleic hydrazide, Ethrel, 2-4-D, *etc.*

v. Alcohol Emasculation

Though this method is more effective than the suction pressure method but not commonly used because of more strictly and sensitive to treatment duration. For a case in alfalfa (*Medicago sativa*), the inflorescence dipped for 10 seconds is sufficient to kill all anthers. Increasing time duration (*i.e.*, > 10 seconds) may damage the gynoecium.

Selfing Techniques in Crop Plants

In plant-like crossing, selfing is also one of the essential procedures in crop plant improvement programs. The exact procedure of selfing merely depends on the floral structure and mode of pollination mechanism in crop species. To ensure self-pollination in crop plants, the selfing is strictly performed in crop breeding programmes as and when required, irrespective of pollination nature of crop plant.

Essentiality of Selfing in Crop Plant

There is certain essentiality behind selfing, which are:

1. To increase the homozygosity.
2. To avoid the natural cross-pollination in both self and cross-pollinated species.
3. To create a true breeding line.
4. To maintain the genetic purity of crop varieties.

Different Techniques of Selfing

1) Natural Method

a. **Space isolation:** Each variety is grown in an isolated plot. Keep minimum isolation distance to avoid cross-pollination. The isolation distance depends on the mode of pollination, pollination mechanisms, pollen viability duration, and mode of pollen dispersal, *i.e.*, wind, insects, birds, animals, *etc.* For instance, castor pollen is dispersed by wind, so such crop species require more isolation distance.

b. **Time isolation:** Each variety is grown in a separate block on different sowing dates. The time of sowing between two varieties adjusts so that flowering periods do not overlap. The time isolation between varieties differs from crop to crop.

Natural methods are suitable for selfing in a large number of plants. When there are a few numbers of plants, artificial methods of selfing are more effective

2) Artificial Method

In the case of artificial methods, the selfing is performed manual basis. The different material and methods are used to protect stigma from foreign pollen according to flower structure, inflorescence type, and pollinator presence. Generally, selfing is done in the early morning or day before the opening of a flower. Before selfing, all unwanted, very young buds, already opened, and disease-pest-infested flowers are removed. The techniques/material and methods are used for selfing are described below:

a. **Bags:** Various paper, butter paper or cloth bags are used. Bags are suitable to protect single flowers or sometimes whole small or medium-sized inflorescences, *e.g.*, sorghum, bajra, wheat, maize, castor, *etc.*
b. **Rubber band/ring:** It is used to avoid opening of flower bud, *e.g.*, cucurbitaceous family
c. **Cages:** A cage is used to cover the whole plant. This method is suitable for crop species pollinated by various insects, *e.g.*, mustard, tomato, onion, and cumin.
d. **Wire ring:** It is used to fix on an unopened flower bud, *e.g.*, okra, cotton, *etc.*
e. **Straw tubes:** Suitable for a crop with a very tiny flower, *e.g.*, pulses.
f. **Mud:** It is smeared on the unopened bud to avoid outcrossing, *e.g.*, cotton, okra, *etc.*

Chapter 4

Crop Breeding Methods: An Overview

Various approaches *viz.*, selection, hybridization, mutation, *etc.* are used for Genetic improvement of crop plants are referred to as plant breeding methods/ approaches/techniques. The choice of breeding methods mainly depends on the mode of pollination; reproduction, gene action, and breeding objectives of crop species.

Classification of Crop Breeding Approaches/Techniques/Methods

Plant breeding methods are generally classified based on application of crop improvement:

A. General Breeding Approaches

1. Plant Introduction
2. Progeny selection
3. Pure line selection
4. Mass selection
5. Pedigree method
6. Bulk method
7. Single seed descent method
8. Back cross method

9. Heterosis breeding
10. Clonal selection
11. Synthetics and composites

B. Special Breeding Approaches

1. Mutation breeding
2. Polyploidy breeding
3. Transgenic breeding
4. Molecular breeding

C. Population Improvement Approaches

1. Recurrent selection
2. Disruptive selection
3. Diallel selective approaches mating system
4. Biparental mating

D. Breeding Approaches used for Self-Pollinated Species

1. Plant Introduction
2. Pureline selection
3. Mass selection
4. Pedigree method
5. Bulk method
6. Single seed descent method
7. Backcross method
8. Heterosis breeding
9. Mutation breeding
10. Polyploidy breeding
11. Distant hybridization
12. Transgenic breeding

E. Breeding Approaches Used for Cross-Pollinated Species

1. Plant Introduction
2. Mass and progeny selection
3. Backcross method
4. Heterosis breeding
5. Synthetic breeding
6. Composite breeding

7. Polyploidy breeding
8. Distant hybridization
9. Transgenic breeding

F. Breeding Approaches Used for Asexually Propagated Species

1. Plant Introduction
2. Clonal selection
3. Mass selection
4. Heterosis breeding
5. Mutation breeding
6. Polyploidy breeding
7. Distant hybridization
8. Transgenic breeding

Plant Introduction

A new plant material (plant genotypes or wild species) is transferred from their original adjusted environments to a new, untested environment. The native environments represent the native area of plant species where it is well adjusted and are being grown. Plant Introduction is considered a simple and quickest method of plant breeding, where plant genetic material is directly introduced into new, untested areas as a new crop or a new cultivar, irrespective of plant breeding forms, pollination behaviours, sexual type, *etc.*

Types of Plant Introduction

Plant Introduction is grouped into two categories (i) Primary and (ii) Secondary Introduction. **Erna Bennet** (1965) described the terms primary and secondary Introduction for the first time. The primary and secondary types are mainly based on the adaptation of plant material in a new area. Each type is briefly described below:

(i) Primary Introduction

The exotic plant material (*i.e.*, cultivated or wild species) is directly introduced into a new environment for commercial cultivation without altering its original genotype. For instance, the introduced plant genotype is directly released as a commercial cultivar after testing yield performance. Hence often termed as a direct Introduction. This type of Introduction is used in rare situations where no alternative cultivar exists or obtains more yield advantage over existing cultivars.

Important Examples

☆ **Wheat:** Semi-dwarf genotypes Lerma Rojo, Sonara 64 introduced from the Mexico.

- ☆ **Rice:** Semi-dwarf varieties IR-8, IR 28.
- ☆ **Tomato:** Sioux, la Bonita from the USA.
- ☆ **Soybean:** Palmetto, Monetta from the USA.
- ☆ **Maize:** Protina from the USA.
- ☆ **Barley:** Sona, Dolma, *etc.*

(ii) Secondary Introduction

The exotic plant material (*i.e.*, cultivated, or wild species) is used as cultivars only after selection or altering its genotype by using it into a hybridization program for desirable gene source, *e.g.*, high yield, disease-pest resistant or quality traits, *etc.*, since introduced material indirectly utilized in cultivar development is often termed as an indirect Introduction. This type of Introduction is most frequently utilized in plant breeding compared to primary Introduction.

Important Examples

- ☆ **Wheat:** Kalyan Sona, Sonalica, Lal Bahadur.
- ☆ **Rice:** Chambal.
- ☆ **Pearl millet:** HB1, HB3, HB4.
- ☆ **Tomato:** Pusa Ruby.
- ☆ **Carrot:** Pusa Kesar.
- ☆ **Brinjal:** Pusa Kranti, *etc.*

Plant Selection

Plant selection has been practised as old as agriculture. It may be an inclusive method in basic plant breeding programs like mass selection, pureline selection, or a part, such as handling of segregating material in each generation. Plant selection is carried out either through nature itself (natural selection) or man's involvement (artificial selection).

Aim of Plant Selection

1. To improve the mean of a trait in a desirable direction
2. To increase homozygosity of traits
3. To improve the stability of traits

Methods of Plant Selection

The plant selection techniques used in crop plants *viz.*, self-pollinated, cross-pollinated, and asexually propagated crops. The methods described are based on direct and indirect applicability of selection.

A. Direct Methods

The selection followed directly on plant population without altering its original genetic makeup.

1. Progeny selection
2. Pureline selection
3. Mass selection
4. Simple recurrent selection
5. Screening methods of biotic and abiotic stress resistance
6. Clonal selection, *etc.*

B. Indirect Methods

The selection is performed on a plant population after altering its original genetic makeup, such as handling of segregating generation.

1. Pedigree selection
2. Mass-pedigree selection
3. Bulk selection
4. Backcross selection
5. Single seed descent selection
6. Recurrent selection, *etc.*

Progeny Selection

It is a natural selection approach where a selection of a superior plant is made from a heterogeneous population based on the progeny test. Here two important termed used; first is 'progeny' means offspring; a plant is grown from the cross seeds produced by sexual means from parental plants. The second is the 'progeny test', which determines an individual's genetic value based on their progeny performance. The progeny selection is primarily used in cross-fertilizing and often cross-fertilizing crop plants. In maize ear to row approach is, the simple way of progeny selection was first used by Hopkins in 1908.

Pureline Selection

The Danish plant physiologist W. Johannsen in 1903 gave the concept of pureline. It refers to '*a progeny of a single, self-pollinated, homozygous (true-breeding) plant*'. The involvement of progeny testing revealed the genotype and phenotype difference in the population by discriminating the environmental role. Pureline selection is the most common approach used to improve self-pollinated crops.

The general procedure of pureline selection consists of five steps.

1. Individual plant selection from heterogeneous/mixed or local landraces population,

2. Phenotypic evaluation of individual plant progenies,
3. Replicated preliminary yield evaluation,
4. Replicated multi-location yield evaluation, and
5. Identification of best line-released/utilized (as per objective) and seed multiplication.

Application of Pureline Selection

1. Direct selection of introduced material.
2. Purify the local (desi) variety.
3. Selection in local (desi) varieties (mixed heterogeneous material).
4. Selection for a typical trait, *i.e.*, biotic or abiotic stress resistance.

Mass Selection

Mass selection is the original and most effortless approach to crop improvement; since the domestication era, mass selection has been practised intentionally or unintentionally without knowing the inheritance patterns of crop plants. This approach is performed in both self and cross-pollinated crops.

The general procedure of mass selection consists of four steps:

1. Selection of plants in mass,
2. Replicated preliminary yield evaluation,
3. Replicated multi-location yield evaluation, and
4. Identification of best line-released/utilized (as per objective) and seed multiplication.

Recurrent Selection

The recurrent selection also refers to cyclic improvement. Involving a simple way of population improvement or through generating new recombinants (population), interbreeding of individuals with repeated cyclic selection over generations. In other words, the genetic recombination generates through repeated cyclic selection over generations with the interbreeding of individuals.

Method of Recurrent Selection

The recurrent selection method is categorised into four different types:

1. Simple recurrent selection (Phenotypic recurrent selection),
2. Recurrent selection for GCA,
3. Recurrent selection for SCA, and
4. Reciprocal recurrent selection for combining ability

Breeding Methods Used to Handle Segregating Generations

I. Pedigree Selection

A method of breeding used to handle segregating generation in self-pollinated crops where individual plants are selected from segregation generation alongside pedigree record is maintained for all selected individual plants. Individual plant selection is performed till the progenies display no segregation means attained homozygosity.

Steps involve in procedure of pedigree method

The standard procedure of a pedigree method of selection consists of four steps

1. Hybridization and raising of the F_1 seeds,
2. Handling of segregating generations/selection from segregating generations,
3. Yield evaluation (preliminary and multilocation trials) of homozygous progenies,
4. Identification of best line-released/utilized (as per objective) and seed multiplication

II. Bulk Selection Methods

The bulk selection refers to a selection procedure of self-pollinated crops where the segregating population continues bulk with or without selection up to F_6 or F_7 generation. After that, the individual plant selection is practised and evaluated individual plant progenies like the pedigree selection method. This method is primarily used in the genetic improvement of self-pollinated crops like cereals, pulses, and minor millets. This method provides the maximum opportunity for natural selection (during the bulk period) followed by artificial selection (advance generation of F_6 or F_7). Moreover, this method can isolate homozygous lines as the most of segregating populations attained homozygosity after the bulking period.

The procedure of a bulk method of breeding

The standard procedure of bulk selection methods of breeding consists of five steps.

1. Hybridization and raising of the F_1 seeds.
2. Handling of segregating generations in a bulk plot.
3. Progeny selection.
4. Yield evaluation (preliminary and multi-location trials) of homozygous progenies.
5. Identification of best line-released/utilized (as per objective) and seed multiplication.

III. Single-Seed Descent (SSD) Method

A single-seed descent (SSD) method is a modified version of a bulk selection method initially suggested by C. H. Goulden in 1941. The term single seed descent was first used by H. W. Johnson and R. L. Bernard in 1962. C. A. Brim in 1966, described detail procedure of SSD in soybean crop. He described it as '*modified pedigree method*'.

In this method, only a single seed is selected randomly from each plant to advance each generation, so-called as single seed descent method of selection. In contrast to a bulk method, only a single random seed is taken from each F_2 plant and bulked to raise the F_3 generation. This procedure is repeated until the maximum level of homozygosity is obtained.

The procedure of a SSD method

The standard procedure of SSD selection methods consists of five steps.

1. Hybridization and raising of F_1 seeds,
2. Handling of segregating generations with the harvesting of a single seed,
3. Progeny selection,
4. Yield evaluation (preliminary and multi-location) of homozygous progenies, and
5. Identification of best line-released/utilized (as per objective) and seed multiplication.

IV. Backcross Method

Backcross refers to a cross between F_1 generation or segregating generation and either parent (dominant or recessive). Generally, the backcross breeding approach has been widely and most successful in plant breeding systems to remove the specific defect or undesirable traits governed by one or two major genes.

The procedure of backcrossing depends on the nature of the gene under transferring. The procedure is quite simple when the dominant and highly heritable gene/allele is under transfer. When gene/allele under transfer is in recessive form, the backcrossing procedure is somewhat complicated and lengthy; however, the primary steps remain the same. The steps of both procedures are listed herewith separately.

A. Procedure for Transfer of a Dominant Gene

The standard procedure of backcross breeding for dominant gene transfer consists of six steps.

1. Crossing and harvesting the F_1 seeds,
2. Raising of the F_1 generation and backcrossing,
3. Backcrossing and generation handling,

4. Self-fertilization,
5. Progeny selection and bulk, and
6. Yield evaluation and released/utilized and seed multiplication

B. Transfer of a Recessive Gene

When genes/alleles under transfer are recessive, the backcrossing procedure is somewhat complicated and lengthy; however, the primary step remains the same. When recessive genes govern the resistant trait, it is challenging to identify phenotypically in the F_1 generation. Identifying the plants carrying resistant recessive gene/allele required selfing of plants after the first backcross and after every two backcrosses.

The standard procedure of backcross breeding for recessive gene transfer consists of five steps.

1. Crossing and harvesting the F_1 seeds.
2. Raising of F_1 plants and backcross.
3. Selfing - backcrossing - selfing and generation handling
4. Progeny selection and bulk.
5. Yield evaluation; released/utilized and seed multiplication.

Heterosis Breeding

Heterosis refers to the superiority of hybrid in terms of overall growth-vigour, yield potential and fertility compared with its parents; it is associated with increasing heterozygosity. Often it is termed hybrid vigour. A German Scientist, **J. G. Koelreuter**, provided the first evidence of hybrid vigour in *Nicotiana* species in 1763. **G. H. Shull** first used the term heterosis in 1914.

Development of Hybrid Cultivars

The exploitation of heterosis in plant breeding is more conveniently by developing hybrid cultivars. The exploitation of heterosis *via* various breeding approaches is called heterosis breeding. Heterosis breeding is considered an alternative plant breeding approach. The heterosis can be revealed in both self and cross-pollinated species. However, heterosis is exploited significantly in cross-pollinated crops like maize, bajra, castor, lucerne, onion, sorghum, sunflower, cotton, vegetable crops, *etc.*

The hybrid variety refers to the F_1 generation obtained from a cross between two or more genetically distinct homozygous parental genotypes. It is directly used for commercial cultivation. The homozygous line comprises purelines (in self-pollinated crops), inbred lines, pistillate lines (cross-pollinated crops), and clones (vegetatively propagated crops).

Types of Hybrids

Conventional hybrids: Hybrids developed from two or more homozygous lines (inbred/pure lines) through conventional means.

Non-conventional hybrids: Hybrids developed from the non-inbred parent.

Steps involved in the development of conventional-hybrid variety in cross-pollinated crops

Step 1 Selection of base material(Genetically heterozygous and heterogeneous)

↓

Step 2 Development of parental lines (inbred lines)

↓

Step 3 Evaluation of inbred lines

↓

Step 4 Crossing between good-inbred lines

↓

Step 5 Production of hybrid seed in bulk after being released commercially

Synthetic Varieties

According to **Lonnquist J. H.** (1961), synthetic variety is an open-pollinated population derived from a intercrossing of selfed plants or lines, maintained through routine mass selection procedures.

Basic Features of a Synthetic Variety

1. Most suitable for cross-pollinated crops.
2. It is a seed mixture of a few selected parents having a good combining ability and combining well with each other.
3. The base material of a synthetic variety is a mixture of inbred lines or clones of known combining ability.
4. The genetic composition is both homozygotes and heterozygotes lines.
5. At the field level, it constitutes a heterogeneous population.
6. It exploited additive gene action.

7. A synthetic cultivar can be restored from a same set of maintained parental (inbred) lines.
8. Stable and high adaptation due to broad genetic base.
9. The yield of a synthetic variety can be predicted in advance based on *per se* performance of inbred (parental) lines and their possible crosses.
10. The maintenance of a synthetic variety is done in open pollination. The improvement is possible via a population improvement approach like recurrent selection.
11. The generations (*viz.*, F_1, F_2, F_3, F_4,.... F_n) of a synthetic variety are designated by Syn_1, Syn_2, Syn_3, Syn_4,... Syn_n.

Composite Varieties

A variety is constituted by a seed mixture of several phenotypically promising open-pollinated varieties, heterogeneous base populations, or germplasm lines that neither tested for combining ability nor focused on inbreeding. A composite cultivar is maintained in open-pollination in isolation.

Basic Features of a Composite Variety

1. Most suitable for cross-pollinated crops.
2. The base material of a composite variety is phenotypically promising open-pollinated varieties, heterogeneous base populations, or germplasm lines
3. It exploited additive gene action.
4. At the field level, it constitutes a heterogeneous population.
5. The genetic composition is both homozygotes and heterozygotes lines.
6. Composite variety can be maintained in open pollination in isolation.
7. Stable and high adaptation due to broad genetic base.

Steps Involved in Synthetic and Composite Varieties Development

The synthetic varieties are synthesized by mixing the seed of all cross combinations of different inbred lines with known combining ability. Hence, the development of synthetic cultivars required base material, *i.e.*, inbred lines/clones or open-pollinated varieties with good general combing ability. The basic steps involved in synthetic variety development are briefly described below:

1. Step 1: Collection of a source material. If suitable, base material not available produce it.
2. Step 2: Evaluate the base material (inbred line) and identify the best lines.
3. Step 3: Make F_1 crosses in all possible combinations or allow random mating in open pollination.

4. Step 4: Composite the F_1 seeds and multiplication.
5. Step 5: Released for commercial cultivation.

Note: *In the case of* ***composite variety development*** *procedure, no combining ability test performed and source population may vary to some extent. The rest procedure remain the same.*

Section-II

Breeding Cereals (Wheat, Rice, Maize, Pearl Millet, Sorghum and Small Millets)

Chapter 5

Breeding Wheat (*Triticum* spp.)

Introduction

Wheat is one of the essential rabi cereal in the world, giving about one-third of the total cereal production, followed closely by rice. It is a staple food of about one-third of the world's population. In India, the wheat crop gives higher per day productivity compared to the global situation due to its short duration (~140 days) against major winter wheat-growing countries (> 300 days). The demand for wheat is increasing as it has been utilized chiefly in the form of floor and floor-based multiple products, bread, and confectionery products, *i.e.*, cakes, cookies, crackers, semolina, *etc.* It is a major source of starch and energy as well as cereal dietary fibre; hence it is considered one of the essential cereal crops for safeguarding food and nutritional security in India and many parts of the world.

Background Information

1.	*Common name*	Gehun, Gau
2.	*Botanical Name*	**Common (Bread) wheat:** *Triticum aestivum* **Macaroni (durum) wheat:** *Triticum durum* Desf
3.	*Family*	*Poaceae*
4.	*Chromosome number*	**Diploid:** 2n=2x=14 **Tetraploid:** 2n=4x=28 **Hexaploid:** 2n=6x=42

5.	*Centre of origin*	**Diploid:** Asia Minor **Tetraploid:** Abyssinia, North Africa **Hexaploid:** Central Asia
6.	*Nature of pollination*	Self-pollination
7.	*Pollination mechanisms*	Bisexuality, Chasmogamy, Cleistogamy, *etc.*
8.	*List of wild relatives*	*Aegilops speltoides, T. aegilopoides, T. boeticum, T. dicoccoides, T. timopheevi, Triticum araraticum, etc.*

Table 5.1. Classification of Wheat According to Ploidy

Ploidy Level	*Species*	*Common name*	*Genome*
Diploid (2n=2x =14)	(*T. aegilopoides*) *T. monococcum*	Einkorn	AA
	T. boeticum	Wild einkorn	AA
	T. urartu	Wild	AA
	Aegilops species	Wild	BB
	Aegilops tauschii	Wild	DD
	Aegilops speltoides	Wild	SS
	T. timopheevi	Wild	GG
Tetraploid (2n=4x =28)	*T. dicoccoides*	Wild Emmer	AABB
	T. durum	Macaroni wheat	AABB
	T. turgidum	Rivet or pollard wheat	AABB
	T. persicum	Persian wheat	AABB
	T. dicoccum	Emmer	AABB
	T. polonicum	Polish wheat	AABB
	T. aethiopicum	Ethiopian wheat	AABB
	T. timopheevi	Zanduri wheat	AAGG
Hexaploid (2n=6x= 42)	*T. aestivum*	Common or bread wheat	AABBDD
	T. sphaerococcum	Indian dwarf wheat	
	T. macha	Macha wheat	
	T. spelta	Dinkel wheat	
	T. compactum	Club wheat	
	T. vavilovii	Yunanense Wheat	AABBGG
	*T. zhukovskyi**	Zhukovsky'swheat	AAAAGG

* Recently originated through hybridization between cultivated *T. timopheevii* and *T. monococcum.*

Distribution

Wheat is grown across the globe except the Antarctica continent. Every month of the year, a wheat crop is harvested worldwide. In India, based on agro-climatic conditions, the total growing area of wheat is broadly divided into six zones *viz.*, NHZ (Northern Hill Zone), NWPZ (North Western Plain Zone), NEPZ (North Eastern Plain Zone), CZ (Central Zone), PZ (Peninsular zone) and SHZ (Southern Hill Zone). Among different cultivated species of wheat, mainly three species *viz.*, *T. aestivum* (common wheat or bread wheat), *T. durum* (Macaroni wheat) and *T. dioccum* (Emmer wheat) are grown and occupies > 90 per cent, < 5 per cent and very marginal area, respectively of a total growing area of the country. The Macaroni wheat is primarily grown in some parts of the central and Peninsular zones, while Emmer wheat is grown in southern parts of the country, *i.e.*, Karnataka and the regions of south Maharashtra.

Floral Biology, Selfing and Crossing Techniques

Inflorescence	:	Racemose type is commonly known as ear (A compound distichous spike). It contains 13-23 spikelets, each arranged on rachis on the alternate side in a zig-zag manner. Each spikelet has 3-6 florets attached alternatively on the rachilla.
Flower	:	The floret is Bracteate (subtended by two empty or sterile glumes arranged alternatively on the contrary side of the rachilla). The flower is zygomorphic, sessile, incomplete (due to perianth), perfect and hypogynous. The uppermost floret is imperfect.
Perianth	:	Each floret comprises an outer perianth called the *lemma,* and a thin two-nerved perianth called the *palea.* The outer lemma with a stiff hair-like structure is called awn. The awn may be variable in length, sometimes absent.
Androecium	:	Three stamens with long filaments, anther–dithecous, versatile, introrse. Pollen is short-lived, and the weather conditions may influence the viability of pollen. The viability lasts 30 to 40 min after pollen shedding.
Gynoecium	:	Single carpels; one superior ovary with two short styles, single ovule and basal placentation; bifurcated feathery stigma.
Fruits	:	Caryopsis
Seed	:	Albuminous or monocots.
Floral formula	:	Br. ⊕ ⚥ $P_{2(Lodicules)} A_3 \underline{G}_{(1)}$

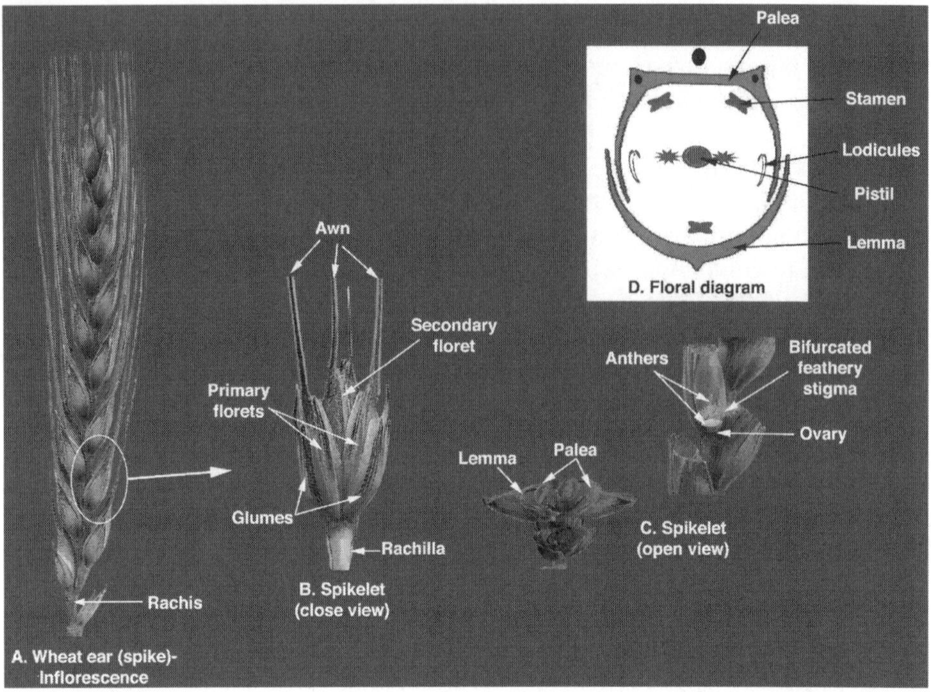

Figure 5.1. Floral Biology and Diagram of Wheat (*Triticum aestivum* L.).

Anthesis: Anthesis begins from the middle third of the spike and continues upwards rapidly compared to downwards. In the spikelet, the basal flowers are open first.

Selfing: The spike is to be covered with a butter paper bag before flowering to avoid the crossing and then labelled.

Crossing: Select a spike in which anthesis will occur one or two days later. Remove awns with the help of scissors so that a small portion of glume, lemma and palea is also cut (Figure 5.2). This will ease the removal of anthers. One-fourth (1/4) of lower and upper florets are removed, keeping 16-20 spikelets only. With the help of fine forceps, remove all three anthers in the evening or early morning. Check the emasculated spike with a magnifying lance so that anthers may not be leftovers in the floret. Bagging and labelling are done.

Pollination is performed the following morning or the second day, depending on weather conditions and stigma receptivity. Select the spike for pollination from the pollen parent and remove it from the plant. Keep the spike in the sunlight for 10-15 minutes. Anthers will burst, and pollen grains will be released. The pollen grains are dusted on all the sides of emasculated spike. Bag the pollinated spike and label it accurately.

Figure 5.2. Prepartion of Wheat Spike for Emasculation Portion of Glume, Lemma and Palea is Cut.

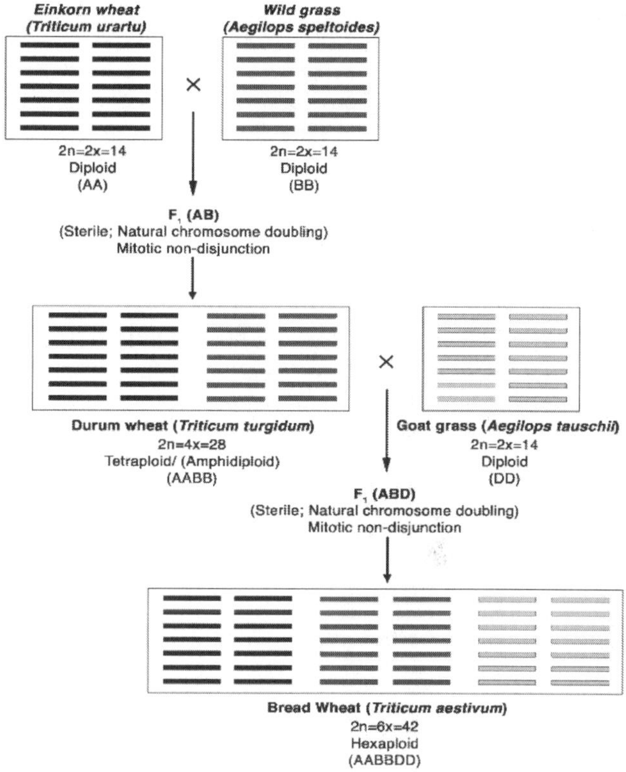

Figure 5.3. Genetic Origin of Common (Hexaploid) Wheat.

Genetic Origin of Common (Hexaploid) Wheat

The common (hexaploid) wheat is believed to be arisen by a natural crossing between tetraploid wheat (2*n* = 4x = 28, AABB) *T. turgidum* ssp. *durum* Desf, and diploid goat grass (2*n* = 2x = 14, DD) *Aegilops tauschii* Coss. (Figure 5.3). **Mc Fadden and Sears** (1946) developed synthetic hexaploid wheat through hybridization between tetraploid wheat (2*n* = 4x = 28, AABB) *T. turgidum* ssp. *durum* Desf and diploid goat grass (2*n* = 2x = 14, DD) *Aegilops tauschii.* This experiment proved the involvement of these two species in the evolution of common wheat. The hexaploid wheat evolution is an excellent model for illustrating structural, functional, and other changes, *i.e.* epigenetics in an allopolyploid chromosome set (genome).

In nature, tetraploid or hexaploid wheat homeologous group chromosomes strictly follow diploid type behaviour, forming 21 bivalents in hexaploid, while 14 bivalents in tetraploid wheat. In other words, they usually perform like diploids due to mutation in the '*ph*' gene of the *Ph1* locus on chromosome 5BL, inhibiting homoeologous chromosomes from pairing at meiosis. Besides the *Ph1* locus, two additional suppressors have been recognized, *i.e.*, Ph2 on 3DS and another suppressor on chromosome 3AS.

The Birth of Semi-Dwarf Wheat: A Fundamental Basis of a Green Revolution

The semi-dwarf wheat varieties resulted due to the equal efforts of Japanese and American scientists. The dwarfing gene-carrying variety '*Norin* 10' was developed through systematic breeding efforts (Figure 5.4). The Japan-origin semi-dwarf wheat variety was used as the primary donor line of the semi-dwarf trait called *Daruma,* which carried a recessive gene (*i.e., rht1* and *rht2*). It was crossed with a high-yielding American winter soft wheat variety, namely '*Fultz*'. The line *Daruma-Fultz* hybridized with another high-yielding American variety, namely '*Turkey Red*'. This breeding resulted in a genotype popularly known as *Norin* 10. Around the 1950s, **Dr. Orville Vogel** at Washington State University used *Norin* 10 line in a breeding program and made several crosses with local American varieties. The resulting line was called '*Gaines*'.

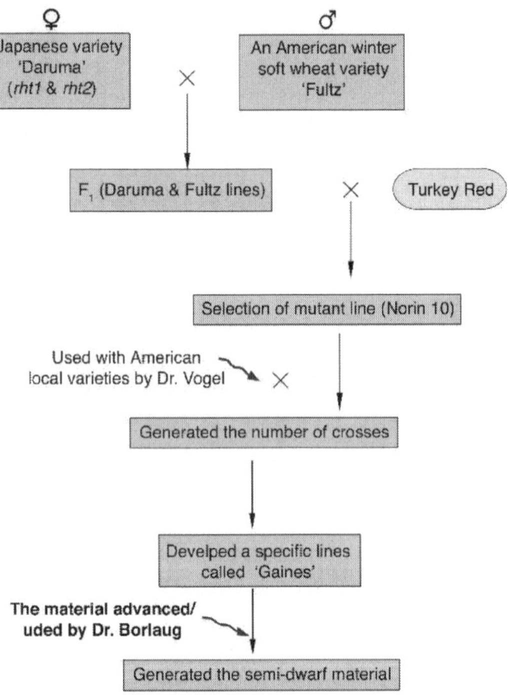

Figure 5.4. Birth of Semi-Dwarf Wheat.

Dr. N.E. Borlaug in Mexico used Gaines's culture to develop semi-dwarf wheat varieties. Dr. Borlaug supplied these semi-dwarf wheat materials in bulk to many developing countries, including India and Pakistan. In mid-1960, the Indian Scientist **Dr. M. S. Swaminathan** played a crucial role in the wheat breeding program (he used the shuttle breeding programme to evaluate received material from Dr. Borlaug), leading to the release of popular varieties called *Sonara* 64 and Lerma Rojo 64A. Later, two important varieties, *Kalyansona* and *Sonalika,* were also selected and became popular until mid 1970. All modern wheat varieties carry these dwarfing genes with critical features: wide adaption, short height, disease resistance, highly fertilizer responsive, and photoperiod insensitivity. Dr. M. S. Swaminathan was honoured as a father of a green revolution in India for his extraordinary work for a green revolution in India.

Triticale

Triticale is credited as the first artificial cereal crop. **W. Rimpau** (1888) made first intergenic hybrid called triticale. The hexaploid triticale (2n=6x=42) is derived from cross between tetraploid wheat (*Triticum turgidum* 2n = 4x = 28; AABB) and rye (*Secale cereale* 2n = 2x = 14: RR). While octaploid triticale (2n=8x =56) derived from cross between hexaploid wheat (*Triticum aestivum* 2n = 6x = 42; AABBDD) and rye (*Secale cereal* 2n = 2x = 14: RR) Figure 5.5.

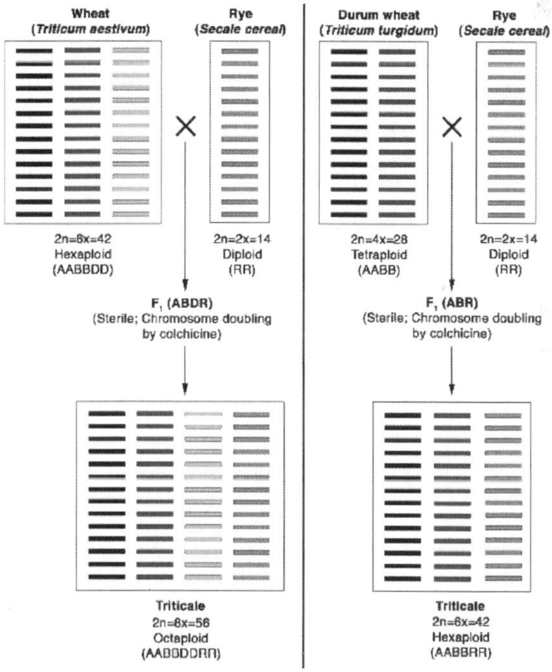

Figure 5.5. A Synthesis of Triticale.

Classification of Triticale

1. **Primary Triticale:** It is obtained directly from the cross of wheat × rye. The crosses are made using rye as a pollen parent due to the slow growth of wheat pollen tubes in rye-style tissues.

2. **Secondary Triticale:** It is obtained by crossing among triticales or crossing triticale with wheat.

Useful Characters of Triticale

i. Winter hardiness,
ii. Hybrid vigour,
iii. Resistance to yellow rust,
iv. High yield even at low fertility level, and
v. High lysine content.

Problems Linked with Triticale Breeding

1. F_1 shows partial sterility and meiotic instability,
2. Shrunken endosperm,
3. Poor germination,
4. Sensitive to day length,
5. Seeds shattering,
6. Plant Lodging,
7. Absence of seed dormancy, and
8. Problems associated with bread and chapatti making.

Breeding Objectives

1. Increase grain yield potential

Improving grain yield in crop breeding is the ultimate objective. In wheat, high grain yield can be improved by focusing on different yield attributing traits, *viz.*, improving numbers of spikes per unit area, number of tillers per plant, increased grain per spike, and bolder grain size. Based on the investigation, it cleared that 1000-kernel weight high and significant positive correlation with yield–selection of yielding components with additive effects. Synthetic wheat increases variability in the primary gene pool *via* the formation of new genetic combinations using different wheat progenitor's *viz., T. monococcum, T. turgidum, Ae. Tauschii, etc.*

2. Breeding for earliness

In wheat breed variety with earliness (early flowering/maturity) is an essential objective. In wheat cultivars, earliness is promoted *via* incorporating vernalization and photoperiod-insensitive genes. Wheat breeders have developed cultivars linking vernalization to stimulate early maturity and increase grain yield potential. Generally, a negative correlation observed between earliness and high yield. In such a situation, breed cultivars with faster growth rates to accumulate enough biomass in a shorter duration to raise grain yield potential in wheat.

3. Breeding Short Height and Lodging Resistance Cultivars

In wheat, dwarfness and logging resistance have a positive correlation. The lodging resistance is possible after identifying the dwarfing gene in the Japanese variety *Norin 10*. The genes *viz., Rht1, Rht2,* Rht-*D1c*, and *Rht8* responsible for dwarfing have been used broadly in wheat breeding programs. The yield potential in wheat significantly improved after developing and using semi-dwarf cultivars.

4. Breeding for Quality; Biofortification

About 46 scientific parameters in wheat can judge grain quality. Majorly categorized into:

1. **Physical parameters** *viz.*, Colors, external appearance, hardness, hectoliter or test weight, *etc.*
2. **Chemical composition** *viz.*, Protein content, gluten content, starch content, lysine content, beta carotene, milling quality, *etc.*

The quantity of gluten content varies according to the types of wheat. Gluten is an indefinite, tough, nitrogenous compound (endosperm protein) retained in wheat flour even after being washed to remove starch. These properties make it suitable for making bread. Different genes and their different forms control the gluten proteins, *viz.*, Glu-1, Glu-3 and Gli-2. The presence or absence of two small proteins called puroindolines determines the grain hardness. The protein puroindolines are encoded by the Pina-D1 and Pinb-D1genes. Hard wheat or bread wheat has higher gluten content than soft wheat. Hence, soft wheat is not suitable for bread making but is ideal for cakes, cookies, pastries, *etc.* On the other hand, the durum wheat lack of gluten and is suitable for making macaroni, pasta and a granular product known as semolina.

Table 5.2. Identified Genotypes/Cultivars for Quality Traits in Wheat

Sl.No.	*Quality Parameters*	*Genotypes/Cultivars*
1.	High protein	HI 6896 (d), HD 2833, HPW 533, PBW 373, PR 31, WH 716 (d), MP 1168, UP 2425, UP 2620, *etc.*
2.	High β-carotene	PBW 295 (d), WH 896 (d), HI 8627 (d), PBW 233 (d), GW 1189 (d), PDW 273 (d), *etc.*
3.	High glutenin	DWR 237, HD 2745, MP 1103, *etc.*
4.	Low gliadin	PDW 275 (d), PDW 1233 (d), MP 01106 (d), Raj 6547 (d), *etc.*

Biofortification is a process of increasing the nutritional level of crop plants by plant breeding, agronomic and transgenic approaches. Landraces are one of the most vital sources of wheat biofortification. *T. dicoccoides* (wild emmer), *T. monococcum* (einkorn), diploid progenitors of common wheat *viz., Aegilops tauschii*, *T. spelta*, and *T. polonicum* are also promising sources of high Fe and Zn in wheat. Improved cultivars with 40 per cent higher Zn concentration than the existing cultivars. In India, two varieties WB-02 (Zn=42.0 ppm and Fe=40.0 ppm) and HPBW-01(Zn=40.0 ppm and Fe=40.6 ppm) have been released.

Table 5.3. Various Quality Products with Suitable Cultivars in Wheat

Sl.No.	*Product*	*Genotypes/Cultivars*
1.	Chapatti (Roti)	LOK 1, GW 173, GW 273, GW 322, HI 1500, HW 2004, HD 2285, PBW 373, PBW 533, SUJATA, *etc.*
2.	Bread	LOK 1, GW 173, GW 190, GW 120, HD 2733, HI 977, HS 240, VL 738, HD 2285, K 9107, *etc.*
3.	Biscuit	PBW 373, Sonalika, HD 2687, UP 2425, WH 542, Raj 3765, DBW 16, *etc.*
4.	Pasta	PBW 233, WH 896, HI 8498, HD 4672, Raj 1555, NP 200, *etc.*

5. Breeding for Biotic Stress Resistance

Rust is the primary disease in wheat. In India, three types of rust are found, *viz.*, Black rust (stem rust), Brown rust (leaf rust) and yellow rust (strip rust). Among these, stem rust and leaf rust are important. Since the rust has different races, the horizontal resistance and multiline breeding approaches are effective. Biotechnological tool marker-assisted backcross breeding is rewarding for developing cultivars by introducing the multi genes for disease resistance in wheat. In India, the cultivars developed *via* resistance genes introgressions with the help of markers-assisted breeding are mentioned in Table 5.4.

Table 5.4. Improved Varieties and Introgressed Resistance Genes

Sl.No.	*Improved Variety*	*Resistance Genes Introgressed*	*Traits Improved*
1.	PBW 723 (Unnat PBW 343)	Lr57/Yr40+Lr37/Yr17	Brown and Yellow rust and Karnal bunt
2.	Unnat PBW 550	Yr15	Foliage resistance to rust
3.	PBW 771	Lr57/Yr40	Brown and Yellow rust
4.	PBW 752	Yr10/Yr36	Yellow rust

6. Breeding for Abiotic Stress Resistance

The yield and quality of wheat decline significantly due to the adverse effect of different abiotic stress. The variety developing may give resistance against different stress like heat, drought, water stress, *etc.* Abiotic stress is merely linked with physiological traits, so the combination of physiological characteristics improvement with favourable alleles provides more opportunity for abiotic stress resistance in wheat.

Breeding Approaches (Methods)

1. **Introduction and germplasm collection:** Introduction of Semi-dwarf wheat from Mexico, *i.e.*, Sonara 63, Sonara 64, Mayo 64, and Lerma Roja 64A.
2. **Pure line and mass selection:** Earlier varieties like P_4, P_6, and P_{12} evolved at Pusa Institute by pure line selection from the local population.
3. **Hybridization selection methods:** These comprises pedigree selection, bulk selection, shuttle breeding, *etc.* In mid-1960, the Indian Scientist **Dr. M. S. Swaminathan** played a crucial role in the wheat breeding program (he used the shuttle breeding programme to evaluate received material from Dr. Borlaug), leading to the release of a popular variety called *Sonara* 64 and Lerma Rojo 64. These were utilized in breeding programmes, and amber colour wheat varieties like Kalyan Sona, Safed Lerma, and Sharbati Sonara were released; these are double-gene dwarfs. Nowadays, in all breeding programmes, these lines or derivatives are used as parental material for improving wheat in India.
4. **Back cross method of breeding:** Mainly used for developing numbers of Isogenic lines (NILs) for developing disease resistance. *e.g.*, rust
5. **Heterosis and hybrid wheat:** The male sterility line was identified by the interaction between the nuclear gene of *T. aestivum* and the cytoplasm from *T. timopheevii*. It is also induced by the use of CHA to produce hybrid wheat. However, limited success could be achieved due to its strictly self-pollinated nature, complex hybridization system, limited heterotic cross, low level of heterosis, high cost of seed production due to lower seed multiplication rate, and poor agronomic performance, *etc.*
6. **Mutation breeding:** Dr. M. S. Swaminathan did extensive breeding work on mutagenic agents, *i.e.*, Gamma rays. He used umber colour grain mutant Lerma Rojo and Sonara-64 varieties as parental material and resultant synthesised variety Pusa Lerma and Sharbati Sonara, respectively.
7. **Development of multiline for disease resistance:** Multiline is a line developed by composite (mixing) of isogenic lines or near-isogenic lines (NILs). **N. E. Borlaug, 1958** presented a systematic approach to multiline variety development using near isogeneic lines for stem rust resistance in wheat. In India, the sonalika multiline variety *viz.*, MLSKA-9 and MLSKA-12 were developed by mixing six and nine isogenic lines, respectively. Each line gives resistance against a particular race of disease.
8. **Biotechnological approaches:** Various biotechnological tools have been utilised to improve yield and resistance breeding.
   - ✰ *Invitro* haploid production, *i.e.*, Androgenesis and Gynogenesis;
   - ✰ Somaclonal variation for improvement of different morphological traits, *viz.*, plant height, awn length, leaf shape, seed shape, size, colour, *etc.*,

- ✰ Genomics Assisted Breeding
  - § Marker-assisted breeding molecular markers techniques-effectively used for tagging the major genes.
  - § Marker-assisted backcross breeding (MABB) - Effective means to accumulate the alleles for the various disease resistance trait.
- ✰ Targeted Mutagenesis: ZFN (Zinc Finger Nuclease,) TALEN (Transcriptional Activator like Effector–Nuclease), ODM (Oligonucleotide Directed Mutagenesis) and MGN (Meganuclease) techniques
- ✰ Genomic Selection: A promising breeding approach to hasten genetic gain in crop plants.
- ✰ Speed Breeding: promising technology for faster and better phenotyping and also fit well with multiple traits selection.
- ✰ Genome editing techniques- CRISPR/Cas9 system.

Breeding Institutes

- ✰ **International level:** International Maize and Wheat Improvement Centre (CIMMYT) -Mexico, USA.
- ✰ **National level**: Indian Institute of Wheat and Barley Research (IIWBR) Karnal, India.
- ✰ **State level:** Regional Research Stations-State Agricultural Universities.

Breeding Achievements

A. Bread Wheat

Timely Sown	*Late Sown*
GW 273, GW 496, GW 322, GW 366, WH-542, HD2687, HD 2781, HD 2967, PBW 343, PBW-502, GW 451, HD 3226, DBW 17, DBW 222 (Karan Narendra), DBW 252 (Karan Shariya), DBW 168, DBW 93, DBW 303, VL Gehun 3010, PBW 872, PBW 826, Pusa Wheat 3369, Unnat HD 2967, Pusa Harsha, Badshah, Karan Bold, Karan Shivangi, GW 513, GW543, GW 547, Pusa Gehun Sharbati, *etc.*	Lok-1, GW 190, GW 173, GW-11, GW 499, DBW 173, DBW 107, DBW 90, DBW 71, *etc.*

B. Durum Wheat

Raj-1555, HI 8498, GW 1139, GDW 1255, GW 1339, PDW-233, WH-896, HD 4728, DDW 47, HI 8713, HI 8759 (Pusa tejas), HI 1605 (Pusa Ujala), Pusa Kirti, Pusa Poshtik, DDW55, Krishidhara Surya, Pusa Gehun Gaurav, *etc.*

Chapter 6

Breeding Rice (*Oryza sativa* L.)

Introduction

Rice is considered the World's most important staple-food crop, feeding more than half the World's population. While in India, it serves as staple food for about 65 per cent of the population. In India, it is grown under a wide range of climatic conditions. Globally, India is the top in terms of area and second in terms of production. Rice was domesticated around 15000-10000 BC. Presently, two cultivated species of the genus *Oryza viz., O. sativa* and *O. glaberrima*. The species *O. sativa* cultivated across Asia is also known as Asian cultivated rice, while *O. glaberrima* is mainly cultivated in western Africa, called African rice.

Background Information

1.	*Botanical Name*	Two cultivated species **Asian Rice:** *Oryza sativa* **African rice:** *O. glaberrima*
2.	*Family*	*Poaceae*
3.	*Chromosome number*	**Diploid:** 2n=2x=24
4.	*Centre of origin*	**Asian Rice:** South East Asia **African rice:** South Africa
5.	*Nature of pollination*	Self-pollination

6.	*Pollination mechanisms*	Bisexuality, Cleistogamy/chasmogamy, *etc.*
7.	*List of wild relatives*	25 wild species: *O. rufipogon, O. nivara, O. breviligulata, O. longistaminata, O. officinalis, O. latifolia, etc.*Representing 11 genomes, namely AA, BB, CC, BBCC, CCDD, EE, FF, GG, HHJJ, HHKK, and KKLL

Classification of *Oryza* Species based on Ploidy Levels and Genomes

Ploidy Level	*Species*	*Genome*
Diploid (2n=2x =24)	*Oryza sativa* (Cultivated)	AA
	O. glaberrima (Cultivated)	AA
	O. nivara	AA
	O. rufipogon	AA
	O. longistaminata	AA
	O. punctata	BB
	O. officinalis, O. rhizomatis and *O. eichingeri*	CC
	O. australiensis	EE
	O. brachyantha	FF
	O. granulate and *O. meyeriana*	GG
Tetraploid (2n=4x =48)	*O. minuta* and *O. schweinfurthiana*	BBCC
	O. alta, O. latifolia and *O. grandiglumis*	CCDD
	O. longiglumis and *O. ridleyi*	HHJJ
	O. schlechteri	HHKK
	O. coarctata	KKLL

A Comparison of Essential Characteristics of Two Forms of Cultivated Rice *viz., O. sativa* (Asian rice) and *O. glaberrima* (African rice)

Sl.No.	*Characteristics*	*Forms of Rice*	
		O. sativa	*O. glaberrima*
1.	Distribution	Across the Globe	Limited to West Africa
2.	Ecotype	Many	Limited
3.	Habit	Basically perennial	Annual
4.	Numbers of branches per panicle	Many	Few
5.	Variability among varieties	High	Limited
6.	Ligule	Soft and long	Hard and short

Gene Pool of Rice

A gene pool of rice is grouped into three categories

Sl.No.	Gene Pool	Species
1.	Gene pool 1 (GP-1)	Comprises the species of *sativa* group
2.	Gene pool 2 (GP-2)	Comprises the species of *Officinalis* group
3.	Gene pool 3 (GP-3)	Includes *Meyeriana, Ridleyi* and *Schlechteri* groups

Distribution

Rice is one of the oldest cultivated crops in India and China. *Oryza* originated ~130 million years before and spread as a wild grass in the past Gondwana land. It disseminated across different areas, *viz.*, Asia, Africa, Australia and America, after the interior drift divided the entire area into these continents. While it was domesticated around 15000-10000 BC. It is a warm-season crop grown extensively in the World's humid tropical and subtropical regions. The major rice-producing countries are China, India, Japan, Korea, Pakistan, Bangladesh and other South-East Asian countries. In India, the leading states are West Bengal, Orissa, U.P, A.P, Karnataka, Tamil Nadu, Panjab, *etc.*

Floral Biology, Selfing and Crossing Techniques

Inflorescence	:	Racemose type is commonly known as panicles (conical shaped). It is born on the uppermost internode of the culm. Partially or fully covered with the flag leaf. The central axis of the panicle is known as a rachis that bears ten or more primary branches. The rachis of a panicle is slightly angular, more or less glabrous, and somewhat hairy at the nodes.
Spikelets-Flower	:	The flower produced on rachis is known as spikelets. The spikelets are born singly at the branches' end or in a cluster of 2 to7. The spikelets number on the panicles varies from 50-500. Spikelet is born on the pedicel (spikelet axis)-pedicelled, generally single. The pedicel apex is enlarged and oblique, termed a facet. On a facet, two glumes occur, one outer and another inner. The flower is bracteate and consists of two large interlocking glumes, a boat-shaped lemma and opposite side palea forming a floret. A flower is zygomorphic, sessile, incomplete (due to perianth), perfect and hypogynous.
Perianth	:	Flower lacking sepals, inside the palea, two broad, thick and fleshy lodicules are present in the whorls. It plays an essential role in the opening of a floret.

Androecium	:	Six stamens arrange in two regular whorls inside the lodicules, anther-dithecous, versatile, slender and free filaments long. Pollen grains are viable only for 5 minutes.
Gynoecium	:	Single carpels; one superior ovary with two long styles, single ovule and basal placentation; bifurcated feathery stigma. Stigma remains receptive for 3–7 days.
Fruits	:	Caryopsis. The mature grain remains bounded in the lemma and palea
Seed	:	Albuminous or monocots.
Floral formula	:	Br. ⊕ ⚥ $P_{2\ (Lodicules)}\ A_{3+3}\ \underline{G}_{(1)}$

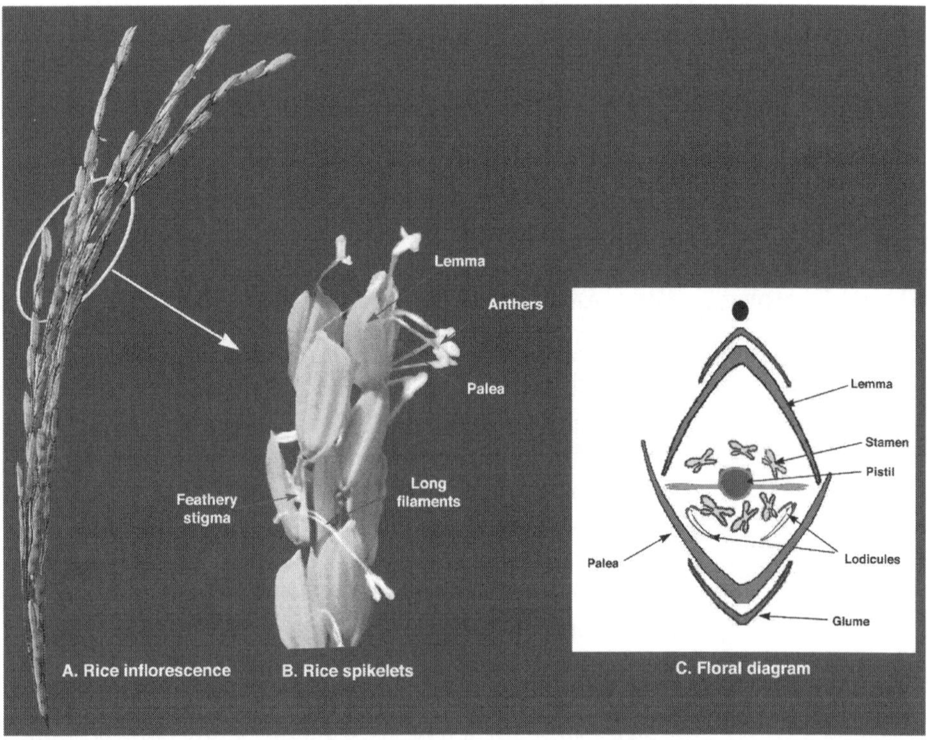

Figure 6.1. Floral Biology and Diagram of Rice (*Oryza sativa* L.).

Anthesis: Anther dehiscence occurs just before or at the time of the opening of the lemma and palea. Anthesis starts in the spikelets of primary branches at the tip of the panicle. Then, it proceeds basipetal towards the bottom of the panicle. Around 7–10 days are required to complete the anthesis of all the spikelets within a panicle. However, it may vary according to weather conditions also. The time of anthesis begins at 8:00 AM and lasts up to 01:00 PM.

Selfing: The spike is to be covered with a butter paper bag before flowering to avoid outcrossing and then labelled.

Emasculation and Crossing: Panicles are selected for emasculation in the second or third days of blooming. Emasculation is finished one day before the commencement of anthesis in the evening. Sometimes it is done early in the morning, about 1-2 hrs before anthesis. Clipping one-third of portions of the desired female parent's top and bottom parts of the panicle using scissors, leaving middle spikelets. In the middle spikelets, the top one-third in each spikelet is clipped off in a slanting position. Emasculation is done by removing six anthers in each spikelet with the help of the forceps/needles. Take care that stigmas should not be injured. The emasculated spikelets are then covered with a butter/parchment paper bag to avoid contamination with the undesired pollen. For effective pollination, the pollen is collected next day (at about 9:00 AM) from the bloomed panicle of the desired male parent. The top portion of the paper bag covering the emasculated panicle in the female parent is cut to expose the panicle. The male parent panicle is inserted in an inverted position into the butter paper bag and turned in both ways to scatter the pollen. After ensuring the abundant pollen distribution, begging with butter paper, and tagging is done with labels.

Origin of Cultivated Rice

There are two concepts about the origin of cultivated rice

1. Polyphyletic origin and 2. Monophyletic origin.

1. **Polyphyletic**: According to this concept, the two cultivated species of rice, *viz.,* Asian rice (*O. sativa*) and African rice (*O. glaberrimal*), have originated individually in their corresponding regions from different species.
2. **Monophyletic:** Based on the monophyletic concept, both cultivated species of rice, *viz.,* Asian rice (*O. sativa*) and African rice (*O. glaberrimal*), originated from the common AA genome ancestor, *O. perennis* in South East Asia and African rice in the upper valley of Niger River in Africa (Western Africa), respectively. This concept is widely accepted among modern rice workers since both species are similar except for glume pubescence, ligules size and colour of a pericarp, *i.e.,* red in African rice. The evolutionary pathway is described in figure 6.2.

Classification of the Cultivated Asian Rice (*Oryza sativa*)

Based on morphological and physiological features, distribution, and geographical adaptation, the cultivated Asian rice (*Oryza* sativa) is differentiated into three subspecies or eco-geographical races, *viz.,* i) *Indica,* ii) *Javanica,* and iii) *Japonica.*

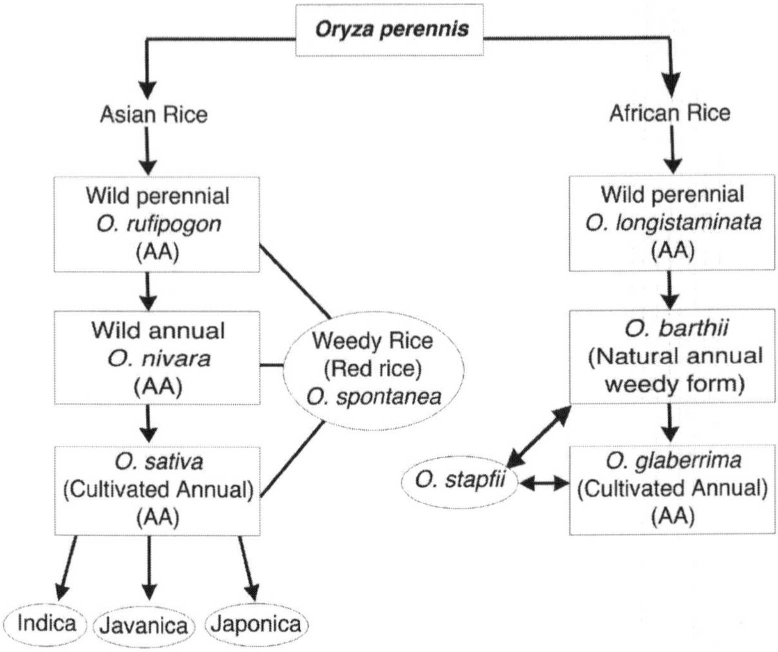

Figure 6.2. Evolutionary Pathway of Asian and African Cultivated Rice Species.

The characteristics of each subspecies/race are presented in the Table 6.1.

Table 6.1. Characteristics of Three Subspecies or Races of Asian Rice (*Oryza sativa*)

Sl.No.	*Characteristics*	*Sub-Species or Races*		
		Indica	*Javanica*	*Japonica*
1.	Geographical adaptation type	Tropical	Intermediate (Equatorial)	Warm temperate to sub-tropical type
2.	Tillering habit	Medium	Few	Many
3.	Plant height	Tall	Tall	Dwarf
4.	Foliage colour	Light green	Light green	Dark green
5.	Lodging	Susceptible	Resistant	Resistant
6.	Flag leaf	Long and narrow	Wide and long	Narrow and short
7.	Awns	Less prominent	More prominent	Absent
8.	Grain shape	Narrow and long	Round- bold and short	Roundish and Short
9.	Grain texture	Non-sticky	Intermediate	Sticky
10.	Seed dormancy	Flexible	Absent	Present

Sl.No.	Characteristics	Sub-Species or Races		
		Indica	Javanica	Japonica
11.	Grain shattering	Easily	Non-shattering	Non-shattering
12.	Threshing	Free	Hard	Hard
13.	Drought response	Resistant	Sensitive	Intermediate
14.	Low temperature	Sensitive	Tolerant	Tolerant
15.	Photoperiod	Very sensitive	Non-sensitive	Non-sensitive

Classification of Asian Rice Varieties

According to Glaszmann (1987), the Asian rice varieties grouped into six genetically distinct groups

Sl.No.	Group	Remarks
1.	Group I	Varieties include typical indicas
2.	Group II	Upland rice, also called *aus*
3.	Group III and IV	Deep water rices belong to India and Bangladesh regions
4.	Group V	Aromatic rices from Indian sub-continent
5.	Group VI	Varieties include *japonicas* and *javanicas*

The Birth of Semi-Dwarf Rice: A Fundamental Basis of a Green Revolution

The variety of key players in the genesis of semi-dwarf rice varieties was a Chinese variety called 'Dee-geo-woo-gen' (DGWG), which contained a recessive gene '*sd1*' responsible for dwarfness in rice. The World's first semi-dwarf japonica rice variety, Taichung Native 1 (TN1), was developed in Taiwan in 1949-1956 using Dee-geo-woo-gen (DGWG), parent. In India, it was introduced in 1960.

The most popular semi-dwarf rice variety IR 8, was developed in 1962 by Peter Jennings, Henry Beachell and S.K. De Datta at International Rice Research Institute (IRRI), Philippines. It was a cross derivative of *Dee-geo-woo-gen* (a Chinese dwarf variety having a stiff strawed) and an Indonesian tall and vigorous variety called *Peta* (Figure 6.3). In India, this variety was introduced in 1966. It became trendy among farmers due to many adaptable attributes, such as high yield potential, semi-dwarf in nature, high fertilizer responsiveness, lodging resistance, and photoperiod insensitivity. Hence, it is popularly termed 'miracle rice'. Both TN 1 and IR 8 and many new semi-dwarf varieties played a significant role in a green revolution.

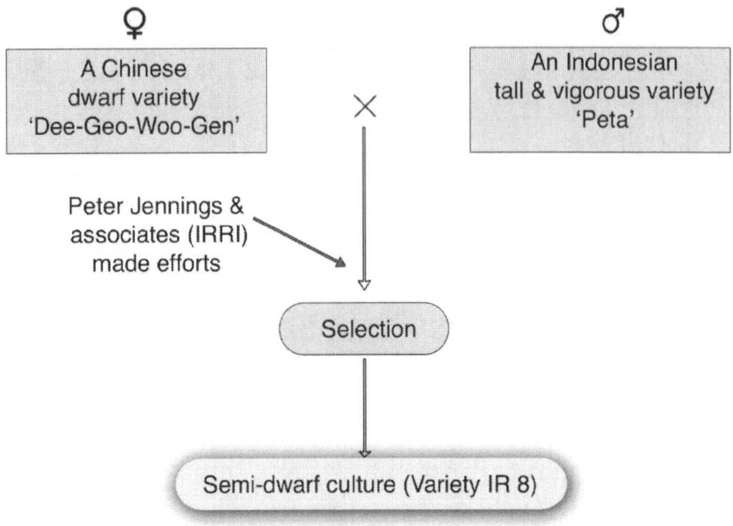

Figure 6.3: Birth of Semi-Dwarf Rice (*i.e.,* IR 8).

Breeding Objectives

The main common breeding objectives of rice breeding are high yield, biotic and abiotic stress resistance and quality improvement/biofortification. Some objectives may according to region-specific and as per consumer-market demand. The essential objectives are briefly discussed below:

1. **High yield ability:** Yielding ability of rice may region specific, and it is controlled by various component traits, *viz.,* length of panicle, grains per spike, number of productive tillers per plant, 1000-grain weight, *etc.*
2. **Breed for stable and wide adaptable variety:** Rice is grown in varying climate conditions in India. Breeding genotypes for high adaptability and stable yield potential over environments is important to maintain yielding ability in many areas.
3. **Breeding for earliness:** The maturity duration may vary according to variety, *i.e.,* early, medium and long: The earliness is ideal that can fit into a multi-cropping system.
4. **Breeding for lodging and shattering resistance:** Both lodging and grain-shattering characters are complex and governed/affected by multiple factors. The ideal plant, like short stature, thick and strong clum, and narrow internode, can serve better to overcome lodging in rice. In the case of shattering, care should be taken while breeding with wild relatives.
5. **Breeding for resistance to biotic and abiotic stress:** The rice crop experienced a number of biotic and abiotic stresses. Some important

biotic stresses are a blast, stem rot, bacterial blight, tungro, sheath blight, stem borer, green leaf hopper, gall midge, brown plant hopper *etc.* while abiotic stress are temperature, salinity, alkalinity, water stress, iron, zinc deficiency, *etc.* Several donors are available for biotic and abiotic stress resistance/tolerance are described in Table 6.2.

Table 6.2. List of Donors of different Biotic and Abiotic Stresses in Rice

Stresses		*Donors*
A. Biotic stresses		
Diseases	Bacterial blight	Xa-4, BJ 1, Sayapha, Boro II, Java 14, DV 85 and TKM-6
	Blast	Ram-tulsi, Tetep, Dawan, Co 4, Carreon, and M-302.
	Rice tungro virus	Ambemohor-159, ARC 10599, ARC 11554, Ptb-18, Ptb-21, ARC 12981, ARC 14320, ARC 14766, IR-20 and Kamod
	Grassy Stunt	IR-28, IR-29, IR-30, and IR-34
Pests	Brown plant hopper	ARC 10313, ARC 6006, ARC 7012, Ptb 8, Babawee, Rathu Henati
	Gall midge	PTB-10, Ptb-18, PTB-21
	Stem Borer	TKM-6, ADT 2, ASD 10, ARC 10598, ARC 10257, Ptb 18, TKM 6, MTU-15, TKM-3, and Ratna
B. Abiotic stresses		
Drought		BR 23, Chanda 2, DJ 5, EB 17, Dhani, Broom gora, Patang, *etc.*
Salinity		Nonasail, Patnai 23, Pokkali, Chettivirappa
Submergence		Boyan, Boku, Bayyahundee, Hb Dw8

6. **Breeding for high grain quality:** The grain quality can be determined by various quality parameters, *viz.*,
   a) Grain shape, size, colour and translucency.
   b) The texture of endosperm and quality of starch in the endosperm.
   c) Aroma and Cooking quality (Basmati type): In rice, the fragrance is governed by two genes *viz.*, *BADH1* and *BADH2*.
   d) Amylose content: A gene Waxy/GBSSI was identified that governs amylose content in rice.
   e) Hulling percentage (per cent): Proportion of de-husked rice grains (brown rice) after de-husking.
   f) Milling quality: Milling refers to a process of removing husk as well as the bran from the rice grains.
   g) Milling recovery: Recovery of unbroken white rice (head rice) determined in percentage.

h) Biofortification is a strategy to improve nutritional values: Increasing bioavailable micronutrient content of food crops through genetic selection *via* various plant breeding approaches. Recent progress in conventional plant breeding has given importance to fortifying essential micronutrients, vitamins, and antioxidants. In many cases, the biotechnological approach is also fruitful in enriching the nutritional level. The best example is golden rice development for reducing vitamin A deficiency. A rice was also biofortified by improving protein quantity and quality with two amino acids, lysine and threonine–Glycemic index, Fe, Zn, lysine, protein content, *etc.* Various genes linked with Fe and Zn content rice grains have been identified *viz.*, OsFER1, OsNRAMP4, OsYSL4, OsNRAMP5, OsYSL6, OsNRAMP6, OsYSL12, OsZIP8 for Fe content and OsZIP10. Whereas various ZIP proteins, *viz.*, OsZIP1, OsIRT1, OsIRT2, OsZIP3, OsZIP4, OsZIP5, OsZIP7 and OsZIP8 for Zn content. This information is vital for advancing breeding programs for biofortification in rice. The various biofortified genotypes/verities identified are listed in Table 6.3.

Table 6.3. List of Important Biofortified Genotypes/Varieties Released in India

Sl.No.	*Name of Cultivar/Hybrid*	*Improved Quality Traits*
1.	DRR Dhan 45	22.6 ppm Zn
2.	DRR Dhan 48	24.0 ppm Zn
3.	DRR Dhan 49	25.2 ppm Zn
4.	CR Dhan 315	24.9 ppm Zn
5.	CRR Dhan 310	10.3 per cent protein
6.	CRR Dhan 311	20.1 ppm Zn and 10.3 per cent protein
7.	Surabhi	22.8 ppm Zn
8.	CGZR 1	22.0 ppm Zn
9.	CGZR 2	25.0 ppm Zn
10.	HPR 2720	20.9 ppm Zn

7. **Breeding for alternate sources of dwarfing genes:** In the past century, the quantum jumps of yield were achieved due to the synthesis of semi-dwarf varieties through the incorporation of *sd1* gene that aided in refining the plant architecture and the harvest index. Now there is a yield plateau and susceptibility of biotic and abiotic stresses by continuous use of a mono source of dwarfing gene. So, it needs to break the yield plateau; more diversification and incorporation of biotic and abiotic stress resistance requires an alternate source of dwarfing gene.
8. **Breeding varieties suitable for export – scented/basmati rice:** In India, basmati rice (Basmati 370) is exported. The basmati varieties are

grown in a limited area, *i.e.*, valleys of the Himalayan range only have export quality compared to other parts of the country. Need to identify/ breed varieties grown in other parts of the country with comparable export quality.

An essential feature of basmati rice

- ✰ It is tall, with light green, dropping leaves, and a weak stem.
- ✰ The traditional varieties are low yielding, photosensitive, long duration and lodging susceptible.
- ✰ Grain is a long slender type, banana shape (slightly twisted toward the tip).
- ✰ Intermediate amylase content (20-22 per cent), intense aroma, low gel consistency, and intermediate gelatinization temperature.
- ✰ After cooking, kernels remain uniform in shape and free from chalkiness.
- ✰ Cooked kernels are sweet, have no bursting, and have an elongation ratio of nearly twice.
- ✰ The cooked kernel is neither too hard nor too soft.

9. **Breeding new plant type (NTP):** The developed plant with ideal traits that directly accelerate the yielding ability. Janoria (1989) was hypothesized the important morphological characteristics of ideotype, *viz.*, tall height, lodging resistance, erect growth habit, fewer but all effective tillers, stiff leaves, heavy panicles, deep broadly branched roots, *etc.*
10. **Breeding super rice:** The super rice variety can produce stable and high yields against minimum agronomical inputs. The variety has high nitrogen use efficiency and is more stable under rainfed conditions, moderate tillering capacity, narrow, erect, V-shaped and thick leaves, heavy and drooping panicles at maturity, drought resistance, and multiple diseases and pest resistances. The super rice breeding programme was initiated by Longping Yuan (1998) in china.
11. **Breeding for the attainment of more heterosis - hybrid:** The hybrid technology in rice is becoming viable and popular in China. The scientist **Yuan Long Ping** had done tremendous work on hybrid rice. At the end of 1990, hybrid rice technology gained popularity in China. Yuan L. Ping was honoured as the Father of Hybrid Rice for his significant contribution to rice hybrid technology. He had shared the 2004 world Food Prize with Dr. Monty Jones, an African rice breeder from the West African Rice Development Centre. In India, systematic research on hybrid rice was initiated in 1989. Though, hybrid technology in rice is not much popular in India due to the following reasons;

- Costly hybrid seed production due to spray of Gibberellic Acid and other inputs.
- Absence of sound seed production technology that increases hybrid seed yield.
- Non- synchronization of male and female flowering.
- Poor grain quality as compared to traditional rice varieties. So poor response by consumers.
- Hybrid varieties failed to satisfy the yield expectation of farmers (*i.e.*, 30-40 per cent).
- Lack of comprehensive extension activities that popularize hybrid technology among farmers.

Breeding Approaches (Methods)

1. **Introduction:** It is the simplest, fastest and oldest method of rice breeding. Various varieties or lines were introduced from Taiwan, *i.e.*, Taichung Native-1, Taichung 65, Kalimpog 1. The lines Dee-Gee-Woo-Gen and IR-8, Palman 579 from IRRI, Philippines, indica/japonica derived Mahsuri from Malaysia. CH 45 and CH 9008 from china; Norin- 18 and Norin 8 from Japan. Various outstanding genotypes/varieties, *viz.*, Jaya, Rasi, Swarna and Padma were also introduced from India to several countries in the World.
2. **Pure line selection:** Pureline selection is the most common approach used to improve self-pollinated crops. Number of varieties were developed, *viz.*, CO1, CO4, Janaki, Kolhapur, T3, Taraori Basmati, GEB 24, Mozhagolukulu (BCP1), Basmati 370, Type 3, Mtu 2, Ar1, S 155, BR3, T1, T3, T 29, Ptb 33, Co-43. T 141, Sabita, Jitendra, AU-1, Terua, Mangala Mahsuri, Ranbir Basmati, *etc.*
3. **Hybridization followed by selection:** It includes different approaches *viz.*, pedigree method of selection, bulk method, SSD, and back cross methods are used in rice breeding. The pedigree method is one of the most important approaches used in rice improvement and varietal development. In 1940, CO14 was the first variety developed by this approach from the cross, CO3 × T275. CO15 (Jada Molakolukulu) was synthesized from the cross between GEB24 × ADT10 with a blast-resistant trait. The varieties were developed between inter varietal crosses, *viz.*, Ponni (Mashuri) (Taichung 65 × ME 80), ADT 27 (Norin 10 × GEB 24), *etc.* In the case of inter-specific crosses Co 31 (*O.perennis* × GEB 24) had drought resistance traits. So far large number of varieties developed by hybridization between tall indica varieties and a semi-dwarf rice variety.
4. **Back cross method of breeding:** A back cross breeding approach is used in different ways in rice breeding programs like transfer of specific gene/s for disease-pest resistance, transfer of male sterility into a new

line, development of RIL and NIL lines, also importantly used in marker-assisted backcross breeding in improving varieties with specific traits.

5. **Mutation breeding:** In the initial era of the rice improvement programme, mutation breeding played a significant role. A mutant is identified from either natural mutation or induced mutation. Such mutant lines are either directly identified as a variety or dominantly used as a parental line in rice improvement programs. For example, a milestone variety GEB24 a spontaneous mutant was identified from Konamani. Likewise, Shyama from Kalimounch, Ranbir Basmati and ADT 41 is a mutant of Basmati 370.

   In the case of induced mutation, several varieties have been identified; some examples; Jagannath from T141; Radhi (CRM-40) from Swarnaprabha; Remanica (MO-15) from MO1; Early Samba from Mutant of BPT5204; Anashwara (PTB 58) from PTB 20; Malaviya Sugandh 4-3 from Lanjhi, *etc.*

6. **Heterosis breeding (Hybrid Rice):** The heterosis was first time reported in rice by Jones (1926) at USA and Ramaiah (1933). Whereas Sampath and Mohanty in 1954 reported the role of cytoplasm in causing malc sterility in rice and its possible usefulness in breeding hybrid rice.

There are three breeding techniques involved in developing a hybrid rice

i. Three-line method (Cytoplasmic Genetic Male Sterility-CGMS)
ii. Two-line method [Environmental sensitive genic (nuclear) male sterility –EGMS]
iii. One-line system (Apomixis)

i. Three-line Method (Cytoplasmic Genetic Male Sterility)

In 1970, after the discovery of wild abortive (WA) source of cytoplasmic male sterility (CMS) came a breakthrough in hybrid breeding research. In the year 1977, China released the first commercial rice hybrid, Nanyou 2. Later, a number of CMS sources were identified (Table 6.4); however, WA cytoplasmic male sterility systems have been found to be most stable in China and other parts of the World since producing nearly complete male sterility. The WA-CMS is the most widely used CMS system in hybrid rice production. Across the World, many hybrids have been released.

Table 6.4: List of Diverse Cytoplasm that Induce Male Sterility and their Sources in Rice

Sl.No.	*Name of Cytoplasms*	*Sources*
1.	BT-CMS	Chinsurah Boro II (indica variety- *Oryza sativa*)
2.	LD-CMS	Lead Rice (*Burmese indica* variety)
3.	WA (Wild Abortive)-CMS*	*Oryza sativa* sp. *spontaneae*

Sl.No.	Name of Cytoplasms	Sources
4.	HL-CMS; CW-CMS; FA-CMS; W18-CMS	*O. rufipogon*
5.	CMS-GA	Gambiaca
6.	Kalinga-I, Lalruma*	*O. sativa*
7.	CMS-D	Dianyu 1 A
8.	K-CMS	K52 (*japonica*)
9.	IR66707A	*O. perennis*
10.	Y-CMS	Yegong (*indica* landrace)
11.	LX-CMS	Luihui rice (*indica*)
12.	ID-CMS	Indonesia paddy rice (*indica*)
13.	GA-CMS	Gambiaca (*indica*)

* Used commercially

In India, the hybrid rice research program was initiated in 1989. The APHR 1 was the first hybrid developed at APRRI, Maruteru, for Andhra Pradesh state, followed by MGR 1, KRH 1 and APHR 2 hybrids developed and released commercially in the year 1994. So far, more than 127 hybrids have been released commercially by the public and private sectors. The important hybrids released in the last two decades is listed in Table 6.5.

Table 6.5. List of Rice Hybrid Varieties Released in India

Name of Hybrids	Cytoplasm Types	Released year
Pusa RH 10,	WA (Wild Abortive)-CMS	2001
Narendra Usar Sankar Dhan 3	WA-CMS	2004
Ajay (CRHR 7), DRRH 2, Sahyadri 2, Sahyadri 3	WA-CMS	2005
Rajalaxmi (CRHR 5)	Kalinga I CMS	2005
CORH-3, HKRH-1	WA-CMS	2006
JRH-4, JRH-5	WA-CMS	2007
Sahyadri – 4	WA-CMS	2008
DRH - 775	WA-CMS	2009
DRRH- 3	WA-CMS	2010
CO–RH-4	WA-CMS	2011
PNPH 24, CR Dhan 701 (CRHR 32)	WA-CMS	2012
KPH 272, NPH 8899	WA-CMS	2017

ii. Two-Line Method [Environmental sensitive genic (nuclear) male sterility -EGMS]

The expression of a genetic male sterility system is also altered due to environmental changes. For instance, changes in temperature or photoperiods directly affect male sterility. This system was first reported in rice and has been well characterized. It is a relatively simple genetic system usually governed by one or two recessive genes. The male sterility expression is more likely to depend on genetic constitution and environmental conditions. This type of male sterility usually originates from a spontaneous mutation in origin.

It has been used in a hybrid seed production system as a two-line breeding approach in rice. Hybrid rice varieties developed by this system have several advantages over the conventional CMS system.

- ☆ The choice of parents for developing heterotic hybrids is significantly broadened.
- ☆ Avoid adverse effects due to sterile cytoplasm.
- ☆ No need for a maintainer line, so multiplication is relatively easy.

It is of two types:

A. **Thermo-Sensitive Genetic Male Sterility (TSGMS):** A range of temperatures markedly influences male sterility is called thermo-sensitive genetic male sterility (TSGMS). In rice, it was first discovered in 1997 in a japonica rice species as a spontaneous mutant [AnnongS-1 (Ans-1)]. In rice, male sterility reverses when a temperature goes below 23°C, and hence male-sterile plant becomes fertile.

B. **Photoperiod-Sensitive Genetic Male Sterility (PSGMS):** A male sterility regulated by photoperiod (day length) is called photoperiod-sensitive genetic male sterility (PSGMS). It was first reported in 1973 in a japonica rice species as a spontaneous mutant [Nongken 58S (NK58S)]. Based on the report, the rice plant remains male-sterile when day length is more extended than 13.75 hours (long day condition) during the anther formation. However, plants become partially or completely fertile when a day length is shorter than 13.50 hours (short day condition).

Complete male sterility is somewhat encouraged through high temperatures (*i.e.*, ~29°C) with long-day conditions. When male sterility responds in dual ways, *i.e.*, based on photoperiod and temperature sensitivity, it is termed photoperiod-temperature-sensitive genetic male sterility (PTGMS)

Table 6.6. List of different Environment-Sensitive Genetic Male Sterility (EGMS) According to their Cytoplasm Types and Source in Rice

Sl.No.	*Types of Cytoplasm*	*Source*
1.	TGMS	Spontaneous mutants of Annong S-1 and Zhu 1S
2.	PGMS	Mutant of *japonica* cultivar Nongken 58
3.	P/TGMS	Nongken 58S
4.	rPGMS	Mutant of *japonica* cultivar 9522

iii. One-Line Method

In one line system of rice hybrid breeding can be developed through:

- Vegetative propagation –Fixation of heterosis.
- Apomictic lines- Fixation of heterosis.
- Tissue culture means *i.e.*, micropropagation, anther culture.

Among the above techniques, apomictic lines and anther culture are cost-effective for large-scale cultivation.

7. Biotechnological Approaches

Rice is gifted with rich of genetic and genomic variability; hence a large number of biotechnological works have been done successfully for yield, stress and quality improvements, *viz.*,

- **Marker-assisted selection (MAS) and Marker-assisted backcross breeding; gene pyramiding:** In Rice, accessibility of genes, a gene linked or gene-based markers and donor germplasm permit the use of marker-assisted breeding. These techniques enable the transfer of specific desirable gene(s)/QTLs of important traits, *viz.*, resistance to diseases, pests, grain quality, and tolerance to drought, salinity, low phosphorus and herbicide, into popular rice varieties. In India, more than 35 improved rice varieties developed through marker-assisted breeding. Among major bacterial blight resistance and blast resistance. The list of some important varieties developed through MAS breeding is listed in Table 6.7.
- **Transgenic technology** – Used for biotic and abiotic stress resistance. Quality improvement - Golden rice- high beta carotene, a source of vitamin A.
- Genomic Selection and Genome editing tools are being used to improve plant architecture concerning yield and agro-morphological, resistance against various stresses and quality parameters *e.g.*, DRR Dhan 100, Pusa DST Rice 1.

Table 6.7. List of Improved Rice Varieties with Targeted Genes of Resistance and Related Traits

Sl.No.	Name of Improved Variety	Name of Traits with Incorporated Genes
1.	Improved Samba Mahsuri	Bacterial blight (*xa5* + *xa13* + *Xa21*)
2.	Swarna sub1	Submergence (Sub1)
3.	Improved Pusa basmati I; Punjab Basmati 3; Punjab Basmati 4; Punjab Basmati 5	Bacterial blight (*xa13* + *Xa21*)
4.	K01, K04	Low-amylose content gene (*Wx-mq*)
5.	Pusa RH10	Rice blast (*Piz5* + *Pi54*)
6.	Improved Tapaswini	Bacterial blight (*Xa4*+*xa5* + *xa13* + *Xa21*)
7.	PR127	Bacterial blight (*Xa45*(t)
8.	DRR Dhan 59	Bacterial blight (*Xa33*)
9.	Pusa Samba 1850	Rice blast (*Pi1*+ *Pi54* +*Pita*)
10.	Pusa Basmati 1847; Pusa Basmati 1885 and Pusa Basmati 1886	Bacterial blight and Rice blast (*xa13* + *Xa21*+ *Pi2*+*Pi54*)
11.	Pusa Basmati 1979 and Pusa Basmati 1985	Herbicide tolerance (Mutant AHAS allele)

Breeding Institutes

- **International Level:** International Rice Research Institute (IRRI), Philippines.
- **National Level**
  - i National Rice Research Institute (NRRI), Cuttack, Odisha.
  - ii Indian Institute of Rice Research (IIRR), Hyderabad.

    [The All India Coordinated Rice Improvement Project (AICRIP) was established in 1965].
- **State Level:** Regional Research Stations-State Agricultural Universities.

Breeding Achievements

A. **Improved varieties:** GR-3, GR-4, GR-5, GR-103, GR-6, GR-7, CSR 36, CSR 30, CSR 27, CSR 23, Gurjari, Jaya, Ranjtih Sub1, Samba Sub1, Swarna Shreya, DRR Dhan 59, DRR Dhan 51, DRR Dhan 310, DRR Dhan 311, Pusa Samba 1850, Pusa 2090, TKM 15, CR Dhan 321, GNR-9 (Lalkada Gold), Shalimar Rice-6, Rajendranagar Vari-5, CR Dhan 211, CR Dhan 329, Surma Dhan, Swarna Purvi Dhan-4, Gujarat Anand Rice 22, GR 24 (Navsari Parimal), GR 25 (Mahatma), Gujarat Aerobic Rice 201, Swarna

Purvi Dhan-5, CR Dhan- 416, VL Bosi Dhan, DRR Dhan- 75, DRR Dhan 81, *etc.*

B. **Hybrid varieties:** APHR-1, APHR-2, KRH-1, KRH-2, DRRH-1, VRH-4, PRS 101, MPH 517, Sava 127, Pusa RH 10, Sahyadri, BIO 799, PAN 2423, Kaveri Rice Hybrid -7299, DR8611, Arize Dhan DT, DRRH-6, HRI- 211, *etc.*

C. **Basmati Varieties:** Basmati – 370, Pusa 150, Pusa-523, Kasturi, DRR-Hyderabad, Pusa basmati 1121, Pusa basmati 1509, Pusa basmati 1692, Pusa basmati 1885, Pusa basmati 1847, Pusa Basmati 1882, Punjab Basmati 7, *etc.*

D. **Genome Edited Varieties:** DRR Dhan 100 (Kamala), Pusa DST Rice 1.

Chapter 7

Breeding of Maize (*Zea mays* L.)

Introduction

Maize (*Zea mays* L.) is one of the world's most important economic cereal crops. It was domesticated over the past 10,000 years from the grass teosinte. The word "*Zea*" (zela) was derived from an old Greek name for food grass. It is an important crop based on cultivated area and production following rice and wheat. Due to the presence of the C_4 photosynthetic pathway, maize is more efficient than the C_3 pathway under high temperatures and dry-land conditions. Maize grains are gaining acceptance and enormous demand in our country due to the nutritionally significant and multiple functions of a traditional farming system. It has diversified uses as food for humans, live stocks, and poultry. In India, it is a pride among the coarse cereal crops. No cereal on the earth has such enormous potential as maize; hence, it resides in a vital place as the '*Queen of Cereals*'.

Background Information

1.	*Botanical Name*	*Zea mays* L.
2.	*Family*	*Poaceae*
3.	*Chromosome number*	2n=2x=20
4.	*Centre of origin*	Central America and Mexico
5.	*Nature of pollination*	Cross-pollination
6.	*Pollination mechanisms*	Monoecism, protandrous and anemophily
7.	*List of wild relatives*	*Zea mexicana, Z. diploperennis, Z. perennis* and *Z. luxurians, etc.*

Theories about the Origin of Maize

There are several theories have been proposed about maize origin

1. **Tripartite theory:** Theory elucidated that maize originated from an unidentified plant that possessed similar ear morphology
2. **Catastrophic sexual transformation theory:** According to this theory, maize originated from teosinte due to sudden sexual transformation.
3. ***Tripsacum* (*Zea diploperennis*) theory:** This theory explains that maize plants evolved from the cross between *Zea diploperennis* and *Tripsacum dactyloides*. The genus *tripsacum* is a close relative of maize descent separate from the genus *Zea*.
4. **Teosinte theory:** This theory is well-accepted and scientifically validated. According to this theory, maize plants arose by acquiring new variations due to mutation in several loci of teosinte. Early people were cultivated teosinte and selected various forms of maize plants. Teosinte is a wild grass belonging to the genus Zea. It is the closest wild relatives of maize having similar flowering habits to maize.

Distribution

The cultivated maize originated in Central America and Mexico from wild grass teosinte. However, European explorers of the overseas period have disseminated it to different continents/regions, *viz.*, North America, Asia, Europe, Australia and Africa. The tremendous variation led to acclimatization in local areas.

Different Groups of Maize

As discussed in maize origin, the maize plant evolved due to several mutations from wild grass teosinte that creates enormous variation, particularly in grain forms. The maize plants are grouped into various types based on grain forms; each type is briefly described below with unique features;

Sl.No.	*Name*	*Species Name*	*Features*
1.	Dent corn	*Zea mays* ssp. *mays* var. *indentata*	Depression (dent-like structure) on top of the grain, made up of hard starch
2.	Flint corn	*Zea mays* ssp. *mays* var. *indurata*	Kernel hard due to the outer portion being made up of hard starch
3.	Flour corn	*Zea mays* ssp. *mays* var. *amylacea*	Kernel is made up of soft starch so easy to grind and make chapati from it
4.	Popcorn	*Zea mays* ssp. *mays* var. *everta*	Extremely flint, kernel size is small, eruption and turns inside out after heating

Sl.No.	Name	Species Name	Features
5.	Pod corn	*Zea mays* ssp. *mays* var. *tunicata*	Each grain is enclosed in pod. This form of maize arose due to a mutation that occurred at *tunicate* locus
6.	Sweet corn	*Zea mays* var. *rugosa* Bonaf	Immature kernels contain high sugar and low starch. Seed becomes wrinkled as it turns mature/dry. The sweetness is due to mutation in genes *viz., sugary1* (*su1*) *and shrunken2* (*sh2*) which regulating sugar metabolism
7.	Waxy Corn	*Zea mays* ssp. *mays* var. *ceratina*	Grains are waxy and become sticky after cooking due to high amylopectin and low amylose content. This form of maize arose due to a mutation that occurred on *waxy1* (*wx1*) locus

Floral Biology, Selfing and Crossing Techniques

Inflorescence	:	☆ Maize is a monoecious plant where male and female inflorescences are borne on a single plant. ☆ The main shoot of the plant ended into male inflorescence, referred to as 'tassel'. It comprises a central axis and lateral branches. The central axis produced many rows of paired spikelets; in contrast, the lateral branches produced only two rows of paired spikelets, one pedicelled and another sessile. ☆ The female inflorescence is called 'silk' that matures into the 'ear' on the shoot in the axil of a leaf. Morphologically the female inflorescence is slightly similar to the main shoot. However, the outer leaves are modified into leaf sheaths of different layers called husk, which protect the ears by covering them. These protective leaf layers are also termed bracts or spathes. The maturity period of both male and female inflorescence is different, *i.e.*, protandrous, in which male inflorescence matures before female.

Flower	:	☆ **Male spikelet:** A pair of glumes present at the base of each spikelet. Each glume comprises two functional florets borne on a short joint rachilla. The flower (floret) is bracteate, composed of pair of thin bract layers outer called the lemma and the inner called the palea. A flower is zygomorphic, sessile, incomplete (due to perianth), and imperfect- unisexual, *i.e.*, male. ☆ **Female spikelet:** Pistillate spikelets are arranged in a pair in a similar fashion on the axis of the cob. Spikelets are sessile and comprise a pair of glumes at the base. The floret is bracteate, composed of couple of thin bract layers, lemma and palea–each spikelet composed of two florets. The upper is fertile, while the lover is sterile. *Thus, kernel rows are always in even numbers on the ear.* The flower (floret) is zygomorphic, sessile, incomplete (due to perianth), and imperfect- unisexual, *i.e.*, female.
Perianth	:	**Male floret:** Present in the form of two fleshy lodicules. **Female floret:** Normally absent, sometimes present as two small, scaly lodicules.
Androecium	:	**Male floret:** Three stamens-one present opposite the palea, while the remaining two opposite to the lemma, anther–bilobed, versatile filaments are short and free. The pollen grains are very small, light in weight and carried away by the wind; they remain viable for 24 hours. **Female floret:** Generally absent (Sometimes may be visible as three rudiments stamens).
Gynoecium	:	**Male floret:** Absent (one rudimentary pistil) **Female floret:** Single carpels (three fused)- one bears an ovule and remains to extend to form silk. One dome-shaped superior ovary with fine long thread-like silk, filiform style, stigma long and hairy. It remains receptive for up to 14 days throughout its entire length. Single ovule and basal placentation
Fruits	:	Caryopsis
Seed	:	Albuminous or monocots.
Floral formula	:	**Male flower:** Br. ⊕ ♂ $P_{2(Lodicules)} A_3 G_0$ **Female flower:** Br. ⊕ ♀ P_0 or $_{2(Lodicules)} A_0 \underline{G}_{(1)}$

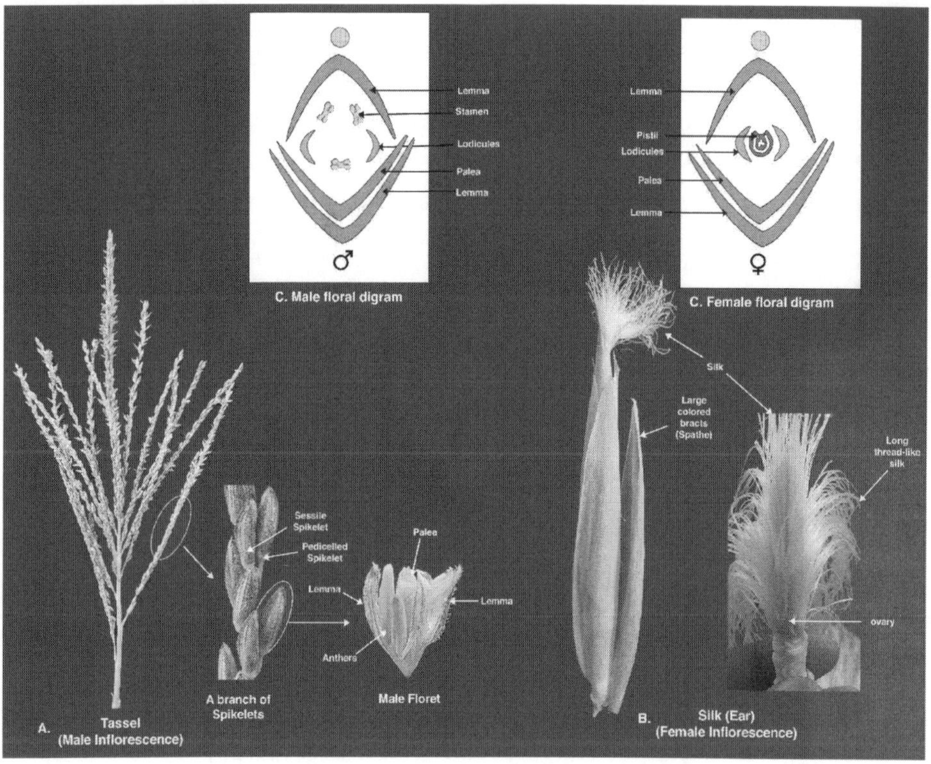

Figure 7.1. Floral Biology and Diagram of Maize (*Zea mays* L.).

Anthesis: Maize plant has a protandrous condition in which male inflorescence matures before the female. The period between male anthesis and female silking is called the anthesis-silking interval. Anthesis begins from the middle third of the spike and continues upwards and downwards for 2-14 days. The silking starts 3-4 days after the beginning of anthesis and continues up to 2-5 days.

Selfing: Select the plant where silk will emerge within 1-2 days. The ears should be cut about 1" (1 inch) below the tip of the husk in the evening and covered with a butter paper bag. The tassel of the same plant is also covered with a brown paper bag for the collection of pollen dust. By the following 1-2 days, the equal size of the silk (1 - 1.5") will grow. Pollen collected in the tassel bag is dusted on silk in the morning. Rebag the silk quickly to avoid contamination with foreign pollens and label the ear shoot.

Crossing/Hybridization techniques: A suitable size of brown paper bag is covered over the tassel of the desired male parent; similarly, the ear shoots of the maize plant are also covered 1 to 2 days before the emergence of silk by butter paper bag to prevent contamination. The top of the ear shoot is cut by a sharp razor on the day before pollination for uniform growth of the silk and covers the bag immediately. Pollination is done the following morning between 10.00 AM to

12.00 PM. The freshly shed pollen grains in the tassel bags are dusted with a jerk on the silk after removing the silk bags. After pollination, the tassel bags and silk bags are replaced to prevent contamination from foreign pollens immediately. New brown paper bags indicate cross number with date. The crossed cobs are duly tagged and harvested after maturity.

Significance of Maize in Context to Genetical and Cytogenetical Studies

At the beginning of the genetics and cytogenetics era, the maize plant was considered a model crop for genetical and cytogenetical studies because of various unique features includes:

- ☆ It is a widely grown crop due to its versatility.
- ☆ Highly cross-pollinated cereal crop.
- ☆ Large numbers of seeds can be obtained from a single cross.
- ☆ Many phenotypic hereditary traits are available for study.
- ☆ Self-pollination and cross-pollination are made easier due to monoecious.
- ☆ Due to cross-pollination in nature, many recessive traits are visible through inbreeding or by induced mutation.
- ☆ Diploid in nature and a less number of chromosomes (*i.e.*, 2n=2x=20).
- ☆ Individual chromosomes can be identified from their length and the presence of typical knobs on some chromosomes (*i.e.*, chromosome number 4, 9, *etc.*).

QPM (Quality Protein of Maize) and Biofortification

In maize protein, two amino acids, namely lysine and tryptophan, are generally deficient proteins; there are several mutants, *viz.*, Opaque-2, Opaque-6, Opaque-7, Flory-2, Flory-3, *etc.*, have been revealed for increasing lysine and tryptophan. In 1964, Dr. Oliver Nelson's team at Purdue University, USA, discovered that the homozygous recessive *o2* allele had substantially higher lysine (+69 per cent) in grain endosperm compared to normal maize hence, Opaque-2 has been considered the important for the improvement of maize quality protein. In maize, an abundant protein stored in grain endosperm is called zein. The fraction is called 'α-zein'. The lysine-tryptophan amino acids content in maize is reversed with zein. The opaque-2 mutant cause to decrease in the production of the alpha-zein protein, which ultimately causes to increase in tryptophan and lysine content.

In maize, nutrition enrichment through the bio-fortification approach has been successfully developed and released in biofortified maize varieties with improved quality protein (QPM) (Table 7.1). The allele opaque-2 has been identified for more tryptophan and lysine content in endospermic protein in maize–a popular single-cross hybrid, Vivek Maize Hybrid 9 developed through marker-assisted backcross breeding. The introgression of an opaque-2 allele in the original QPM variety, the improved QPM hybrid (Vivek QPM 9), showed a 41per cent increase in tryptophan content and a 30 per cent increase in lysine content over the original hybrid. Furthermore, efforts have been made to develop the β-*carotene*-rich (provitamin A) maize hybrids through marker-assisted introgression of β-*carotene hydroxylase* allele (*crtRB*1), *lycopene-ε-cyclase* (*lcyE*) and opaque 2(o2) genes. The various hybrids have been released with dual quality parameters *viz.*, rich in provitamin A and high lysine and tryptophan content (QPM), are listed in Table 7.1.

Table 7.1. List of Important High-Quality Protein Maize (QPM)/ Biofortified Cultivars Released in India

Sl.No.	*Name of Cultivar/Hybrid*	*Improved Quality Traits*	*Year of Released*
1.	Shaktiman 1	QPM -1.01 per cent tryptophan	2001
2.	Shaktiman 2	QPM -1.04 per cent tryptophan	2004
3.	Shaktiman 3	QPM -0.73 per cent tryptophan	2006
4.	Shaktiman 4	QPM -0.93 per cent tryptophan	2006
5.	HQPM 1	HQPM	2006
6.	HQPM 5	HQPM	2007
7.	HQPM 7	HQPM	2008
8.	Vivek QPM 9	HQPM	2008
9.	Pusa HM-4 Improved	QPM- tryptophan 0.91 per cent and Lysine 3.62 per cent	2017
10	Pusa Vivek QPM-9 Improved	Provitamin-A rich maize QPM hybrid Provitamin-A 8.15 ppm	2017
11.	Shalimar QPMH-1	QPM-tryptophan 0.81 per cent.	2019
12.	LQMH 1	QPM- Tryptophan (0.75-0.80 per cent) and Lysine content (3.0-3.5 per cent)	2020
13.	Pusa HQPM 5 improved	Provitamin-A rich QPM Provitamin-A 6.77 ppm; Tryptophan (0.94 per cent) and Lysine content (4.25 per cent)	2020
14.	Pusa HQPM-7 Improved	Provitamin-A rich QPM Provitamin-A 7.10 ppm; Tryptophan (0.93 per cent) and Lysine content (4.19 per cent)	2020
15.	Pusa Vivek Hybrid-27 Improved	Provitamin-A rich maize hybrid Provitamin-A 5.49 ppm	2020
16.	IQMH202	QPM- Tryptophan (0.66 per cent) and Lysine content (3.05 per cent)	2021
17.	IQMH203	QPM- Tryptophan (0.77 per cent) and Lysine content (3.48 per cent)	2021
18.	PMH1-LP	36 per cent low phytic acid	2022

Source: ICAR–IIMR, Ludhiana and ICAR–IARI, New Delhi.

Xenia Effects

This unique effect is reported in maize, where the effect of pollen sources on certain seed characters like kernel colour and kernel weight, *i.e.*, light yellow kernel results when pollen of yellow corn fertilizes to ovule of white corn.

Breeding Objectives

1. Breeding for Improve Grain Yield

Yield is a complex trait controlled by polygenes. It can be improved by selecting different yield-attributing traits like ear length, girth, kernel rows, numbers, size, weight, *etc.* Recently two important QTLs (KRN4 and KRN1) related to kernel row number and qHKW1 for kernel size and weight have been identified with the help of biotechnological tools. These QTLs will help to improve the grain yield in maize.

2. Breeding for Earliness

Earliness is desirable in any crop. In maize, the flowering days are complex and vary according to climatic region. Three QTLs, *viz.*, *Vgt1*, ZmCCT9 and *ZmCCT10* have been recognized, cloned and mapped too. Among these QTLs, *ZmCCT9* is important for days to anthesis. The number of days for anthesis can be reduced by muting the function of *ZmCCT9*.

3. Breeding for Biotic Stress Resistance

In maize, different biotic stresses occur due to various pests and diseases, and they cause considerable yield losses in India. The stem borer, pink borer, shoofly, Heliothis, *etc.*, are major pests of maize. DMR 7, CM 502 are reported resistant sources of pink borer. In the case of diseases mydis leaf blight, downy mildew, leaf sheath blight, rust are important. Generally, Mexican varieties and Taiwan lines are exhibited resistance to downy mildew, while BML-5, BML-6, BML-7, BML 15, *etc.* are resistant sources of leaf blight. There are several genes-specific diseases have been identified with the help of biotechnological tools. This approach helps to identify resistant lines or introgression of resistant sources into susceptible but agronomically desirable varieties.

4. Breeding for Abiotic Stress like Drought Resistance

Drought is major abiotic stress in maize. Major morpho-phycological traits like earliness, stomatal conductance, leaf water content, ABA concentration, and root morphological characters are important for breeding maize drought-resistant genotypes. Besides drought, heat stress also plays a significant role in yield reduction.

5. Breeding for Grain Quality

A quality of maize can be judged with various parameters such as:

- ✰ High crude protein content and quality balanced protein: The quality protein includes a proportional number of amino acids, which can be improved by increasing the size of the embryo since 20 per cent of protein present in the embryo is balanced while the rest of 80 per cent of protein present in endosperm commonly known as zein that deficient in lysine and tryptophan so it is imbalance.

- Oil content: The embryo size is a positive association with oil content in maize.
- Grain colour, size, shape, and external appearance according to the utility of maize are also important parameters for fetching higher prices in the market.
- Sweet corn: High sugar content in immature grain (milky stage); Popcorn: popping percentage and popping volume; Waxy corn: High amylopectin; Baby corn: high prolificacy and the maximum number of ears.
- High provitamin-A content.
- Low antinutritional factors *i.e.*, Phytic acid

6. Breeding for Specialty Corn

Besides the common type of maize, the various specialty maize gets vast consumer demand in agriculture. Therefore, breeding specialty maize is also an important objective. Such types include sweet corn, popcorn, baby corn, waxy corn, high amylose content maize. Now that genetic composition and inheritance are known, it is easy to breed such cultivars. The special types of maize and its regulating genes/QTLs are mentioned in Table 7.2

Table 7.2. The Special Types of Maize and its Regulating Genes/QTLs

Sl.No.	*Types of Special Corn*	*Genetic Architecture/Regulating Genes*
1.	Sweet corn	☆ Two genes *viz.*, *su1* on chromosome 4 and *sh2* on chromosome 3, brittle2 (*bt2*) on chromosome 4.
2.	Baby corn	☆ A gene teosinte branched1 (*tb1*) on chromosome 1; QTL prol1.1 on chromosome 1 are important for prolificacy. ☆ A new QTL 'qProl-SP-8.05' identified on chromosome 8 in prolific maize land races.
3.	Pop corn	☆ Three QTLs *viz.*, metaQTL1_1, metaQTL1_5 and metaQTL1_7 located on chromosome 1, and one QTL metaQTL6_2 on chromosome 6 revealed significant for popping traits.
4.	Waxy corn	☆ Identified and mapped mutant gene *wx1* on the short arm of chromosome 9.
5.	High amylose maize	☆ A recessive amylose extender1 (*ae1*) mutant on chromosome 5. ☆ Modifier genes *Sbela* and *mae1* also important.

6. Develop Hybrids with Alternate Sources of Cytoplasm

In maize, three types of male-sterile cytoplasm *viz.*, CMS – T, S and C are reported. The cytoplasm CMS – T (Texas cytoplasm) is susceptible to maydis leaf blight diseases (MLB). Therefore, hybrids must be developed using a leaf blight-resistant source of alternative cytoplasm like CMS-C and S.

Breeding Approaches (Methods)

1. **Maize Introduction:** Various Mexican origin lines were first introduced during 16th century by the Portuguese, *e.g.*, Protina
2. **Population Improvement:** The different approaches have been used:
   A. **Recurrent selection-** Recurrent selection also refers to cyclical improvement involving a simple population improvement or generating new recombinants (population), interbreeding of individuals with repeated cyclic selection over generations. In other words, genetic recombination generates through repeated cyclic selection over generations with the interbreeding of individuals. The detailed procedure of recurrent selection was explained in maize by **M. T. Jenkins** in 1940. It is performed in two ways 1. Simple recurrent selection (Phenotypic recurrent selection) and 2. Selection based on combining ability.

   B. **Synthetic and Composite varieties:** According to **Lonnquist J. H.** (1961), a synthetic variety is an open-pollinated population derived from an intercrossing of selfed inbred, maintained through routine mass selection procedures. Therefore, breeding synthetic varieties is worthwhile in cross-pollinated crops like maize.

   The cultivar developed by population approaches are Kisan, Jawahar, Sona, Vikram, Amber, Vijay, *etc.*
3. **Heterosis breeding/Hybrid technology:** The hybrid variety is derived from two diverse parents and F_1 seed commercially cultivated at farmers' fields. In 1908 **G. H. Shull** introduced the phenomena of heterosis in maize and did a pioneer work behind hybrid maize development. The world first hybrid variety of maize called '**Burr Leaming Dent**' was commercially released in 1922 in the USA. Initially, it was not much popular, but around the 1950s, it changed maize cultivation by replacing the open-pollinated varieties. In India, the first hybrid technology was used in maize in 1960. In 1960-61 double cross hybrids *viz.*, Ganga 1 & 2, Ganga 101, VL 54, Deccan and Ranjit were released for commercial cultivation. Afterward many single cross hybrids have been released commercially by the public and private sectors. The improved hybrid cultivars of maize resultant a significant increment in maize production from 1.73 mt (1950-51) to the top level of > 35.0 mt (2023-24).

Different Types of Hybrids Produced in Maize

A. Single or Simple Cross (Hybrid)

A hybrid generated by a cross between two parental lines (inbred lines). The single cross hybrid is popular in both self and cross-pollinated crops. For instance, Vivek hybrid 9 is produced by a cross between CM 214 × CM 145. The number of all possible single crosses among selected inbred lines is easily counted by the formula

$$= \frac{n(n-1)}{2}$$

where n = number of inbred lines used in crossing.

Single cross hybrids are superior than double cross hybrids due to high uniform and yield potential and more productive. It is more vigorous than parents.

B. Double-Cross (Hybrid)

A hybrid generated by a cross between two single crosses. In other words, this cross comprises of a total four parental genotypes (inbred lines). D.F. Jones (1918) proposed concept of double cross hybrid. The number of all possible double-crosses among selected inbred lines is easily counted by the formula

$$= \frac{n(n-1)(n-2)(n-3)}{8}$$

where n = number of inbred lines used in crossing.

Initially, it was more prevalent in maize. It is represented by [(A × B) × (C × D]. For example, the double-cross maize hybrid Ganga 2 was developed by four inbred lines, *i.e.,* [(CM 109 × CM 110) × (CM 202 × CM 111].

C. Three-Way Cross (Hybrid)

A hybrid generated by a cross between a single cross and an inbred line. In other words, this cross/hybrid involved three parental genotypes (inbred lines). Initially, it was also popular in maize. It is represented by [(A × B) × D]. For example, the three-way cross maize hybrid Ganga 5 was developed by three inbred lines, *i.e.,* [(CM 202 × CM 111) × CM 500]. The number of all possible three-way-crosses among selected inbred lines is easily counted by the formula

$$= \frac{n(n-1)(n-2)}{8}$$

where n = number of inbred lines used in crossing.

All these methods were done by detasseling method, after 1950, introduced CGMS system (Texas cytoplasm CMS-T) replaced these conventional methods of hybrid seed production. But during 1970, the Texas cytoplasm of maize became susceptible to leaf blight caused yield drop up to 90 per cent. Therefore, hybrid seed production method by conventional- detasseling has been substituted over CGMS system.

Inbred Development and Evaluation

- **Development:** A homozygous line developed by continuous selfing (inbreeding) in a cross-pollinated crop is referred to as **inbred**. Generally, 6-7 generations of continuous selfing or sib mating are required to develop a homozygous line. In cross-pollinated crops, inbreeding may cause a reduction of vigour and overall productivity of lines. So alternative generation sib mating is desirable to maintain vigour and productivity. While inbreeding generations, the natural cross-pollination is strictly barred.
- **Evaluation:** The genetic worth can be evaluated based on the performance and productivity of inbred lines in single and double-cross combinations. The genetic analysis of inbred is performed based on combining ability. An inbred can produce superior performance in a hybrid combination with other inbred lines. It is of two types (1) General Combining Ability (GCA) and (2) Specific Combining Ability (SCA).
- General Combining Ability (GCA): An average performance of inbred lines in a set of crosses or hybrid combinations. It represents the features of parents.
- Specific Combining Ability (SCA): Deviation of better or poor performance of specific hybrid/cross from the average performance of the parental inbred lines. It represents the features of hybrids or crosses (Sprague and Tatum, 1942).

7. Biotechnological Approaches

- **Double haploid techniques:** In vivo haploid induction is used to produce complete homozygous lines quickly and efficiently.
- **Genomics-Assisted Breeding:** Various approaches includes, marker-assisted selection, marker-assisted backcross breeding (MABB), genomic selection (GS), forward breeding (FB) are considered to be effective tools for hastening genetic gain by increasing selection intensity, shorten the selection cycle, introgression of desirable genes/QTLs for quality improvement, biotic and abiotic stress resistance/tolerance in maize.
- **Transgenic breeding:** Many cultivars has been released world-wide specific to herbicide tolerance, insect resistance, abiotic stress resistance modified quality, *etc.*
- **Gene Editing Technology:** Various tools such as TALEN, ZFN, CRISPR/Cas9 promisingly used for the modification of various traits such as herbicide tolerance, drought tolerance, biotic stress tolerance, grain composition, male sterility, lignin biosynthesis, secondary metabolism, superior plant types, *etc.*

Breeding Institutes

- **International level:** The International Maize and Wheat Improvement Centre (CIMMYT)-Mexico, USA
- **National level:** Indian Institute of Maize Research (IIMR), PAU Campus, Ludhiana.
- Winter Nursery, Rajendra Nagar, Hyderabad.
- Regional Research Stations-State Agricultural Universities.

Breeding Achievements

A. **Synthetic and composite varieties:** Shakti, Ratan, Vijay, Kisan, Prabhat, Kiran, Jawahar, Sona, Ambar, Surya, Arun, Vikram, Parvati, Pusa Composite-4, Pusa Composite-3, Him Palam Maize Composite 2, Pant Composite Maize 4, Shalimar Maize Composite-8, SKOMC-3 and 4, *etc.*

B. **Varieties developed:** GM-1, GM-2, GM-3, GM-4, GM-6, Aravali Makka-1, Jawahar Makka-216 (JM-216), Sheetal, Megha Maize-1, Megha Maize-2, HT 5110, Shalimar Kishenganga -3, *etc.*

C. **Hybrids:** GAYMH 3, GAWMH-2 (White), GAYMH 1, Ganga-1. Ganga-3, Ganga safed-2, Ganga-5, Ganga-9, Ganga-11, Deccan-103, DMRH1301, DMRH1308, DMRH1305, Vivek Maize Hybrid-5, Vivek Maize Hybrid-9, Vivek Maize Hybrid 47, Vivek Maize Hybrid 45, VLQPM Hybrid 45, Pusa Biofortified Maize Hybrid-3, VLQPM Hybrid 63, NMH 4140, KMH 8322, Gujarat Dantiwada Yellow Maize Hybrid 101, Shalimar Maize Hybrid-5, Pusa Jawahar Hybrid Maize-2, COH (M) 11, Pratap Hybrid-6, DHM 206, IMH 228, VL Madhurima, VNR- 4597, Pusa Bio fortified Maize Hybrid – 4, Pusa Biofortified Maize Hybrid-5, 6, 7, and 8 VL Shikhar, *etc.*

D. **Baby corn:** Baby corn GAYMH 1, IMHB1539, IMHB1532, LBCH3 (DMRHB1305), Vivek Hybrid 27 (Central Maize VL Baby Corn 2), Central Maize VL Sweet Corn 1 (FSCH18), HM-4, COBC-1, BCH-402, Rajendra Baby Corn- 1, Punjab Baby Corn 1, Pusa male sterile baby corn hybrid-3, *etc.*

E. **Sweet corn:** Madhuri, GASCH 11, Shalimar Sweet Corn-1 (KDM-1263SC), Hi-Brix 53 (ADVSW-2), Hi-brix-39 (ADVSW-1), Hy-brix 53 (ADVSW-2), CANDY (KSCH-333), NSCH-12 (Misthi), Pusa super sweet corn-3, *etc.*

F. **Popcorn:** VL Amber, GAPCH 21, LPCH 3 (IMHP1540), LPCH 2 (IMHP1535), DMRHP1402, Shalimar Pop Corn-1 (KDPC-2), BPCH-6, Pearl Popcorn, Pusa Popcorn Hybrid-2, Rajendra Pop Corn 1, Mizo Popcorn 1, *etc.*

G. **QPM and Biofortified cultivars:** As per Table 7.1.

H. **Waxy corn:** Pusa Waxy maize hybrid-1.

Chapter 8

Breeding Pearl Millet (*Pennisetum glaucum* L.)

Introduction

Pearl millet (*Pennisetum glaucum* L.) also known as bajra. The genus *Pennisetum* is derived two Latin words *Penna* meaning 'feather' and *seta,* meaning 'bristles'. It is an important millet crop of traditional farming systems in tropical and subtropical Asia and sub-Saharan Africa. It was domesticated 3500- 5000 years ago in Africa along with the Southern margins of the Saharan central highlands, and it was introduced to India about 2000 B.C, now, it is the fourth most cultivated food crop after rice, wheat and maize. India is a global leader in the context of the area and production of bajra. In India, it used as dual purposes *i.e.,* food and fodder. Pearl millet has unique features like rapid growth under minimum inputs, high productive efficiency in dry and hot conditions due to the C_4 pathway, high tolerance against adverse climate conditions and balanced nutritional properties; hence, it is also called 'super cereal'. It is adaptable in dry areas, low rainfall or drought-prone regions and is popular among poor farmers due to drought tolerance. Pearl millet is the staple food grain with a high nutritional value, with more protein (11.6%), fat (5%), and mineral content (2.3%). It can provide food and nutritional security even in the toughest condition, hence considering the potential to deliver food and nutritional security of the country. It contains unique nutritional, climate-resilient traits, the powerhouse of nutrients and health benefit properties. The Central Government declared millets consisting of Pearl Millets (Bajra) as a 'Nutri-Cereals' for production, consumption and trade point of view.

Background Information

1.	*Common name*	Bajra, Bajari, Bulrush millet, Cattail millet, Babala
2.	*Botanical Name*	*Pennisetum glaucum* (L.) R. Br Previous name: *P. americanum* (L.)
3.	*Family*	Poaceae
4.	*Chromosome number*	2n=2x=14
5.	*Centre of origin*	West Africa (Ethiopia)
6.	*Nature of pollination*	Cross-pollination (> 80 per cent cross-pollination)
7.	*Pollination mechanisms*	Protogynous and anemophily
8.	*List of wild relatives*	*Pennisetum monodii, P. ramosun, P. purpureum, P. squamulatum, P. tetrastachyum, etc.*

The genus *Pennisetum* has more than 140 species. Based on morphological features, the genus *Pennisetum* was classified into five groups: *Brevivalvula* Döll, *Eupennisetum*, *Gymnothrix*, *Heterostachya Schumach* and *Penicillaria*. Pearl millet belongs to the *penicillaria* group.

The Gene Pool of the Genus *Pennisetum*

According to Harlan and de-Wet (1971), the genus *Pennisetum* is classified into three gene pools (GP-1, GP-2 and GP-3) according to the cross-compatibility relationship between cultivated pearl millet and its wild relatives (Table 8.1).

Table 8.1: Classification of Important Species of Genus *Pennisetum* According to Gene Pools

Gene Pool	*Species*	*Common Name/ Types*	*Chromosomes/ Ploidy Level*
GP-1	*P. glaucum* ssp. *glaucum* (L.) R. Br	Pearl millet Cultivated type	2n = 2x = 14
	P. monodii (Maire) Brunken	Wild	2n = 2x = 14
	P. purpureum Schum.	Napier/elephant grass	2n = 4x = 28
GP-2	*P. squamulatum* Fresen	-	2n = 6x = 54 or 2n = 8x = 56
GP-3	*P. alopecuroides* (L.) Spreng.	Swamp foxtail grass	2n = 2x = 18
	P. pedicellatum Trin.	Deenanath grass	2n = 4x = 36
	P. polystachion L. Schult.	Mission grass	2n = 6x = 54
	P. macrourum Trin.	Needle grass	2n = 4x = 36
	P. villosum Fresen	Feather top grass	2n = 5x = 45

- *P. purpureum* Schum: Most closely associated with pearl millet, Napier grass used in developing perennial fodder varieties.
- *P. squamulatum* Fresen: It has been used in breeding programmes to develop draught and cold-resistant cultivars.

Distribution

Pearl millet was domesticated over 4000 years ago in the West African Sahel, and later distributed to East Africa and India. It is being cultivated worldwide; however, majorly grown in Africa and Asia. In India, it is widely cultivated in the states of Rajasthan, Maharashtra, Gujarat, Uttar Pradesh, and Haryana.

Nutritional Significance and Variability

- It contains a good quantity of dietary fibre and protein.
- It is rich in fat content with better fat digestibility.
- It is rich in unsaturated fatty acids with higher content of nutritionally essential omega-3 fatty acids, *viz.*, oleic acids, linoleic acid, and linolenic acid, than other cereal.
- Good source of vital amino acids like methionine, tryptophan, lysine, leucine, isoleucine, *etc.*
- Grain is a gluten-free and low prolamin fraction.
- It is a good source of dietary minerals, such as calcium, phosphorus, iron, magnesium, zinc and copper.
- It can be the finest food for constipation, diabetics, celiac diseases and obesity due to its low glycaemic index score and rich impervious starch.
- It has a relatively high concentration of both Fe and Zn. As per ICRISAT studies, the range of variability for grain Fe and Zn densities in *Iniari* germplasm accessions (51–121 mg/kg Fe; 46–87 mg/kg Zn), population progenies (18.0–135.0 mg/kg Fe; 22.0–92.0 mg/kg Zn), inbred parents (30.3–102.0 mg/kg Fe; 27.4 mg to 84.0 mg/kg Zn), hybrids derived from diverse inbreds (25.8–80.0 mg/kg Fe; 22.0–70 mg/kg Zn) and commercial hybrids (31.0–61.0 mg/kg Fe, 32.0–54.0 mg/kg).

Floral Biology, Selfing and Crossing Techniques

Inflorescence	:	☆ The Inflorescence is referred to as panicle or ear. It is terminal, cylindrical to conical in shape, very compact, unbending and can grow up to 46 cm long with a perimeter of 7-9 cm. ☆ Inflorescence consists of a central rachis covered with soft/short hairs and bears fascicles on rachillae. Each fascicle contains 1–9 spikelets surrounded by a wall of bristles (*i.e.*, involucre). The whorl consists of 30 to 40 bristles.

Flower	:	☆ A spikelet comprises 2-4 flowers or florets, but mainly two. Each spikelet has two types of flowers, imperfect (staminate) and perfect (bisexual/hermaphrodite). ☆ Staminate floret is bracteate, sessile, zygomorphic, incomplete (due to perianth) with a single lemma. ☆ Bisexual (hermaphrodite) floret is bracteate, zygomorphic, incomplete with short pedicel.
Perianth	:	In the case of staminate floret, lodicules and palea are absent, while in bisexual floret, two fleshy lodicules are present.
Androecium	:	The staminate floret has only three stamens; similarly, the bisexual floret also has 3 stamens surrounded between the broad lemma and palea. Anther–bilobed, versatile filaments are short and free. The pollen grains are very small, light in weight and simply carried away by wind, it remains viable for 24 hours.
Gynoecium	:	Single carpel with feathery style divided into two branches, One unilocular superior ovary single ovule and basal placentation. Stigma remains receptive for 1-2 days.
Fruits	:	Caryopsis
Seed	:	Albuminous or monocots.
Floral formula	:	**Staminate flower:** Br. ⊕ ♂ $P_0 A_3 G_0$ **Bisexual flower:** Br. ⊕ ♀ $P_{2(Lodicules)} A_3 \underline{G}_{(1)}$

Anthesis: The style begins to appear 2 to 3 days after the emergence of the ear. From the top portion and gradually proceed downward. Anther emerges after styles dry up. The anthers of bisexual flowers appear 2 to 3 days before those of staminate flowers. The maximum appearance of style is occurred between 9.00 AM to 3.00 PM, while the maximum opening of the flowers occurs between 10.00 AM to 2.00 PM and the minimum between 3.00 and 6.00 PM.

Selfing: The panicle is to be covered with a butter paper bag before the emergence of stigmas to avoid crossing and then labelled it.

Crossing/Hybridization techniques: In pearl millet, emasculation is laborious and challenging due to the small size of flowers and the late development of anthers concerning stigma. Being protogynous in nature, the flower can be crossed without emasculation. Due to the overlap of the female flower with the male phase at the tip of spike, about four-fifths of the upper portion of the ear is removed and the rest is bagged with butter paper bag before the styles appear to prevent contamination. Pollination should always be done in the morning after

2 to 3 days when styles appear, fresh pollen grains collected from the protected desired male parent and dusted over the stigma of the bagged ear heads and re-bag instantly and labelled the crossed heads. The pollination may be repeated 2-3 times till the anthers of the same region display their presence.

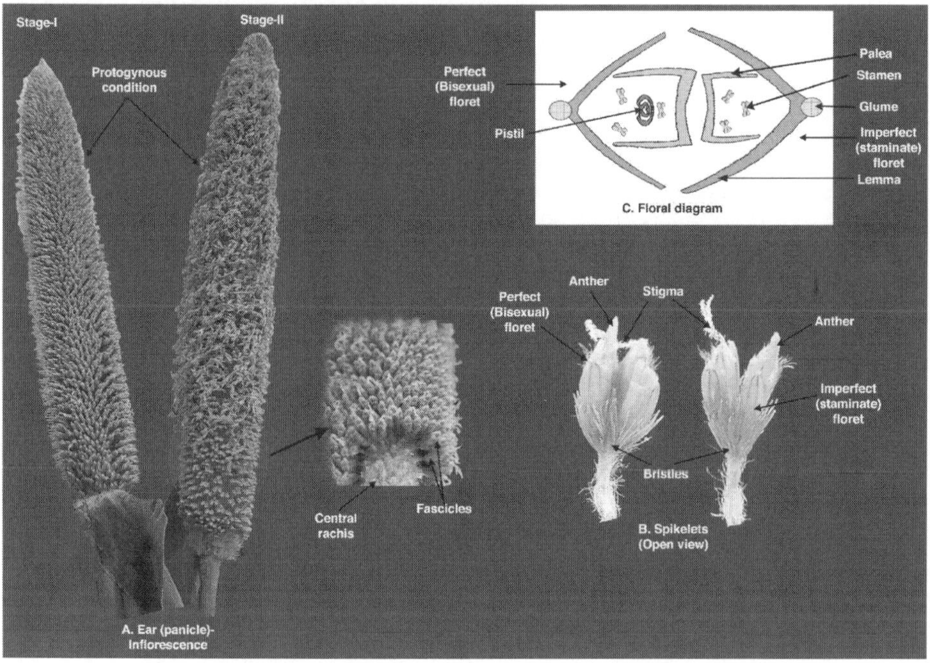

Figure 8.1. Floral Biology and Diagram of Pearl Millet (*Pennisetum glaucum* L.).

Breeding Objectives

1. **Breeding for higher yield:** Being a C_4 plant, pearl millet has more yield potential. In India, dual-purpose bajra is cultivated. Accordingly, both grain and fodder yield improvement are essential. The yield may be improved by focusing on different yield component traits *viz.*, number of tillers per plant, panicle long, well-filled and compact, bolder grain size, synchronous maturity, and high harvest index. Low tillering is the ideal selection criteria for higher grain yield per panicle. In contrast, for higher dry fodder yield, more tillers per plant is an important criterion.
2. **Breeding for lodging resistance:** The dwarfness is positively associated with lodging resistance, but good stover yield is essential. The dwarfing genes have been reported to improve forage nutritive value by reducing stem percentage and increasing leafiness. There are four recessive dwarfing genes, *viz.*, *d1*, *d2*, *d3*, and *d4* have been reported in pearl millet that control plant height, among gene *d2* widely used in breeding programs.

3. **Breeding for early maturity:** Earliness is essential in low and erratic rainfall areas, drought-prone areas, and terminal water stress conditions. The ideal plant type with a stay-green trait can mature within 70-75 days in available soil moisture. It is also suitable in multi-cropping sequences and experiences the least bird damage. The most widely used sources of earliness are the *Iniadi*-type landraces from western Africa.
4. **Breeding for quality**
   - ✰ **Grain quality:** The bajra is an important staple food in arid and semi-arid regions. The quality can be determined by various parameters, such as:
     - ❖ Grain colour: light or white grain colour is more preferable.
     - ❖ Grain size: Should be bold and lustrous.
     - ❖ Protein content: Grain contains 9 to 15 per cent protein. A positive correlation was reported between white endosperm and high protein content.
     - ❖ Mineral nutrients, amino acids contents: In recent breeding, the major focus on the improvement of Fe and Zn, the approach referred as biofortification (more detail are discussed in the biofortification section). Grain content 2 to 7 per cent, mineral matter, also rich in vitamins A and B, thiamin, riboflavin. There is a correlation between yellow endosperm and vitamin A content.
   - ✰ **Fodder quality:** The fodder quality is determined through the number of tillers per plant, deep green colour, the juiciness of stem, number of leaves with more breadth suitable for multi-cut and the nutrient value of fodder, including high sugar content, *invitro* dry matter, digestibility are desirable traits for fodder yield and quality.
5. **Breeding for drought tolerance:** Most of pearl millet growing areas fall under arid and semi-arid regions, low and erratic distribution of rainfall is common phenomena, these factors are responsible for the drought. The genotype breeds with early flowering, early vigour, high tillering and desynchronizing of tillering contribute more to adaptation to drought stress during the vegetative growth phase. Therefore, emphasis is given to traits like high water potential, more adventitious root, low tillering and large panicles while breeding drought-tolerant genotypes in pearl millet. The hybrids GHB 538, GHB 719, and GHB 757 are drought resistant. The good drought-tolerant genotypes are CZP 9802, 863B, ICMP 83720, ICMV 9413, and ICMV 94472.
6. **Breeding for salinity tolerance:** Up to 33 per cent yield reduction have been reported due to salinity stress in bajra. A shoot-biomass ratio associated with salt tolerance and shoot Na^{+} concentration may be used a one of the selection criteria while screening genotypes against salinity tolerance in pearl millet. Some identified lines for salinity tolerance are

KAT/PM-2, Kitui local; 10,876, 10,878, 18,406, 18,570; IP 3732, IP 3757. Recently one variety HASHAKI 1 has been identified for saline areas of Uzbekistan.

7. **Breeding for disease resistance:** Downey mildew, blast, ergot are major diseases in bajra. Downey mildew (DM) also known as green ear disease, utmost distressing disease of pearl millet responsible for maximum yield loss in India. There are various conventional and biotechnological work has been done at ICRISAT. The study revealed that DM susceptibility is not associated with the cytoplasm–the resistance related to nuclear genes. Therefore, selection for DM resistance has been a top priority in hybrid parental line development. The resistant line 7042S was identified from susceptible landraces, male-sterile line 841A was reselected from susceptible male-sterile line 5141A. Several putative QTLs have been identified. With the use of conventional and marker-assisted back cross-breeding resistance alleles at two DM QTLs, one each on linkage groups 1 and 4, were incorporate into a male parent (H 77/833-2) of one of a widely grown hybrid, HHB 67 from elite resistance donor parent ICMP 451. The DM-resistant variety released as a '**HHB 67 improved**' in Rajasthan, Haryana and Gujarat in 2005. Further research is in progress to transfer gene into a widely grown susceptible varieties.

   Another important foliar disease is a blast. It is a serious threat in both A_1 and A zone of India in early to medium-maturing varieties. Two resistant genotypes, IP 21544 and IP 21720, have been used as a donor line in blast-resistant breeding programs.

   The other diseases are smut, rust and ergot also important, there are a number of resistant lines have been identified and are used in a resistant breeding program. The various identified resistant genotypes for various diseases in pearl millet are depicted in Table 8.2.

Table 8.2. Important Diseases and their Available Resistant Genotypes in Pearl Millet

Name of Diseases	*Resistant Lines/Accessions*
Downy mildew (DM)	ICML series (*i.e.*, 1, 2, 3, 4, 5, 6, 7, 8, 9, 10), ICMPE 13-6-30, ICMA 92666, ICMB 92666, ICMA 91333, 91444, 91555, WGI 52, GI148, ICMA 91555, ICMR 09999, ICMB 89111-P6, ICMB 0111-P6, ICMA, 88006A and 88006B.
Blast	IP 11010, IP 13636, IP 21525, IP 21531, IP 21536, IP 21540, IP 21594, IP 21610, 21,640, IP 21706, IP 21711, IP 21716, IP 21719, IP 21720, IP 21721, IP 21724, IP 21987, IP 21988, IP 22160, IP 21544, IP 21720, IP 22269, IP 21544, IP 11353, IP 22423, IP 7941, *etc.*
Rust	I Tift 65, P 16438, IP16762; P310-17, P1449- 3; IP18292, IP18293, IP700651; ICML 12 to 16, 22; ICMP 312, 423, 85410; 7042S; 841A; landraces like Desi Bajri-Chomu, Dhodsar local, Ardi-Beniya Ka Bas, 81B-P6, ICMP 451-P8, IP 21629, 21645, 21658, 21660, 21662, *etc.*

Name of Diseases	*Resistant Lines/Accessions*
Smut	ICML 5, 6, 7, 8, and 10; ICML 17, 18, 19, 20 and 21; Tif leaf 3; Tift 3 (PI 547035) and Tift 4 (PI 547036); Tift 65, ICMA 91555, ICMA 88006A and 88006B.
Ergot	ExB 46-1-2-S-2, ExB 112-1-S-1-1, ICMV 8282, 8283; ICMA 91555, ICMA 88006A and 88006B ICMA 91333 to 91555; ICML 5 to 10; SSC FS 252-S-4, ICMPS 100-5-1, 700-1-5-4, 900-1-4-1, 900-3-1, 900-9-3, 1300- 2-1-2, 1400-1-6-2, 1500-7-3-2, 1600-2-4, 1800-3-1-2, 2000-5-2; ICI 7517-S-1, ExB 132-2-S-5-2-DM-1, P-489-S-3; SSC 46-2-2-1, SC 77-7-2-3-1, SSC 18-7-3-1.

Source: ICRISAT and AICRP on Pearl millet, Mandor, Jodhpur, Rajasthan.

8. **Breeding for bird resistance:** Variety with long profuse bristle provides more resistance against bird damage.
9. **Breeding Pearl Millet for Better Nutrient Use Efficiency:** The arid and semi-arid region soil is generally deficient in nutrients, often termed poor soil. The application of excess fertilizer is economically not viable for farmers. Therefore, breeding a desirable genotype gives maximum yield potential under low input or has better efficiency in nutrient use. Recently ICRISAT scientists identified various nitrogen-efficient genotypes, *viz.*, IP 10820, ICMV 221, ICMB 01222-P1, ICMB 90111-P2, ICMB 89111-P2, IP 17720, IP 10379, IP 8069, and ICMV-IS 89305.
10. **Breeding for diversification of cytoplasmic source in male sterile lines:** Only a single source of CMS (A_1 cytoplasm) has been used in the pearl millet hybrid breeding program. For example, the unique cytoplasmic line Tift 23 is highly susceptible to downy mildew since all these hybrids of HB series have been left out from cultivation due to vulnerable of DM susceptibility. Fears with the possible vulnerability of hybrids to diseases and insect-pests associated with the deployment of a single cytoplasmic background. A prompted research effort is required to find alternative CMS sources in hybrid breeding. The research led to the identification of some alternative CMS sources in India and other countries. Currently, five types of cytoplasm have been used (Table 8.3). The 732 A_4 (background Bellary cytoplasm), Lines Pb111, Pb 101A developed at Ludhiana possess a high level of resistance toward downy mildew.
11. **Breeding for better end-use consumers accessibility:** Although pearl millet is a powerhouse of nutrient but their bioavailability is poor, due to the presence of certain anti-nutritional factors like typical grey colour, phytate and polyphenol content; poor milling quality- flour bitterness or rancidity, poor flour rheological properties. The variety breed with low antinutritional factors and with improved keeping flour quality is important for better end-use consumer accessibility in pearl millet.

Table 8.3. List of Diverse Cytoplasm that Induces Male Sterility and their Source in Pearl Millet

Sl.No.	*Name of Cytoplasm*	*Source*
1.	A_1-CMS (Tifton)	Tifton 23A (Inbred 556)
2.	A_2-CMS (Ludhiana)	IP 189
3.	A_3-CMS (Ludhiana)	Amber grain stock
4.	A_v and A_4-CMS (Tifton)	*Pennisetum glaucum* (L.) R. Br. subsp. Monodii (*violaceum*)
5.	A_5-CMS (ICRISAT)	Half-sib progeny of the Early Gene Pool (EGP 261)-large-seeded gene pool (LSGP)

Breeding Approaches (Methods)

1. **Introduction:** *Iniadi* germplasm from West Africa, male sterile line Tift 23 A_1 from USA to India; Hybrid bajra from Punjab to other states of India.
2. **Single plant/Mass selection**: In the pre-hybrid development phase, the pearl millet cultivars improvement was made with simple mass selection, inbred selection several OPVs were released *viz.*, Vansari, Kopargaon Local, N 28–15–1, Co 1, Co 2, Co3, AKP 1, AKP 2, K 1, RSJ, RSK, T 55, *etc.*
3. **Interspecific hybridization and selection:** The best example is the Napier hybrid [a sterile vigorous triploid hybrid ($2n = 3x = 21$)], developed by a cross between *Pennisetum glaucum* ($2n = 2x = 14$) × *P. purpureum* ($2n = 4x = 28$). Another example is developing an ornamental *Pennisetum* cultivars by cross between *Pennisetum purpureum* Schmuach and *P. squamulatum* Fresen.
4. **Heterosis breeding: Hybrid bajra:** Burton G.W. (1958), in Tifton City, Georgia, United States of America (USA) was discovered a cytoplasmic genic male sterile line known as a Tift 23A_1. The development and release of Tift 23A_1 made the production of commercial single crosses hybrid in pearl millet possible. Tift 23A_1 have short stature, profuse tillering, uniform flowering and good combining ability. This system was introduced in India and extensively utilized by Indian scientists from PAU, Ludhiana and IARI, New Delhi. This laid the foundation of pearl millet hybrid breeding in India. The first hybrid HB 1 was released in 1965, followed by a number of hybrids *viz.*, HB 2, HB 3, HB 4, HB 5 were developed using this A_1-CMS source. So far, more than 180 hybrids, including public and private sectors, have been released for commercial cultivation in various agroecological zones of the country. These hybrids occupied over 60 per cent of the country's total pearl millet cultivation area. The hybrid technology leads to an increase crop productivity more than 1300 kg/ha (2023-24) against only 305 kg/ha (1951-55).

5. **Population improvement:** There are various approached, *viz.*, recurrent selection, synthetic and composite varieties, are used to develop open-pollinated varieties (OPVs) in pearl millet. In the era of hybrid technology, the OPVs still exist in farmer's field due to the repeated use of seeds, better resilience toward biotic and abiotic stresses. The major weakness about OPVs is partial exploitation of heterosis. Many OPVs were developed, *viz.*, WC-C75, ICTP 8203, ICMV 155, ICMV 221, Raj 171, HC 4, JBV 2, JBV 3, HC 20, RCB 2, CZP 9802 and Pusa 266; ICMS 7703 (synthetic variety). Pusa composites-383, 443, 612, 701. A biofortified variety 'Dhanshakti'.
6. **Mutation breeding:** A downy mildew-resistant line 5071A was developed from Tift 23A to treat gamma rays at IARI. Three hybrids NHB 3, NHB 4, and NHB 5, were developed from this mutant line.
7. **Biotechnological approaches/Genomics-Assisted Breeding:** Various approaches includes, marker-assisted selection, marker-assisted backcross breeding (MABB), genomic selection (GS), forward breeding (FB) are considered to be effective tools for hastening genetic gain by increasing selection intensity shorten the selection cycle, introgression of desirable genes/QTLs for quality improvement, biotic and abiotic stress resistance/tolerance in pearl millet.
8. **Biofortification:** The pearl millet biofortification breeding program at ICRISAT initiated with the support of HarvestPlus. The program is taken up in a three-pronged breeding phase-I, II and III. The first phase is a short-term approach dealing with traits genetics, germplasm screening and creating genetic variability. The second phase is the medium-term approach of validating identified high-Fe and Zn breeding lines and hybrid parents from the regular breeding program to develop fast-track biofortified variety/hybrids. Finally, the third phase consisted of long-term objective- development high-Fe/Zn breeding lines and hybrid parents and its genetic diversification through steady mainstreaming micronutrient traits at ICRISAT and NARS breeding programs.

   A world-first high-Fe pearl millet variety, '**Dhanashakti**' was developed by utilizing the intra-population variability within ICTP 8203, an early-maturing, large-seeded, disease resistant and high-yielding open-pollinated variety that has been under cultivation in India since 1990. Dhanashakti is officially released and notified by Central Variety Releasing Committee in 2014 for the cultivation in all pearl millet-growing states of India. Then after number of biofortified hybrids have been released from ICRISAT and other research institutes are mentioned in Table 8.4.

Table 8.4. List of Important Biofortified Cultivars Released in India

Sl.No.	Name of Cultivars	Improved Biofortified Traits		Year of Released/ Notified	Name of Breeding Station
		Grain Fe Content (ppm)	Grain Zn Content (ppm)		
1.	Dhanshakti	81	43	2014	ICRISAT, India and MPKV, Dhule
2.	HHB 299	73	41	2018	CCS HAU, Hisar
3.	AHB 1200	77	39	2018	NARP, Aurangabad
4.	AHB 1269	91	43	2019	NARP, Aurangabad
5.	HHB 311	83	39	2019	CCS HAU, Hisar
6.	RHB 233	83	46	2019	SKNAU, Jobner
7.	RHB 234	84	41	2019	SKNAU, Jobner
8.	GHB 1305	77	40	2025	JAU, Junagadh

Source: ICRISAT and AICRP on Pearl millet, Mandor, Jodhpur, Rajasthan.

Breeding Institutes

- **International level:** International Crops Research Institute for Semi-Arid Tropics (ICRISAT), Hyderabad, India.
- **National level:** All India Pearl Millet Improvement Project (AIPIP), Jodhpur (Rajasthan), India.
- Regional Research Stations-State Agricultural Universities.

Breeding Achievements

Table 8.5. List of some Important Hybrids Released in India

Name of Hybrid	Released Year
Hybrid Bajra 1 (HB 1)	1965
HHB 67	1990
Pusa 444	1995
ProAgro 9443, ProAgro 9445	2001
GHB 558, GHB 526	2003
HHB 67 improved (improved version of HHB 67)-First molecular marker-assisted cultivar	2005
MH 1234	2006
Nandi-52	2008
Nandi 65, HHB 223, Nandi 61	2010
Shine (MH-1578), 86M86	2012

Name of Hybrid	*Released Year*
Kaveri super boss (MH 1553), Nandi 72	2013
MPMH 17 (MH 1663), Pratap (MH 1642)	2014
86M88, NBH 5767	2015
Bio 8145, 86M82, 86M01, Nandi 75	2016
Pusa 1201, MH 1743	2018
MH 2192, MH 2224, MH 2228	2019
MSH 346 (MP 7366)	2020
MH2545 (*HHB 67 improved 2-7), MH 2423, MH 2439	2021
MH 2474, MH 2480, MSH 361, MSH 363, 86M80, SVPMH-101, MPMH 35, BLPMH 109	2022
GHB 538 (EDV:DM) (Maru Sona), ADV 955	2023
GHB 538 Improved (Maru Sona), Banas Nayan (GHB 1351), MPMH 42 (Shree Anna Bajri 42), GHB 1294, 86M22, PA 9898	2024
MP7214, PA9010	2025

The General Procedure of Hybrid Seed Production in Pearl Millet

- ✰ In bajra hybrid seed production is done by a three-line breeding system (Cytoplasmic genetic male sterility). This system uses three inbred lines *viz.,* A, B and R lines.
- ✰ Cytoplasmic line developed by introducing a cytoplasm into desired inbred line by backcross breeding method. [CMS line × Inbred line; act as a recurrent parent]→ A new male sterile (A-line).

Maintenance of Three Lines

- ✰ The maintenance of the A-line is done by crossing with its counterpart (B-line, a male fertile). The B-line maintain by selfing or sibbing. Similarly, the restores line (R-line) also maintains same way of B-line. Both B- line and R-line are maintained in separate plots.
- ✰ The breeder has to maintain the nucleus seeds of these lines. The progeny of the nucleus seed, known as the foundation seed, are multiplied on government farms with an isolation distance of 1000 m. The male sterile A-line with B-line are planted in the plot in the ratio of 4:2. The complete plot is surrounded by 2-4 border rows of B-line. Seeds harvested from A-line are used for hybrid seed production programmes.

Certified Seed Production Programme in Pearl Millet

- ✰ Generally, this programme is taken on the farmer's field with supervision as well as gives certification from state level seed certification agency. An isolation distance of 200 m is maintained. Line A and Line R are planted in the ratio of 4:2, 2-4 pollinator lines should be grown surrounding the field.

- ☆ **Rouging**: The rouging is done at different stages of the crop *viz.*, 30 DAS, commencement of flowering, at the time of maturity, *etc.*

Seeds obtained on line-A are hybrid seed, which should be given to farmers for general cultivation. *e.g.*, ICMA 99222 (A-line) × J-2565 (R-line)→ Hybrid seed of GHB-1129.

Chapter 9

Breeding Sorghum [*Sorghum bicolor* (L.) Moench]

Introduction

The word "sorghum" typically refers to cultivated sorghum [*Sorghum bicolor* (L.) Moench subsp. *bicolor*]. It is the fifth most important cereal crop, next to wheat, rice, maize, and barley worldwide. While in India, it is the fourth most important cereal crop. Sorghum is considered a staple food for millions of poor people in the semi-arid tropics of Africa and Asia, where sorghum is cultivated dual-purpose crop, as grain and fodder are highly valued for human and animal consumption, respectively. Sorghum has high biomass production potential, making it a climate-resilient crop due to its short-day habit and C_4 photosynthetic pathway. It is mainly cultivated in the arid and semi-arid tropics because of better adaptation to drought, heat, salinity, and flooding. Regarding nutritional quality, it is more highly palatable and digestible than maize and pearl millet. Considering the potential to deliver food and nutritional security of the country, contains unique nutritional, climate-resilient traits, the powerhouse of nutrients and health benefit properties, like bajra the Central Government also declared sorghum (Jowar) as a 'Nutri-Cereals' for production, consumption, and trade point of view.

Background Information

1.	*Common name*	Juwar, Jowari, Great millet, Indian millet
2.	*Botanical Name*	*Sorghum bicolor* (L.) Moench
3.	*Family*	*Poaceae*
4.	*Chromosome number*	2n=2x=20
5.	*Centre of origin*	North-eastern Africa
6.	*Nature of pollination*	Often Cross-pollination (average outcrossing 6 per cent)
7.	*Pollination mechanisms*	Bisexuality, Entomophily, and anemophily; protogyny in some florets
8.	*List of wild relatives*	*Sorghum halepense, S. sudanense, S. miliaceum, S. propinquum, S. sudanense, S. vulgare, S. almum, etc.*

Sorghum Progenitor and Wild Species

Cultivated sorghum is the only single member of the genus *Sorghum*, consisting of 25 species and divided into five taxonomic units: *Chaetosorghum, Heterosorghum, Parasorghum, Stiposorghum*, and *Eusorghum*. Among these, the *Eusorghum* species are agronomically important, which comprises of cultivated sorghum and its wild species. Some important are

- ☆ *Sorghum verticilliflorum* [2n=2x=20]: Annual wild progenitor of cultivated sorghum, known as Sudan grass.
- ☆ *S. propinquum* [2n=2x=20]; wild perennial.
- ☆ *S. halepanse* [2n=4x=40]; Perennial, tetraploid developed by a cross between *S. arundinaceum* × *S. propinquum*; commonly known as Johanson grass.
- ☆ *S. almum* [2n=4x=40]; Cross between *S. bicolor* × *S. halepense*; commonly known as Columbus grass.
- ☆ *S. sudanense* [2n=2x=10]: Sudan grass.

A. The Classification of Commercial Sorghum

Based on utility, the cultivated sorghum is grouped into different types

1. Grain sorghum- it also classified into various class, *viz., kaffir, milo, feterita, durra, shulllu* and *kaoling.*
2. Forage sorghum: Sudan grass (annual) and Johnson (perennial) grass.
3. Sweet (Sorgos) sorghum.
4. Miscellaneous or special purpose sorghum *viz.*, Broom corn, pop sorghum, waxy sorghum, hurda sorghum, *etc.*

B. Races of Sorghum

Based on spikelet types and grain enclosed by glume, Harlan and De Wet (1972) were subdivided cultivated sorghum (*Sorghum bicolour*) into five primary inter-fertile races (Figure 9.1) and 10 intermediate races (based on floral morphology). This classification has been broadly accepted

Sl.No.	*Race*	*Basic Features of Grain and Spikelets*
1.	*Bicolor* (B)	Glumes clasping the grain, which may be covered entirely or 1/4th part exposed at the tip, grain elongate, sometimes slightly obovate; spikelets persistent
2.	*Caudatum* (C)	Asymmetrical type of grain, concave, the opposite side rounded and bulging; glume generally total half-length of grain or less
3.	*Durra* (D)	Obovate rounded shape grain, wedge-shaped at the base and broadest slightly above the middle; glumes very wide.
4.	*Guinea* (G)	Grain flattened dorso-ventrally, sub-lenticular in outline, twisting 90° at maturity between gaping involutes glumes that are nearly as long to longer than grain
5.	*Kafir* (K)	Grain almost spherical, not twisting, glume clasping and varying in length

Intermediate Races

1.	Race *Caudatum-Bicolor* (CB)	6.	Race *Guinea-Durra* (GD)
2.	Race *Durra-Bicolor* (DB)	7.	Race *Guinea-Kafir* (GK)
3.	Race *Durra-Caudatum* (DC)	8.	Race *Kafir-Bicolor* (KB)
4.	Race *Guinea-Bicolor* (GB)	9.	Race *Kafir-Caudatum* (KC)
5.	Race *Guinea-Caudatum* (GC)	10.	Race *Kafir-Durra* (KD)

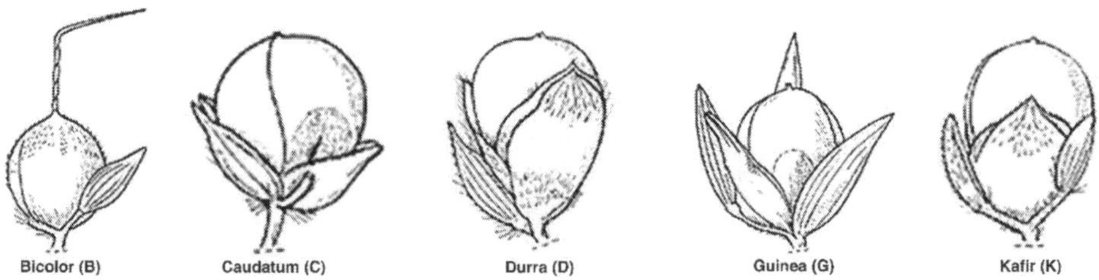

Figure 9.1. Primary Inter-Fertilize Races of Sorghum.

Distribution

Based on the hypothesis, the sorghum crop originated and was domesticated in north-eastern Africa or at the Egyptian–Sudanese border around 5000–8000 BC. Although the major diversity of cultivated and wild sorghum is reported in these regions of Africa, the spreading of sorghum took place from this early domesticated site to other parts of Africa and ultimately to Asia, including India, the Middle East, and China. Among the five cultivated races, durra types currently spread from Ethiopia along the Nile to the Near East and beyond to Thailand and across India. Although, in India, sorghum occupies the area of arid and semi-arid, cultivated as a grain, fodder and dual purposes, some special types of sorghum also cultivated in the middle to southern parts of the country.

Nutritional Significance and Variability

- ✰ Sorghum and millets are nutritionally superior as it grains contain high amounts of protein, essential amino acids except lysine, minerals, and vitamins.
- ✰ The macronutrient composition of sorghum grain is comparable to wheat and maize, on average, 70 per cent carbohydrate, 12 per cent protein and 3 per cent fat; grain sorghum has a higher percentage of fat than other cereals.
- ✰ It is a good source of vitamin B complex.
- ✰ Antinutrients like phytates, polyphenols, and tannins reduce the bioavailability of nutrients.
- ✰ In the global germplasm of sorghum, the protein content varied from 8.1 to 18.8 per cent, fat content ranged from 1.0 to 4.3 per cent while starch content varied from 61.7 to 71.1 per cent.
- ✰ The grain Fe and Zn concentration in elite parents generally ranges between 20 to 40 and 20 to 30 ppm, respectively.
- ✰ The *bicolor-durra* and *durra* sorghum races had the highest protein and fat content *i.e.,* 12.6 and 3 per cent, respectively, while the lowest starch content (66.1 per cent). Whereas *kafir* sorghum had the lowest protein content and the highest starch content *i.e.,* 10.9 and 67.6 per cent, respectively.

Floral Biology, Selfing and Crossing Techniques

Inflorescence	:	✰ The inflorescence is referred to as a panicle. It is a short and compact or loose and open or spreading type. It may be erect or curved shaped, depending on the genotype.

		☆ Inflorescence consists of a central rachis that differs greatly in length and shape, completely or partly hidden by dense panicle branches. It may be striated, hairy or glabrous. A rachis bears a several primary branches from different nodes. The secondary branches produced from primary branches and it bear spikelets.
Spikelet and floret	:	☆ Spikelets are usually found in pairs on lateral branches but may be three at the tip. In paired spikelets, one spikelet is always sessile (perfect) and the another pedicellate (staminate). In case of three spikelets, one spikelet is sessile (perfect) and the other two are pedicellate (staminate). ☆ **Sessile (perfect) spikelet:** Without pedicel, varied shape, lanceolate to almost round and ovate and sometimes depressed in the middle. Possess two glumes, green in colour, but vary in colour as it matures according to genotypes. It is hairy or hairless, Glumes are hard. **Floret:** Each spikelet possesses two florets within two glumes, lower is sterile and the upper one is perfect. There are two lemmas, each a delicate white tissue. The lower lemma is elliptic or oblong about equal in length to the glume. The upper lemma is shorter, more ovate, and may be awned, broad, and membranous. Palea is small and thin. The perfect floret is sessile, bracteate, zygomorphic, and incomplete (due to perianth). The lower floret with empty lemma, no palea. ☆ **Pedicellate (staminate) spikelet:** It is narrower and longer than sessile spikelet. **Floret:** Upper floret possess only stamens with only lemmas which much reduced in size and rarely an awn which is present on the upper lemma, palea absent. Floret is bracteate, zygomorphic, and incomplete (due to perianth). The lower floret with an empty lemma is represented by a lemma only.
Perianth	:	Two short and broad fleshy lodicules are adjacent to the fertile lemma on either side of the ovary at its base.
Androecium	:	Yellowish three stamens with long and free filaments, anther–bilobed, versatile. The pollen grains are very small, lightweight, it remains viable for several hours.

Gynoecium	:	Single carpel with two long pistils, two feathery protrude stigmas, One unilocular superior ovary, single ovule, and basal placentation. Stigma remains receptive for 8-16 hours.
Fruits	:	Caryopsis
Seed	:	Albuminous or monocots.
Floral formula	:	**Staminate flower:** Br. ⊕ ♂ $P_{2(Lodicules)} A_3 G_0$ **Bisexual flower:** Br. ⊕ ⚥ $P_{2(Lodicules)} A_3 \underline{G}_{(1)}$

Anthesis: Anthesis begins with the exertion of the whole panicle. Flowers begin to open 2–3 days after the complete emergence of the panicle. The anthesis starts at the tip of the panicle and progresses downwards and is complete within 7-10 days. Anthesis occurs first in the sessile spikelets from top to bottom of an inflorescence. Maximum flowering is observed in the morning between 6:00 and 9:00 hours. Flowers remain open for 30–90 minutes.

Selfing: Head bagging is efficient for selfing the ear heads. All heads in a row should be covered once the decision to bag heads is made. If a head has already begun to flower, the flowering portion should be cut off. During head bagging, the boot leaf of the plant is usually removed before placing the bag. After bagging label it.

Selfing of Sorghum.

Crossing/Hybridization techniques: The hand emasculation is carried out during evening hours between 4:00 to 7:00 PM on well-developed about 50 florets of female parents, which likely open to the following day is selected for emasculation. The remaining panicle portion is clipped away. During clipping, the flowered tip and the lower panicle branches are removed. Next, the needle is inserted at the middle of the floret and moved across the glumes. The needle rotated at 90° and three anthers lifted out. After removing the already open and very immature buds, the emasculated panicles of female parents are covered with white butter paper bags (Figure 9.3).

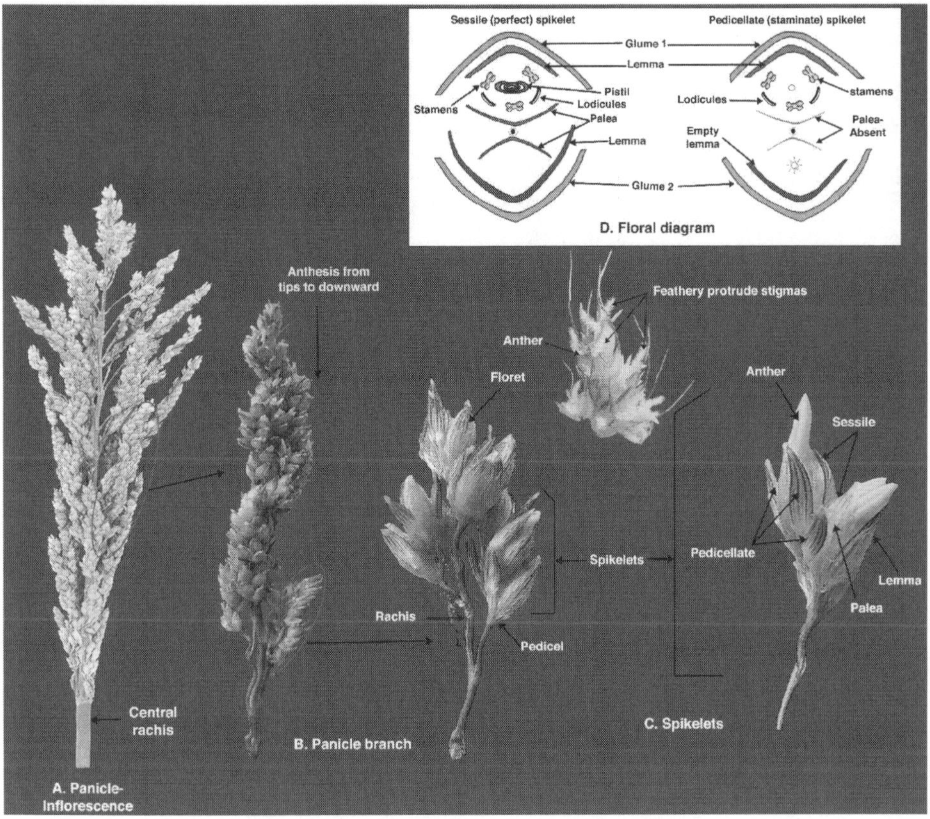

Figure 9.2. Floral Biology and Diagram of Sorghum (*Sorghum bicolor* L.).

Pollination: The pollination is carried out the following day (between 7.00 to 10.00 AM) on emasculated panicles by collecting the pollen grains from the protected flowers of parents. The pollen grains are applied on a receptive stigma by brush, re-covered with a suitable paper bag, and labelled. The selfed and crossed panicle must harvest separately at the time of maturity. The ear head should be dried under the sunlight, and seeds should be collected separately.

Breeding Objectives

1. **Breeding for higher grain or forage yield or both purposes:** The maximum yield advantages can be achieved by exploiting heterosis by developing high-yielding hybrids. In sorghum, commercial exploitation of heterosis has been possible due to the availability of a stable and heritable CMS mechanism, allowing large-scale seed production with minimum cost and the maximum extent of heterosis in yield-attributing traits over a wide range of environments.

Figure 9.3: Hybridization Technique in Sorghum.

2. **Breeding for non-lodging type:** It is positive associated with strong culm with more stilt roots.

3. **Breeding for early maturity:** Earliness is important for escaping pest like stem borer and abiotic stress like drought and fit in a multi-cropping system. The best example of early cultivars are, CSV 1, CSV 2, CSV 3 and CSV 17 (grain sorghum); GFS 4 (forage sorghum).

4. **Breeding for wider adaptability and resistance to abiotic stresses like drought, salinity, temperature and water stress.**

   ✰ Although sorghum crop is drought hardiness, post-flowering *rabi*-sorghum experience significant drought and cause extensive yield losses. Stay greenness is considered as an important trait ascribing drought tolerance in sorghum. Major QTLs contributing to stay-green trait has been identified and efforts are in progress to introgress into superior cultivars.

5. **Breeding for diseases resistance:** In sorghum, important diseases are grain mold, downy mildew, leaf blight, anthracnose, and ergot.

   ✰ Grain mold: it is a major disease in *Kharif* sorghum. Multiple mechanisms of resistance govern grain mold resistance and many agronomically undesirable traits are associated with the resistance. A significant and positive correlation found between seed traits and grain mold. Harder grain, higher levels of seed phenols, and darker glumes contributed to grain mold resistance.

6. **Breeding for insect-pest resistance:** In sorghum, important insect pests are shoot fly, stem borer and gall midge.

   ✰ The shoot fly is the most important pest in sorghum due to the complex nature of inheritance and the limited resistance sources available in sorghum. The important resistant genotypes are IS 18551, IS 2205, IS 2312, *etc.* The major QTLs of shoot fly resistance have been identified; efforts are in progress to transfer in parental lines.

   ✰ In the case of stem borer, *Cry1b* gene is important for resistance. The morphological traits, such as early panicle initiation, faster internode elongation, shoot length, plant growth rate, seedling vigour, leaf glossiness, and ligular hairs, are linked with resistance. Based on studies revealed that additive type of gene action is important for governing resistance to spotted stem borer.

   ✰ Sorghum gall midge resistance is a polygenically inherited trait and is governed by both additive and non-additive gene action. Resistant sources have been identified. A midge resistance variety DSV 3 (selection from ICSC 745) has been released for Karnataka.

7. **Breeding for grain quality-biofortified sorghum:** The quality comprises of various parameters such as;

   ✰ Physical parameters; Grain colour, size, shape, texture and luster.

- ✰ Roti making quality: Pearly white grain colour, its flour having a higher water-holding capacity and good organoleptic taste. The good roti-making varieties are M 35-1 (Maldandi), CSV 22, Phule Vasudha, Phule Chitra, Nizer goti, *etc.*
- ✰ Other quality parameter includes bread making quality, chemical composition, nutritional value (Fe and Zn, amino acid content), low tannin content, high protein, sugar, and lysine content, puffing quality, *etc.*
- ✰ Inheritance studies suggested that the increased amount of lysine in each line is controlled by a single recessive gene which can be easily transferred by standard plant breeding methods.
- ✰ A *rabi* sorghum grain quality is superior as compared to *kharif* sorghum.
- ✰ The first biofortified sorghum cultivar' **Parbhani Shakti'** (ICSR 14001) was released in India in 2008 in collaboration with ICRISAT and Vasantrao Naik Marathwada Krishi Vidyapeeth, Parbhani (Maharashtra). The improved variety has high Fe (45 ppm) and Zn (32 ppm). Besides, it has higher protein content (11.9 per cent) and higher bioavailability of improved nutrients due to low phytates (4.1 mg/100g) (https://www.icrisat.org/).

8. **Breeding for fodder quality:** The fodder quality parameters include more number of tillers, vigorous growth, retaining green colour, leafiness, stalk sweetness and juiciness, high sugar (Brix) content, low HCN (hydro cyanic) content, high palatability and *invitro* digestibility. The genotype with the brown midrib trait (BMR) can improve digestibility.

   In dryland conditions, before flowering, forage sorghum stem possess high HCN content, it may be poisonous when young-aged sorghum is used for animal feeds. So, it is essential to breed a variety with low HCN content. Low HCN content exhibits a partial dominance reaction. More than one gene governs this trait.

9. **Breeding for sweet sorghum:** Breeding sweet sorghum is one strategy for alternate uses of sorghum. It has great potential for ethanol production due to high sugar concentration in the stem with Brix value of up to 24 per cent. From the stem juice, ethanol can be produced, a renewable energy source. In India, sweet sorghum has arisen as a potential substitute crop to sugarcane for bio-ethanol production. The ethanol from sweet sorghum has good quality due to less sulphur content and high-octane rating. Important cultivars are CSV-19-SS, CSV-24-SS, CSH-22-SS (Hybrid) and CSV 49SS (Jaicar Raseela).

There are two types of sweet sorghum.

- ☆ Syrup varieties - Syrup for table purposes can be produced from this. It is also suitable for ethanol production.
- ☆ Sugar varieties: Contents of more sugars and less inflammable organics. Hence less suitable for ethanol production compared to syrup varieties.

10. **Breeding for specialty sorghum**: Papad making, pop sorghum, hurda (roasted) purpose, colour sorghum, flanking purpose, *etc.* Various cultivars identified for special purposes sorghum *viz.*, SMJ 1, RSJ 1, PKV Ashwini, RSSGV 46 and Phule madhuri (hurda purpose), RPASV 3 (papad making), AKJ 1 (flaking purpose) and KMJ 1 (popping purpose), CSV50R, CSV60 and CSV65 (yellow colour sorghum).

11. **Breeding resistance to parasitic weed, *e.g., Striga***: Host plant resistance is the most applied and economically viable means for reducing crop losses due to *Striga*. Early released variety CSV 5 showed resistance to *Striga*.

Breeding Approaches (Methods)

1. Introduction

Introduction of *kafir-milo* sources from USA to India and used these sources in conversion programs into local varieties for traits like photo insensitive and short duration. The First two hybrids were developed *i.e.*, CSH-1 and CSH-2 using the introduced Combined Kafir (CK) 60A CMS line. The cultivar CSV 1 was released from the direct Introduction of line IS 3924.

2. Selection

The varieties were developed from local races through pure line, mass selection approaches *e.g.*, PC 6 [Scl. from IS 5977 (*Durra*)], PC 9 [Sel. From IS 4870 (*Durra*)], M35-1 (Landrace selection from local Maldandi), CSV 1 (selection from IS3924), Sel.3 (Sel. from Bidar rabi local), CSV 216R selection from landrace *rabi* germplasm dhulia, *etc.*

3. Hybridization and Selection

a) Inter-Varietal and Inter-Specific Hybridization

Developing forage sorghum genotypes and new grass species is essential. For example, a multi-cut variety, Co (FS)29, was released from Coimbatore in 2001 through an interspecific cross.

- ☆ *S. halepanse* [2n=4x=40]; Perennial, tetraploid developed by a cross between *S. bicolor* × *S. propinquum*; commonly known as Johanson grass.

- ☆ *S. almum* [2n=4x=40]; Cross between *S. bicolor* × *S. halepense*; commonly known as Columbus grass.

b) Backcross Method

The conversion program was initiated by the backcross method of breeding to transfer genes of male sterility, sterile cytoplasm, photo insensitive and short duration from sources like *kafir-milo* into local material.

c) Heterosis Breeding

The Introduction of *milo* cytoplasm with *kafir* nucleus was discovered by Stephens and Holland (1954). That led to developed Cytoplasmic Genetic Male sterility. Initially, the line was developed known as *Kafir*-60 (cytoplasmic male sterile), the cytoplasm of A_1 (milo). Though various cytoplasm have been known in sorghum (Table 9.1), including Indian sources of CMS. The A_1 (milo) cytoplasm has been widely exploited in the entire hybrid program in India.

Table 9.1: The List of known Cytoplasm and their Source in Sorghum

<table>
<tr><th colspan="2">Name of Cytoplasm</th><th>Source</th></tr>
<tr><td colspan="2">A_1 (milo)</td><td>Milo (Durra race)</td></tr>
<tr><td colspan="2">A_2 and A_4</td><td rowspan="2">Guinea race</td></tr>
<tr><td colspan="2">9E</td></tr>
<tr><td colspan="2">A_3 (IS1112C)</td><td>Caudatum</td></tr>
<tr><td colspan="2">A_5</td><td>Bicolor</td></tr>
<tr><td colspan="2">A_6</td><td>Durra</td></tr>
<tr><td colspan="3">Indian sources of cytoplasm</td></tr>
<tr><td>Maldandi (M 31-2A; M 35-1)</td><td rowspan="3">Indian A_4 type</td><td rowspan="3">Durra based</td></tr>
<tr><td>VZM 2A (Vijayanagaram)</td></tr>
<tr><td>G1A (Guntur)</td></tr>
</table>

The productivity improved in sorghum due to the development and cultivation of hybrid varieties. Because of the well-characterized CMS system (millo A_1), many hybrids have been released in Asia, Africa, and other countries. More than 50 per cent yield increments have been recorded in grain and forage sorghum only due to hybrid technology. In India, CSH 1 was the first successful commercial hybrid released in 1964, followed by the release of more hybrids, *viz.*, CSH 5 and early hybrid CSH 6 were released in 1975 and 1976, respectively. The best-selling hybrid CSH 9 was released in the early 1980s with an 18-20 per cent yield increment over the previous hybrids CSH 5 and CSH6. After many hybrids have been released for *Kharif* and *Rabi* seasons according to the utility of sorghum, *i.e.*, grain, forage, sweet sorghum, *etc.* (Table 9.2). **Dr. Neelamraju G. P. Rao** is popularly known as the '*Father of Hybrid Sorghum*' because of his significant contribution to applied sorghum research and the popularization of hybrid sorghum technology in India.

d) Pedigree Method

The pedigree selection method is vigorously used in sorghum crop improvement programs. There are number of high-yielding varieties have been released for grain, forage, dual purposes, sweet –high biomass sorghum from AICSIP and state-level projects. The important HYVs are listed in Table 9.2.

4. Mutation Breeding

The variety CSV 5 has developed by X-ray, Co 19 is a natural mutant from Co 2; likewise, CSV 7 was developed from mutant CS35 41 (tall mutant). The high-yielding multi-cut forage variety CSV 33MF was developed from EMS mutant of COFS 29 in 2016. A chemically induced high lysine mutant P721 was developed at Purdue University has been used in various sorghum breeding programmes.

5. Population Improvement

Recurrent selection in open pollination or sib mating with use of local races as pollinators (male parent) and CMS line as a seed parent (female line).

6. Use of Apomictic Lines

The apomictic lines identified by vegetative propagation and these can be utilized in breeding programme and also fix up heterosis, *e.g.* R4-73 from Hyderabad.

7. Biotechnological Approach

The various biotechnological tools, including markers-assisted breeding, genomic selection, and genome-wide association mapping, are very promising for the improvement of sorghum in the context of yield components, biotic, abiotic stress resistance and quality improvement (biofortification).

Che *et al*, 2016 has developed transgenic sorghum lines that significantly augment pro-vitamin A levels to improve vitamin A deficiency. The gene *dh-dps-raec*1 related to the first enzyme called dihydropicolinate synthase of the lysine-specific pathway has been successfully introduced in the sorghum genome to improve sorghum with high lysine content and enhanced nutritional quality (Tadesse *et al*, 2003).

Breeding Institutes

- ☆ **International level:** International Crops Research Institute for Semi-Arid Tropics (ICRISAT), Hyderabad, India
- ☆ **National level:** Indian Institute of Millets Research (IIMR), Hyderabad.
- ☆ **State level:** Regional Research Stations-State Agricultural Universities.

Breeding Achievements

Table 9.2. List of Important Cultivars Released in India (National level)

Type of Sorghum/ Purpose	*Varieties*		*Hybrids*	
	Kharif	*Rabi*	*Kharif*	*Rabi*
Grain	CSV 1, CSV 2, CSV 3, CSV 4, CSV 5, CSV 6, CSV 7, CSV 10, CSV 11, CSV 13, CSV 15, CSV 17, CSV 20, CSV 23, CSV 27, CSV 28, CSV 31, CSV 34, CSV 36, CSV 39, CSV 43, CSV 45, CSV 48, GJ 44, Phule Purva, PJ 2510, CO 34, CSV 51, CSV 55 (Gujarat Goti), CSV62	CSV 7R, CSV 8R, CSV 14R, CSV 21R, CSV 18R, CSV 22R, CSV 26R, CSV 29R, GJ-102 (SuratGoti)	CSH 1, CSH 2, CSH 3, CSH 4, CSH 5, CSH 9, CSH 10, CSH 11, CSH 14, CSH 18 CSH 21 CSH 23, CSH 26, CSH 27, CSH 28, CSH 29, CSH 30, CSH 32, CSH 33, CSH 34, CSH 35, CSH 37, CSH 41, CSH 42, CSH 45, CSH 48, HJH1513, CSH-49 (DSH-6), CSH-55	CSH 7R, CSH 8R, CSH 12R, CSH 13R, CSH 15R, CSH 19R, CSH 31R and CSH 39R
Forage	Pant Chari-5, CSV 21F, SSG 59-3, Pusa Chari-23, CSV 30F, CSV 32F, CSV 33MF, CSV 35F, CSV 38F, CSV 40F, CSV 44F, CSV 46F and CSV 47F, CSV 53F, Pant Chari 14, Pant Chari 15, CSV 56F, CSV 57F, CNFS-1, CSV 59 BMR, CSV63F	-	PCH 106, CSH 13, CSH 20MF, CSH 24MF, CSH 36F, CSH 40F, CSH 43 MF, and CSH 46MF, CSH 50F, Jumbo Super (CSH 52MF), Jumbo Diamond (CSH 53MF), CSH 54BMR	-
Dual	CSV 15, CSV 20, CSV 23, CSV 27, CSV 41, GJ 43, GJ 45, *etc.*	-	-	-
Sweet/ High biomass	CSV 19 SS, CSV 24 SS, CSV 49 SS, CSV 52SS, CSV54HB, CSV6655	-	CSH 22SS, CSV 58SS,	-
Speciality	Phule Madhuri, CSV 50 R, CSV 60 Yellow, GJ 103 (Surati Ponk 1), CSV65 yellow, *etc.*	-	-	-

The steps involve in sorghum hybrid seed production

1. Entire program is taken up through State Seed Certification Agency.
2. Sowing of male sterile and pollinator lines in 4:2 or 6:2, or 12:4 rows ratio and entire field boundaries should be sown with four rows of pollinator lines.
3. Roguing should be carried at
   (i) 30 days after sowing
   (ii) At the time of flowering
   (iii) Before harvesting a crop
4. Keep proper isolation distance, *i.e.*, 400m (Breeder seed production) and 200m (Certified hybrid seed production).

The general model for producing hybrid sorghum by using cytoplasmic male sterility and fertility restoring line:

A line Male sterile	× ↓ **A line**	**B line** Male fertile	**Maintenance of A-line:** A line is crossed with maintainer line (B-line). Seed produced from male sterile A line seed and it used for a hybrid seed production programme.
A line Male sterile	× ↓ F_1 (Hybrid Seed)	**R line** Restorer gene	**Hybrid seed production:** A line is crossed with the Restorer line (R line). Seed produced from male sterile A line is a hybrid seed and used for commercial cultivation as a hybrid seed.

Chapter 10

Breeding Small Millets

Introduction

'Millets' is a common term to categorize small-seeded grasses belonging to the Poaceae family. They often termed by different names: small millets, minor millets, Nutri-cereals, dryland- cereals, *etc.* The word "millet" has originated from the French word "*Mile*" meaning thousand, which suggests a handful of millet contains thousands of grains. It is thought that they were under cultivation for more than 8000 years. Millets share specific common characteristics which make them socially essential crops. Millets are highly resilient and require few external inputs. They can be grown under harsh conditions in arid and semi-arid environments, requiring less water than many other cereals, and can often cope with poor soils. Millets are highly nutritious or called a storehouse of nutrients, with diversifying uses in human food, animal feed-fodder, pharmaceutical use, and commercial starch production. Because of such vital properties, they are often calling '*miracle grains*' or '*crops of the future* or '*climate-resilient crops*' and '*powerhouse of nutrients*'. Millets are broadly characterized into two major groups (A) **major millets**, *viz.*, pearl millet [*Pannisetum glaucum* (L.)] and sorghum [*Sorghum bicolor* (L.)]; (B) **minor or small millets**, *viz.*, finger millet [*Eleusine coracana* (L.) Gaertn.], foxtail millet [*Setaria italica* (L.) Beauv.], proso millet [*Panicum miliaceum* (L.)], barnyard millet (*Echinochloa* spp.), kodo millet [*Paspalum scrobiculatum* (L.)], and little millet [*Panicum sumatrense* Roth ex. Roem. and Schult.] cultivated in India.

Background Information

Sl. No.	*Crop Name*	*Common Name*	*Botanical Name*	*Family*	*Chromosome Number*	*Centre of Origin*	*Nature of Pollination*	*Pollination Mechanisms*	*List of Wild Relatives*
1.	Finger millet	Ragi, Bavato, Nagali, Wimbdi	*Eleusine coracana* (L.)	*Poaceae*	2n= 4x=36	East Africa	Self-pollination	Bisexuality, Chasmogamy, Cleistogamy	*E. indica, E. africana, E. kigeziensis, E. multiflora*
2.	Foxtail/ Italian millet	Kang, Italian millet, foxtail bristle millet	*Setaria italica* (L.)	*Poaceae*	2n= 2x=18	Central Asia/ China	Self-pollination	Bisexuality	*S. viridis, S. pumila*
3.	Proso millet	Cheno, broom millet, panic millet, hog millet, common millet	*Panicum miliaceum* (L.)	*Poaceae*	2n= 4x=36	Central Asia	Self-pollination	Bisexuality, Chasmogamy	*P. ruderale, P. halli, P. virgatum*
4.	Barnyard millet	Banti, Indian barnyard, Samo, sawa millet,	*Echinochloa frumantacea* (L.)	*Poaceae*	2n= 6x=54	Japan	Self-pollination	Bisexuality	*E. colona, E. crus-galli, E. pyramidalis, E. stagnina*
5.	Kodo millet	Kodra, Water couch, creeping paspalum	*Paspalum scrobiculatum* (L.)	*Poaceae*	2n= 4x=40	India	Self-pollination	Bisexuality, Cleistogamy	*P. sanguinale, P. orbiculare, P. setaceum*
6.	Little millet	Morauyo, Vari, Kuree, Saga, Gajro, blue panic	*Panicum sumatrense* (L.)	*Poaceae*	2n= 4x=36	India (Eastern Ghat of India)	Self-pollination	Bisexuality, Chasmogamy	*P. psilopodium*

Floral Biology

In poaceae (grass) family have compound inflorescences (*i.e.*, spike, panicle, ear, *etc.*) with numbers of spikelet arranged alternately in two vertical rows on opposite sides of an axis. Spikelet comprises two small bracts (glumes) with enclosed floret. Floret consists of lemma, palea and flower (perianth, stamens and carpels).

Particular	*Finger Millet*	*Foxtail/ Italian Millet*	*Proso Millet*	*Barnyard Millet*	*Kodo Millet*	*Little Millet*
Inflorescences	The inflore-scence consists of 4-5 digitated straight or slightly curved spikes, a whorl form a finger like structure. The spike rachis is flat.	The inflore-scence is a terminal spike, drooping, dense and cylindrical, borne on a thin and very short pedicel. Main stalk with shortened side branches bearing spikes and bristles.	The inflore-scence is a drooping panicle and looks like a broom. Primary branch is spreading or ascending or appressed.	The inflore-scence is a varying shape. Having terminal panicle 10–25 cm long. Contains numerous compact racemes. Arranged on one side, two sides or around the rachis.	The inflore-scence is consists 3–5 spike-like racemes (3-15 cm) consecutively arranged on a short to extended primary axis. Racemes (spike) composed of sub-sessile spikelets arranged in 2–3 rows along one side of flattened rachis.	The inflore-scence is contracted, oblong, nodding and is about 5–40 cm long and 1–5 cm wide panicle. The primary branches are closed and flats. The panicle branches are scabrous and drooping.

Particular	*Finger Millet*	*Foxtail/ Italian Millet*	*Proso Millet*	*Barnyard Millet*	*Kodo Millet*	*Little Millet*
Spikelet, Floret and Flower	Sessile spikelets are arranged alternatively in two rows on one side of the rachis. The terminal floret is sterile. Each spikelet comprises of 3–7 flowers enclosed by joint glumes. Two small scaly lodicules are present at the base of ovary.	Each spikelet comprises a pair of glumes holding two-minute flowers, the upper one bisexual and the lower one sterile. The spikelets are glabrous, elliptic to obovate with one to three bristles at base. The two lodicules are also present. With thickened lemma and palea.	Spikelets is singly, comprised of two glumes and two lemmas. The upper lemma is shorter with palea and has a fertile floret. The lower lemma is a long reduced palea and has a sterile floret. Glume length is unequal.	Spikelets are arranged tightly or loosely in four irregular rows on the triquetrous rachis. Each spikelet comprises of two florets supported by two glumes; upper bisexual floret contains two unequal glumes, lemma and palea, and lower floret is neuter (sterile) with lemma and minor palea.	Spikelets are arranged alternatively in two series (long and short pedicelled) with two florets. The upper glume is as long as the spikelet, while lower glume is absent. The lower lemma is flat more or less membranaceous and without palea.	Spikelets are solitary, the fertile spikelets are pedicellate with two minute florets; basal one is sterile or staminate and upper one is fertile without rachilaa extension. Both the florets possess one lemma and one palea of each. Glumes cover top of florets, thinner than the fertile lemma.

Particular	*Finger Millet*	*Foxtail/ Italian Millet*	*Proso Millet*	*Barnyard Millet*	*Kodo Millet*	*Little Millet*
Androecium	Stamens are three with long filaments and short oblong anthers. Pollen viability is short about 10–15 minutes	The bisexual flower consists three stamens	The fertile floret consists three stamens	The bisexual flower comprises three stamens	The fertile floret consists three stamens	The fertile floret consist three stamens
Gynoecium	Single carpel, feathery branched stigma, monolocular superior ovary. The stigma is receptive for about 5 min after emergence from the glumes.	Single carpel, two style with feathery stigmas, monolocular superior smooth ovary.	Single superior ovary with bifid style and feathery stigma	Single superior ovary with two separate styles with plumose stigmas	Monocarpellary, superior ovary with single ovule, two separate styles with feathery stigmas	Monocarpellary, superior ovary with single ovule

Particular	*Finger Millet*	*Foxtail/ Italian Millet*	*Proso Millet*	*Barnyard Millet*	*Kodo Millet*	*Little Millet*
Fruits	Caryopsis	Caryopsis; oval shape tightly enclosed within the lemma and palea	Caryopsis	Caryopsis; oval shape tightly enclosed in white lemma and palea	Caryopsis; enclosed within the shiny hard lemma and palea that is difficult to separate from grain	The caryopsis is with attached pericarp
Seed	Monocots	Monocots	Monocots	Monocots	Monocots	Monocots
Floral formula	Br. ⊕ ⚥ $P_{2(Lodicules)} A_3 \underline{G}_{(1)}$	Br. ⊕ ⚥ $P_{2(Lodicules)} A_3 \underline{G}_{(1)}$	Br. ⊕ ⚥ $P_{2(Lodicules)} A_3 \underline{G}_{(1)}$	Br. ⊕ ⚥ $P_{2(Lodicules)} A_3 \underline{G}_{(1)}$	Br. ⊕ ⚥ $P_{2(Lodicules)} A_3 \underline{G}_{(1)}$	Br. ⊕ ⚥ $P_{2(Lodicules)} A_3 \underline{G}_{(1)}$
Anthesis	Anthesis begins from top spikelets and proceeds to downwards. The maximum number of flowers opens on the third to fourth day.	Anthesis begins from top spikelets and proceeds to downwards. Single flower open for 30 minutes	Anthesis begins from top spikelets and proceeds to downwards. It usually takes 10–12 days for complete flowering within an inflorescence. Single flower open for 10-15 minutes	Anthesis begins from the top of the inflorescence and proceeds downwards, finishing in 10–15 days. Flowers open from 5 to 10 am, with a maximum number of flowers opening between 6 and 7 AM	Anthesis begins from the middle and proceeds to both sides of the inflorescence	Anthesis begins from the top of the inflorescence and proceeds downwards, finishing in 12–15 days.

Figure 10.1: Finger Millet Floral Biology.

Figure 10.2: Foxtail Millet Floral Biology.

Recognition of Millets at the National and International Level

The year 2018 was declared as 'National Year of Millets'

India is the largest producer and consumer of millet in the world. Considering the potential in delivering food and nutritional security of the country, contains unique nutritional, climate-resilient traits, the powerhouse of nutrients and health benefit properties, the Central Government declared millets consist of Sorghum (Jowar), Pearl Millets (Bajra), Finger Millets (Ragi/Mandua) and Minor Millets *i.e.* Foxtail millets (Kangani/kakun), Proso millet (Cheena), Kodo Millet (Kodo), Barnyard millet (Sanwa/Jhangora), Little millet (Kutki) and two Pseudo Millets (Buck wheat-kuttu, Ameranthus-chaulai) as a 'Nutri-Cereals' for production, consumption and trade point of view [Ministry of Agricultural and Farmers Welfare Gazette Notification No. 133 Dated 13-4-2018]. Promoting millets in the country, the Government initiated the millets mission under the National Food Security Mission (NFSM) launched in the year 2007.

Figure 10.3: Proso Millet Inflorescence.

Figure 10.4: Barnyard Millet Floral Biology.

Figure 10.5: Kodo Millet Floral Biology.

Figure 10.6: Little Millet Inflorescence.

The Year 2023 is declared as 'International Year of Millets'

Understanding food security has broadened beyond the availability and access to 'just' food to encompass nutrition. The UN General Assembly adopted by consent a resolution sponsored by India declared 2023 as the International Year of Millets, aiming to raise awareness about the grains' health benefits and suitability for cultivation under changing climatic conditions.

Why do Millets Hold Promise for Next-Generation Agriculture?

Millets are ancient crops domesticated and cultivated in the world. It is a superior grain with unique characteristics that promise next-generation agriculture. Therefore, millets are considered a promising crop that would address important issues like food, fodder and fuel; malnutrition, health problem, climate alteration, the reservoir of genetic variability, *etc.*, in future agriculture, the key points are listed below:

A. The diversified uses of Millets (*i.e.*, food, fodder and fuel)

The millets are used in diversified manners *viz.*, food, feed for animals and fuel purposes. It is considered the staple food for people living in various parts of Asia, Africa, the Indian subcontinent, and East Asia. People have utilized millets in various forms such as wholegrain foods, meal or flour-based foods, alcoholic and nonalcoholic beverages.

B. Nutritional Benefits of Millets

The millets are more correctly called 'Nutri-cereals' because of the storehouse of nutrients and unique properties that make them capable of hidden hunger. They comprise unique features like

- ☆ High-quality gluten-free protein
- ☆ Low glycemic-index
- ☆ Very easy to digest due to the lower carbohydrate content and non-glutinous
- ☆ Richness in dietary fibres, amino acids, vitamins, antioxidants properties, phytochemicals
- ☆ Content, high concentration of micronutrients, and bioactive flavonoids and small peptides

C. Health Benefit of Millets

Millets are a good alternative for the major cereals and a tonic to overcome health problems like diabetes and obesity due to their superior nutritional values and some unique dietary properties. Posses' anti-diabetic and antioxidant activities, anti-tumorigenic, atherosclerogenic effects, and antimicrobial properties. The consumption of millet improves the digestive system, increases energy levels, detoxifies the body, lowers the risk of cancer, increases immunity in respiratory health, and improves neural and muscular systems.

D. Resilient to Climate Changes

The millets have extraordinary morphological, physiological and biochemical characteristics that make them promising and excellent alternative crops to challenge the hurdles of climate change. Some of the unique traits are

- ☆ Minimum vulnerable to biotic and abiotic stresses
- ☆ Better growth and productivity under low nutrient input conditions, less irrigation conditions
- ☆ High dry matter production capacity and a high degree of tolerance to drought and heat
- ☆ Superior photosynthetic efficiency due to C_4 type
- ☆ Short duration: In specific, the life cycle of millets completes in 12-14 weeks

E. Reservoir of Genetic Variability

Millets have a vast genetic base and genetic diversity in crop cultivars. This diminishes the risk of vulnerability resulting in crop failure due to infestation by insect pests and diseases or unpredictable climatic effects.

Breeding Consequences of Small Millets Improvement (Breeding Objectives and Approaches)

Minor millets received minimal attention concerning crop breeding and production except for finger millet. The crop-wise breeding consequences, important objectives and major breeding approaches are mentioned below:

1. Finger Millet

Finger millet is the important staple food of Africa and Southern Asia. In India the grown mainly grown in southern regions and some parts of the middle.

Breeding Objectives

1. Improve yield potential: Grain and Fodder
2. Breeding non-lodging tall type cultivar
3. Breeding for early maturity
4. Breeding tolerance or resistance to high temperature and drought
5. Breeding tolerance or resistance to blast disease
6. Breeding for improved quality includes protein (10 per cent), calcium (>450 mg/kg), grain Fe (40mg/kg) and Zn (24mg/kg), popping (> 70 per cent), *etc.*
7. Breeding white grain colour variety: Rich in protein (~12 per cent)

Breeding Methods

Because of the self-pollination rule, various breeding methods are used for finger millet improvement

1. Plant Introduction
2. Pure line selection
3. Mass selection
4. Interspecific hybridization
5. Pedigree selection
6. Mutation breeding

In the early finger millet improvement program, breeding work was initiated with the Introduction of Indo-African crosses of finger millet and released various varieties under Indaf series by the late **Dr. C. H. Lakshmanaiah**, who is known popularly a "Ragi Brahma".

2. Foxtail Millet

Foxtail millet is one of the oldest crops among cultivated millets, it is a short duration used for food, feed and fodder. Cultivated in various countries of Asia, Africa and America. In India, currently, it is cultivated on a very limited area, mostly in the state of Andhra Pradesh, Karnataka, Telangana, Tamil Nadu, Maharashtra, Rajasthan, Madhya Pradesh and north-eastern states.

Breeding Objectives

1. Improve yield potential: Target traits include grain and high biomass and harvest index
2. Breeding non-lodging dwarf-type cultivar
3. Breeding for early and synchronous maturity
4. Breeding drought resistant,
5. Breeding for sheath blight resistance and shoot fly tolerance
6. Breeding variety with high water and nitrogen use efficiency

Breeding Methods

In foxtail millet, an autogamy rule, difficulties in emasculation and hybridization, presently the simple selection breeding methods are used.

1. Plant Introduction
2. Pure line selection
3. Mass selection, and
4. Pedigree selection

3. Proso Millet

It is one of the oldest domesticated crops. It has a very short duration (70–80 days), very high water–use efficiency among all cereals. Green plants are used as fodder for animals, while the dehusked grain is generally cooked like rice, it is highly nutritious.

Breeding Objectives

1. Improve yield potential: Grain and high biomass
2. Breeding non-lodging
3. Breeding for early maturity
4. Breeding for non-shattering type
5. Breeding for grain quality: Uniform shape, bold size, waxy types (A waxy form is preferred in the food industry for glutinous nature appropriate for beverage manufacturing for their high fermentation efficacy)
6. Breeding for disease and pest resistance: Important diseases are blast, leaf blight, sheath blight, rust and grain smut. While shoot fly is a major pest

Breeding Methods

The conventional and non-conventional approaches have been used in proso millet improvements are

1. Pure line selection
2. Mass selection
3. Hybridization methods: Pedigree selection and back cross-breeding; inter-specific hybridization
4. Mutation breeding
5. Biotechnological approach: Tissue culture technique; Molecular tools to identify QTLs for yield and resistant traits

4. Barnyard Millet

Worldwide, India is the largest barnyard millet producer, both in relation to area and production. It is a good climate-resilient crop grown for human food and fodder. Due to its unique nutritional properties, it is the best food for patients with dietary-based health defects like diabetes, celiac diseases and heart-related diseases.

Breeding Objectives

1. Improve yield potential: Grain and fodder, high biomass
2. Breeding dwarfing and non-lodging
3. Breeding for early maturity
4. Breeding for non-shattering type
5. Breeding variety with bolder grain that easily dehulling (more dehulling percentage)
6. Breeding for disease: Important diseases are grain smut and leaf blight
7. Identification of male sterile line

Breeding Methods

In barnyard millet an autogamy rule, many small number of flowers per raceme, flowering occurs in early hours, pollen viability is too short are major hampers in emasculation and artificial hybridization. The basic conventional approaches have been used are

1. Pure line selection
2. Mass selection
3. Hybridization methods: Pedigree selection, SSD method, and
4. Mutation breeding

5. Kodo Millet

Kodo millet is one of the toughest crops among the small millets. It is an indigenous cereal of India, regularly cultivated in the states of Madhya Pradesh, Uttar Pradesh, Chhattisgarh, Tamil Nadu, Gujarat, Maharashtra and Karnataka. The grains are bolder than foxtail and little millet grains, it is a balanced and principal food grain of tribal and poor populations.

Breeding Objectives

1. Breeding for improve yield potential
2. Breeding for wider adaptability and yield stability
3. Breeding for early maturity
4. Breeding for disease and pest resistance: Head smut and shoot fly
5. Breeding for drought and salinity tolerance

Breeding Methods

The basic conventional approaches have been used are

1. Introduction and selection
2. Pure line selection
3. Mass selection
4. Hybridization method: Pedigree selection
5. Mutation breeding

6. Little Millet

Little millet is important minor millet which originated in the Indian subcontinent. It is extensively cultivated as a minor cereal across the country, mainly in eastern India, where its systems are part of tribal agriculture. It is tolerant in moisture deficit conditions and gives reasonable yield even in poor soil fertility. The grains of little millets can be conserved for 10 years or more without much deterioration, so it is also known as good reserve food.

Breeding Objectives

1. Improve yield potential: Grain and stover
2. Breeding for disease and pest resistance

Breeding Methods

Little millet is a highly self-pollinated crop, many small spikelets on bristle pedicel are major hampers in hybridization techniques. The basic conventional approaches have been used are

1. Pure line selection

2. Mass selection, and
3. Mutation breeding

Breeding Institutes

- ☆ **International level:** International Crops Research Institute for Semi-Arid Tropics (ICRISAT), Hyderabad, India.
- ☆ **National level:** Indian Institute of Millets Research (IIMR), Hyderabad.
- ☆ **State level:** Regional Research Stations-State Agricultural Universities.

Breeding Achievements

The latest cultivars of small millets are released in India are mentioned in Table 10.1.

Table 10.1. List of Important Cultivars of Small millets Released in India

Sl.No.	*Name of Crop*	*Cultivars*
A.	Finger millet	VL-382, CFMV-2, VR-988, GN 8, Vegavathi, FMV-1102, VL-379, GNN-7, Dapoli-2, GN-5, GNN-6, VL 352, GPU 67, Co 15, Co1, Co 3, Gossaigaon Marua Dhan, Phule Kasari, Gosthani, CFMV 4, VL Mandua 400 (CFMV 5), Shreeratna, VL Mandua 408 (CFMV 6), CFMV 7, VL Mandua- 402, VL Mandua-409, ATL 2, Hagari Ragi-13, *etc.*
B.	Foxtail millet	ATL-1, Hagari Navane-46, SiA 3223, SiA 3156, Surya Nandi, SiA 3085, PS 4, PRK 1, K3, Co 4, Co 5, Co 7, Mahanandi, GPUF 3, CFXMV-1, PDKV Yashashree, ATL 2, *etc.*
C.	Kodo millet	GAK-3, KMV 545, KMV-543, Jawahar Kodo 137, RK 390-25, TNAU-86, JK 98, Indira Kodo 1, JK 13, Gujarat Kodra 2, Gujarat Kodra 2, GK4 (Dahod Kodra 4), Dahod Kodo 1 (CKMV 3), JK- 9-1, CKMV4, CKMV5, *etc.*
D.	Barnyard millet	DHBM-23-3, MDU-1, DHB-93-2, DHBM 93-3, Co (KV) 2, Gujarat Banti 1, Phule Barti-1, ATL 1, CBYMV-1, VL Madira -54, GB 2 (Sabari), *etc.*
E.	Little millet	Jaicar Sama 1, LMV513, GNV-3, DHLM-14-1, JK 4, GV-2, BL 6, JK 36, Co 4, OLM 217, OLM 208, Kalinga Suan 18 (CLMV-2), GV-4 (Ambika), GPUL 6, DHLM-28-4, Konkon Satwik, GPUL-11, GV 5 (Mavli), *etc.*
F.	Proso millet	PMV-442, DHPM-2769, TNPm-230, TNAU 202, Pratap Chenna-1(PR-18), TNAU 164, TNAU 151, CO(PV) 5, TNAU 145, Hagari Barugu-1(HB-1), GPUP 28, ATL 2, *etc.*

Section-III

Breeding Pulses (Chickpea, Pigeonpea, Green Gram, Black Gram, Cowpea and Guar)

Chapter 11

Breeding Chickpea (*Cicer arietinum* L.)

Introduction

Chickpea is an essential rabi pulse crop of the semi-arid tropics, also known as Bengal gram. In India, both desi and Kabuli types of cultivars are cultivated. India is the largest producer, consumer and also leading exporter of chickpea. Nutritionally, it contains high protein (20-22 per cent), fibre, minerals and β-carotene. It is considered as a cheap source of protein and energy in the developing world. In India, chickpea is versatile food use, cooked and eaten as salads, cooked in stews, milled into flour called gram flour (besan) used for making different sweet stuffs.

Background Information

1.	*Common name*	Gram, Chana
2.	*Botanical Name*	*Cicer arietinum* L.
3.	*Family*	*Fabaceae* (*Leguminosae*)
4.	*Chromosome number*	2n=2x=16
5.	*Centre of origin*	Southeastern Turkey and Syria
6.	*Nature of pollination*	Self-pollination
7.	*Pollination mechanisms*	Bisexuality, Cleistogamy, Keel petal
8.	*List of wild relatives*	*C.bijugum, C. reticulatum, C. pinnatifidum, C. echinospermum*

Types of Chickpeas

Chiefly chickpeas are divided into two major types.

1. **Desi:** Also known as '*microsperma*'. Seeds are smaller, dark brown in colour, with a thick rough coat. Plants contain anthocyanin pigmentation, cultivated in India and its subcontinent, Ethiopia, Mexico and Iran.
2. **Kabuli:** Also known as '*macrosperma*'. Seeds are bolder, lighter in colour, larger, with a thin smoother coat. Tall plant stature, pods and leaves size are larger. Plants lack anthocyanin pigmentation. Evolved from the smaller seeded desi types through natural mutation. It is mainly cultivated in Southern Europe, Northern Africa, Afghanistan and Chile. It might moved from the Mediterranean region to India and the subcontinent around the 18th century.

Gene Pool

A gene pool of chickpeas is grouped into three categories

Sl.No.	*Gene pool*	*Species*
1.	Gene pool 1 (GP-1)	*Cicer arietinum, C. echinospermum, C. reticulatum*
2.	Gene pool 2 (GP-2)	*C. bijugum, C. judaicum, C. pinnatifidum*
3.	Gene pool 3 (GP-3)	Rest all species of genus *Cicer*

Distribution

It is mainly cultivated in India, Pakistan, Mexico, Turkey, Ethiopia and Burma. In India, major cultivated states are Madhya Pradesh, U.P., Rajasthan, Haryana, Maharashtra, Bihar, West Bengal and Andhra Pradesh.

Floral Biology, Selfing and Crossing Techniques

Inflorescences	:	Racemose type; solitary axillary
Flower	:	Pedicellate, zygomorphic, irregular, complete, perfect, pentamerous except pistil, slightly perigynous, papilionaceous.
Calyx	:	Five sepals, persistent, gampsepalous, calyx tube oblique and sepaloid.
Corolla	:	Five petals [1+2+(2)], polypetalous, of unequal size, vexillary aestivation, **Papilionaceous** [one posterior largest petal (standard vexillum) that encloses two lateral petals (wings or alae) and two fused forming a unique structure like a keel of a boat, commonly referred to as a keel].

Androecium	:	Diadelphous stamens – (9) +1; one posterior stamen is free, but the remaining 9 are fused to form an incomplete stamina tube that encloses ovary, anther-dithecous, basifixed and dorsifixed. The bursting of anther longitudinally.
Gynoecium	:	Monocarpellary, ovary- superior, unilocular, single ovule, Placentation- marginal, style – short and curved, feathery stigma.
Fruits	:	Legume or pod (dehiscence formed by marginal placentation) Pod is pubescent, covered with glandular and non-glandular hairs bearing 1–2 seeds.
Seed	:	Ex-albuminous with two cotyledons.
Floral formula	:	⊕ ⚥ $K_{(5)}\ C_{1+2+(2)}\ A_{(9)+1}\ \underline{G}_{1}$

Anthesis

Anthesis occurs in the bud stage (1 or 2 days before the flower's opening). Anthesis takes place throughout the day. The flowers of chickpeas pass through different stages of development *viz.*, Close bud, hooded bud, half-open flower, fully open flower and fading flower. The stigma is receptive in the hooded bud to the fully open flower stage, while pollen matures and dehisce of anthers during the half-open flower stage.

Artificial Crossing and Pollination

The emasculation and hybridization are tedious due to the small flower structure and cleistogamous condition. However, artificial hybridization with and without emasculation has been followed in chickpea. In the case of artificial hybridization, select the flower bud (hooded stage), owing to open the next day. The bud is held at the base between the thumb and forefinger. The anterior sepal is exposed with the help of forceps, and the keel is pushed down or a small cut at the base of the keel petal and makes a cut along the central axis down to the distal end. The exposed anthers are then softly removed from the filaments. Tagging of emasculated bud by tying around the internodes below the emasculated bud with coloured cotton thread or a tag.

For a better seed setting, the pollination is generally performed the next morning in the case of Northern India while immediately after emasculation in southern India. Half-opened flower stage is the best source of matured pollen. Care should be taken that no any mechanical damage to the other floral parts (stigma, ovary) at the time of emasculation and pollination.

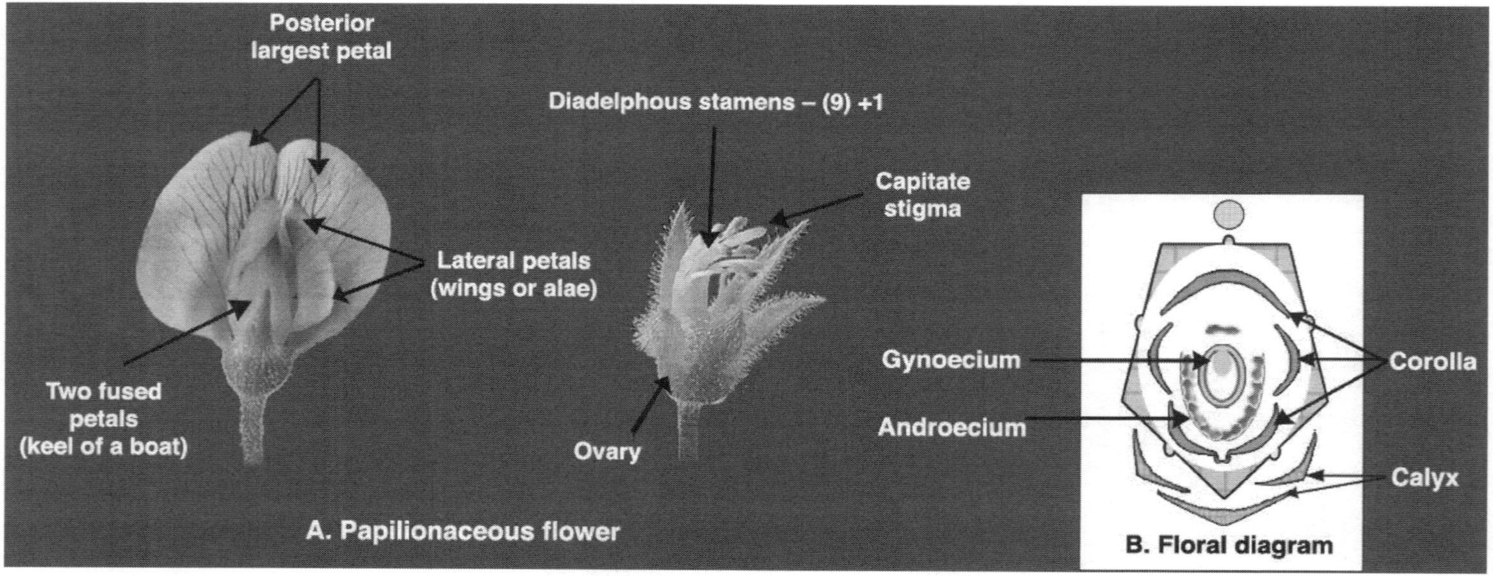

Figure 11.1: Floral Biology and Diagram of Chickpea (*Cicer arietinum* L.)

Breeding Objectives

1. **Breeding for increased seed yield:** Due to Photo and thermo sensitivity, and indeterminate growth habit results in a low harvest index. The breeding cultivar that has high biomass, tall, erect and compact type, the optimum number of primary and secondary and minimum number of tertiary branches with more responsive to fertilizer and irrigation are important.
2. **Broadening of a genetic base:** The lesser genetic gain is mainly due to a narrow genetic base. Increase variability by incorporating wild relatives in a breeding program. Wild species are considered a significant source of genes that give resistance to a diverse range of biotic and abiotic stresses.
3. **Breeding for early type:** Important for suitable in the short growing season and fit in the multi-cropping system and escape biotic and abiotic stresses. The important varieties for short-duration are ICCV 2 (Sweta), JG 16, JG 11, Vijay, Vikas, Vishal, JGK 1, KAK 2, ICCV 10, ICCV 96029 and ICCV 96030.
4. **Breeding variety suitable for mechanical harvesting:** To minimize the cost of cultivation, and due to labour scarcity, the demand of varieties harvested by combined harvesters are increasing. The ideal characteristics of a variety are tall (> 55cm), erect-semi erect growth habit (>60° branch angle from the soil surface) and no pod setting up to 25 cm height (ground clearance). Many cultivars developed in central and south India are NBeG 47, GBM 2, Phule Vikram, Phule G 08108, JG 20016-24, and BG 3062.
5. **Wider adaptability:** Essential for yield stability over the environment, *i.e.*, diverse environments, *e.g.*, varieties Chaffa and C 235.
6. **Breeding for resistance/tolerance to insect pests** *i.e.*, Pod borer and leaf minor, are important pests in chickpea. The yield potential can be maintained by breeding resistant cultivars.
7. **Breeding for diseases resistance**: The major diseases like *Ascochyta blight, Fusarium wilt,* root rot, botrytis grey mould, stunt, root knot nematodes are responsible for yield reduction in chickpea. Therefore, breeding resistance/tolerance cultivars in chickpea is essential criterion. The important cultivars/genotypes identified for biotic stresses are given in Table 11.1.
8. **Breeding for tolerance to abiotic stress** *i.e.*, Drought, cold, frost, heat, salinity and alkalinity are major abiotic stresses. The chickpea production become uncertain due to climate changes and erratic abiotic stresses. The important cultivars/genotypes identified for abiotic stresses are given in Table 11.2.

Table 11.1. The Important Cultivars/Genotypes Identified for Biotic Stresses

Biotic Stress	*Cultivars/Genotypes*
Ascochyta blight	HOO-108, GL 92024, PBG1, PBG 5, GNG 469, GL 88016, Gaurav, ICC 1069, PBG 7, BG 267, and GNG 2171, GNG 469, BG 362, H03-45, ILC 3279, GNG 1581, IPC 79, IPC 129, *etc.*
Fusarium wilt	RKGK 13-499, Pusa Parvati, H 82-2, KWR 108, GPF 2, Vijay, JG 11, Vishal, GNG 663, JG-16, KPG 59, Digvijay, Rajas, BGM 547, BGD 128, GNG 1581, Phule G16111, GJG1603, *RKG* 13-55, NBeG 776, PG 209, JG 74315-2, IPC 08-11, PG 211, JG 2017-50, Phule G 0819, BCP 60, JG 2017-49, GJG 0922, GNG 2325, GNG 2391, GL 13037, IPC 07-28, NBeG 779, H 12-63, SCGP-WR 28, GJG 0814, IPCK 10-134, IPC 17-28, GJG 0921, GJG 1010, SCGP-WR 32, GJG 904, IPC 08-69, CSJK 96, Gujarat Gram 1, Gujarat Gram 2, Pant Gram 5, Gujarat Gram 5, GJG 6, *etc.*
Botrytis grey mould	IPC 15-95, IPC 15-202, IPC 15-48, IPC 15-113, IPC 15-185, GCP 101, CSJ 556, RVG 202, ICC 1069, GNG 1581, Dhanush, IC 12483, ICCW 92, HK 94-134, ICCV 41, CSJK 72, GLW 91, GL 10006, GLW 69, *etc.*
Pod borer (*Helicoverpa armigera*)	ICCL 86102, ICCL 86111, ICCX 730020, H 12-55, RVSSG 63

Table 11.2. The Important Cultivars/Genotypes Identified for Abiotic Stresses

Abiotic Stress	*Cultivars/Genotypes*
Drought tolerance	Vijay, RSG 888, ICC 4958, ICC 8261, Super Annigeri-1
Salinity tolerance	CSG 8962, ICCV 10, JG 62
Cold tolerance	GL 26018, GL 28202
Heat tolerance	JG 14, JSC 55, ICC 15614, JSC 56, ILWC 115, ILWC 21, EC 556270

9. **Development of herbicide-tolerance varieties:** In present agriculture, manual weeding is a major problem due to labour shortage, time consuming and expensive. Cultivars need to develop that tolerance to post-emergence herbicides like imazethapyr and carfentrazone-ethyl. The genotypes have been identified are GLK 10103, NDG 11-24 and GL 22044.

10. **Breeding for quality improvement/Biofortification:** The quality of chickpea includes market/consumers preference, milling quality and nutritional composition.
    - ✰ Market/consumer preference:
        - ❖ Grain colour- yellow to light brown seed coat colour;
        - ❖ Seed size: small to medium size seeds for the desi type and bold size for the kabuli type. The important cultivars were developed in

desi type Pusa 256, JG 11, Phule G 5, Samrat, Vishal and BGM 547, while in kabuli type BG 1003, BG 1053, Haryana Kabuli Chana 1, Haryana Kabuli Chana 2, JGK 1, KAK 2, Virat and Vihar. The extra-large seeded (50 g/100) cultivars Phule G 0517, PKV 4–1 and MNK-1 are developed in kabuli chickpea.

- ❖ Grain shape; Seed coat thickness: Seed coat thickness is an essential parameter for processing desi seeds with thicker coats compared to the kabuli type. Seed mass, cooking time, and seed volume expansion before and after soaking are also essential quality parameters.

- ✰ High dal recovery (milling efficiency)
- ✰ Nutritional quality-High protein content, Sulphur containing-amino acids methionine and cystine. The transgenic breeding approach is being used to enhance the level of such deficient amino acids; β-carotene, Fe and Zn content.

11. **Development of variety for vegetable purpose:** Keep in view the high demand of green pod for vegetable purpose in market, the appropriate breeding program required to develop high-yielding chickpea (desi) varieties suitable for green immature grain for vegetable.
12. **Development of male sterile line in chickpea:** The exploitation of heterosis in chickpea is a hurdle due to the hybridization problem, seed production in a large scale, high seed rate, *etc.* The exploitations of heterosis may be possible by developing a stable and useful male sterile system in chickpea.

Breeding Approaches (Methods)

1. **Introduction:** The introduced genotype is directly or indirectly used in chickpea improvement. *e.g.*, G109-1, a bruchid-resistant line was selected and introduced from the Turkish variety. Russian tall line was used to breed the tall variety Pusa 261.
2. **Pure line selection:** The important varieties were developed by selection from local landraces, and exotic germplasm. For examples Chaffa, Jyoti, Annegri-1, Pb7, S26, Co1, Co2, JG 62, JG 315, BDN9-3, Dahod yellow, GNG 146, *etc.*
3. **Hybridization followed by selection:**
   - ✰ **Pedigree selection:** Breeding for biotic and abiotic stresses resistance cultivars developed by three-way and multiple crosses.
     - ❖ *Single cross:* Two parents involved in hybridization
     - ❖ *Three-way crosses:* Three parents used for specific traits, *e.g.*, BG 244, JG 130, JG 11, BGD 72, H 208, BG 1003, GPF 2, *etc.*

- ❖ *Multiple crosses:* More than three parents involved, crossed in various way, *e.g.*, BG 256, ICCV2, Phule G 95311, SG2, *etc.*

☆ **Back cross method:** Used to transfer disease resistance traits in the case of interspecific hybridization.

☆ **Interspecific hybridization:** The various wild species *viz., Cicer reticulatum, C. echinospermum, C. pinnatifidum, C. judaicum etc.* have been used in chickpea breeding program. For examples variety Pusa 1088 was first variety developed through inter-specific hybridization. A high-yielding variety PBG 8 from an interspecific cross of *C. arietinum* × *C. judaicu*. The cultivar Pusa 1103 was developed by wide cross between *C. arietinum* and *C. reticulatum*.

4. **Mutation breeding:** Generally used to improve specific traits and adaptation. Mostly gamma rays, DMS and fast neutrons have been used in chickpea. The varieties have been developed are given in Table 11.3.

Table 11.3. List of Chickpea Cultivars Released through Mutation Breeding in India

Sl.No.	*Cultivars*	*Year*	*Source of Mutagen*	*Characters Improved*
1.	V16 (Amba)	1981	DMS	High yield, resistance to fungal and bacterial diseases
2.	V37 (Shreshtha)	1981	DMS	High yield, high vegetative growth, suitable for fodder
3.	V38 (Swarna)	1981	DMS	High yield, early maturity, synchronous flowering, better quality pods and grains, resistance to diseases
4.	V240	1984	DMS	High yield, resistance to fungal, viral and bacterial diseases
5.	Co 5	1984	Gamma rays (300 Gy)	More nutritive forage cowpea, high yield (16 per cent), comparable for intercropping with fodder cereals
6.	Kiran	1984	Neutrons	Erect plant type, increased pod number, high yield, early maturity and salt tolerance
7.	Pusa 408 (Ajay)	1985	Gamma rays (600 Gy)	High yield, blight resistance, semi-erect, 140-155 days to maturity and improved plant architecture
8.	Pusa 413 (Atul)	1985	Gamma rays (600 Gy)	High yield, wilt resistance, moderate blight resistance, resistance to stunt virus, foot root, root rot, semi-erect, the higher number of branching, *etc.*

Sl.No.	Cultivars	Year	Source of Mutagen	Characters Improved
9.	Pusa 417 (Girnar)	1985	Gamma rays (600 Gy)	High yield, short, semi-erect, profuse branched, high pod number, wilt resistance, moderate resistance to stunt virus, collar root, foot rot, root rot, low pod borer and nematode damage
10.	BGM 547	2005	Gamma rays	High yield, bold grain size, attractive golden brown colour and moderate resistance to wilt, root rot, stunt and *Helicoverpa armigera*
11.	Pusa 547	2006	Gamma rays (600 Gy)	High yield, good cooking quality, tolerance to *Fusarium* wilt, stunt virus and root rot

5. Biotechnological Approaches

- Double haploid technique
- Transgenic breeding
- Genomic assisted breeding: Identified several QTLs for biotic and abiotic stress. Some of the QTLs are introgressed in popular genetic backgrounds.
- Marker-assisted breeding/selection: With the use of MAS, success achived in the development of stress tolerance genotypes, namely drought tolerant cultivar Pusa Chickpea 10216 and *Fusarium* wilt resistance cultivar Super Annigeri 1, Pusa Chickpea 20211 (Pusa Chickpea Manav), have been developed.

Breeding Institutes

- **International Level**

  i. International Crop Research Institute for Semi-Arid Tropics (ICRISAT), Hyderabad, India.

  ii. International Centre for Agricultural Research in the Dry Areas (ICARDA), Beirut, Lebanon, Syria.

- **National level:** Indian Institute of Pulses Research (IIPR), Kanpur, (Uttar Pradesh).
- **State level:** Regional Research Stations-State Agricultural Universities.

Breeding Achievements

- **Desi type:** Dahod yellow, Chaffa, ICCC 4, Gujarat Gram 1(GG 1), Gujarat Gram 2 (GG 2), Gujarat Gram 3 (GG 3), Gujarat Gram 4 (GG 4), Gujarat

Gram 5, Gujarat Junagadh Gram 3, GJG 6, Teej, RVG 202, Pusa Chickpea 20211, RVG 204, Pusa Parvati, Pusa Chickpea 10216, Super Annigeri 1, Raj Vijay Kabuli Gram 2020, PBG 9, Kota Kabuli Channa 2, HC 6, L 558, Pusa JG 16, Gujarat Gram 7, Jawahar Gram 14, GNG 2461, ADVIKA (NC 7), Gujarat Gram 8, Saatvik, Kundan, Pant Gram 10, Nandyal Gram- 1267, *etc.*

- **Kabuli type:** KAK 2, JGK 1, PG 0517, BDNGK 798, IPCK 2002-29, Karan Kabuli 4, Kota Kabuli Chana 3, Vallabh Kabuli Chana-1, PKV Kabuli 4, IPCK 2013-163 (Madhav), Gujarat Kabuli Gram 1, Gujarat Kabuli Gram 2, *etc.*

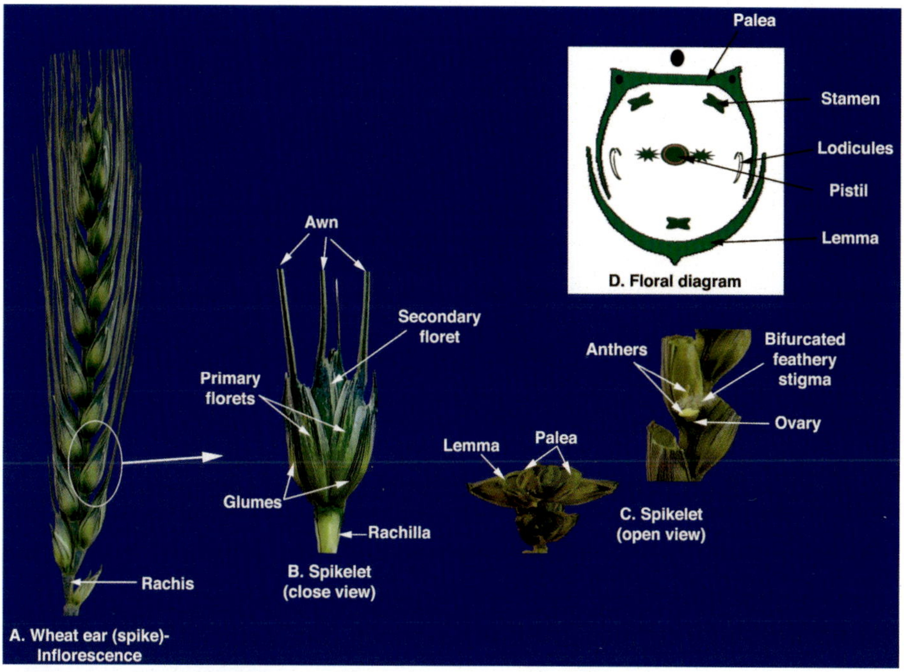

Figure 5.1. Floral Biology and Diagram of Wheat (*Triticum aestivum* L.). (pg. 36)

Figure 5.2. Prepartion of Wheat Spike for Emasculation Portion of Glume, Lemma and Palea is Cut. (pg. 37)

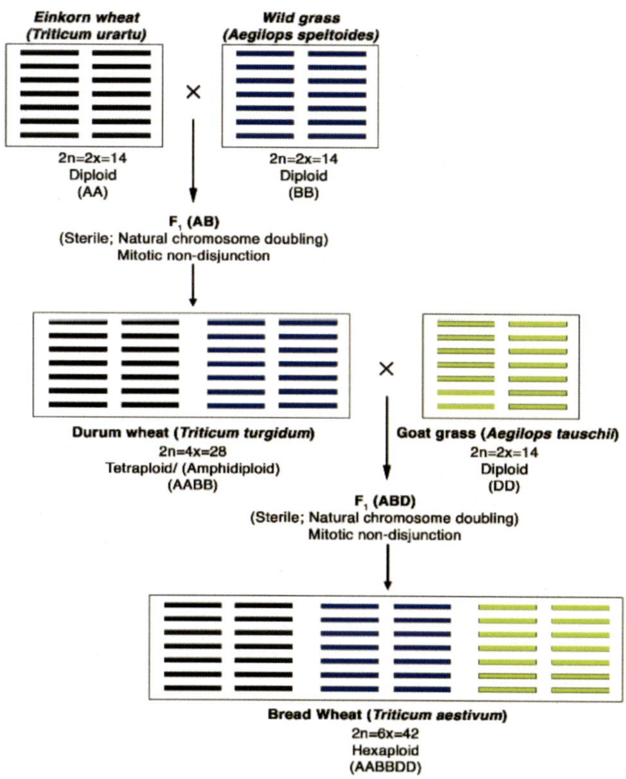

Figure 5.3. Genetic Origin of Common (Hexaploid) Wheat. (pg. 37)

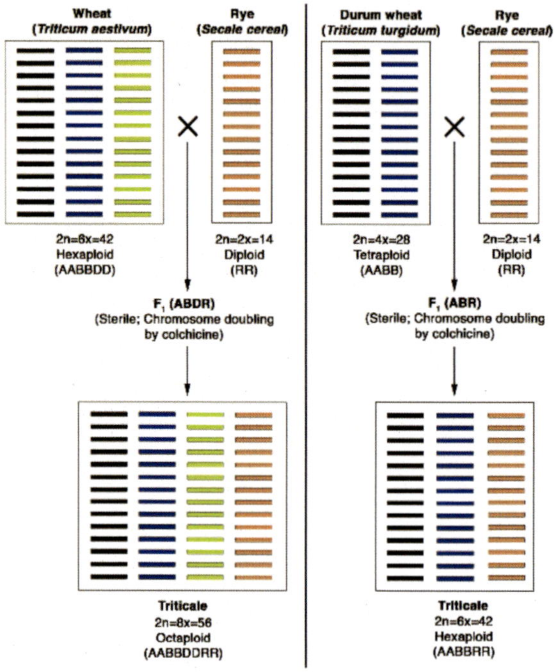

Figure 5.5. A Synthesis of Triticale.(pg. 39)

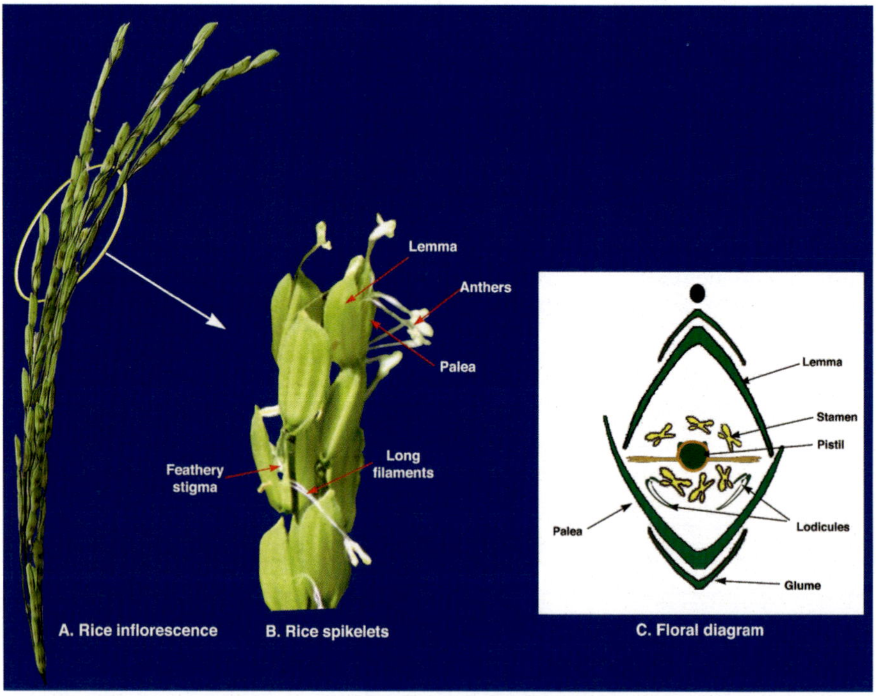

Figure 6.1. Floral Biology and Diagram of Rice (*Oryza sativa* L.). (pg. 48)

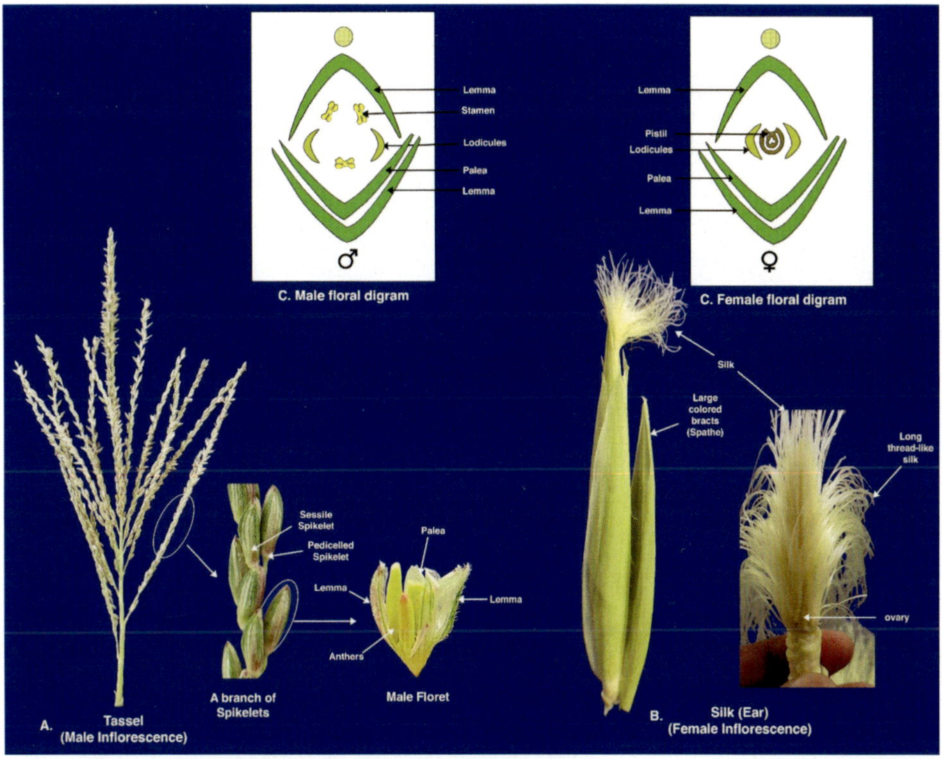

Figure 7.1. Floral Biology and Diagram of Maize (*Zea mays* L.). (pg. 67)

Figure 8.1. Floral Biology and Diagram of Pearl Millet (*Pennisetum glaucum* L.). (pg. 81)

Selfing of Sorghum. (pg. 96)

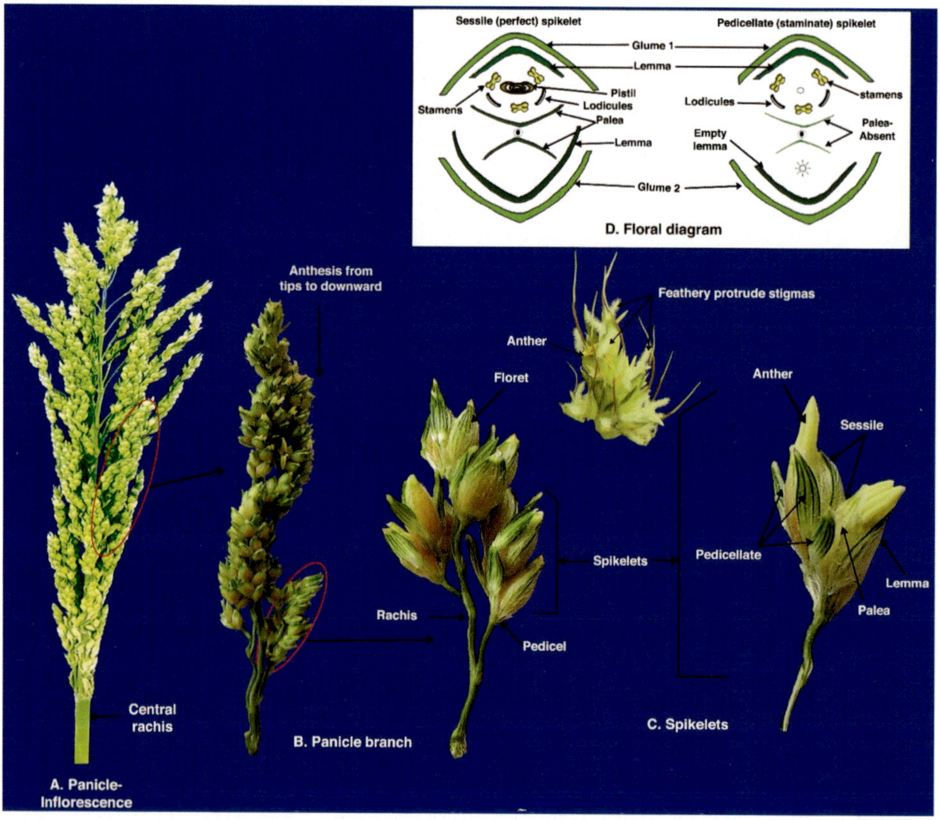

Figure 9.2. Floral Biology and Diagram of Sorghum (*Sorghum bicolor* L.). (pg. 97)

Figure 9.3: Hybridization Technique in Sorghum. (pg. 98)

Figure 10.1: Finger Millet Floral Biology. (pg. 113)

Figure 10.2: Foxtail Millet Floral Biology. (pg. 113)

Figure 10.3: Proso Millet Inflorescence. (pg. 114)

Figure 10.4: Barnyard Millet Floral Biology. (pg. 114)

Figure 10.5: Kodo Millet Floral Biology. (pg. 114)

Figure 10.6: Little Millet Inflorescence. (pg. 115)

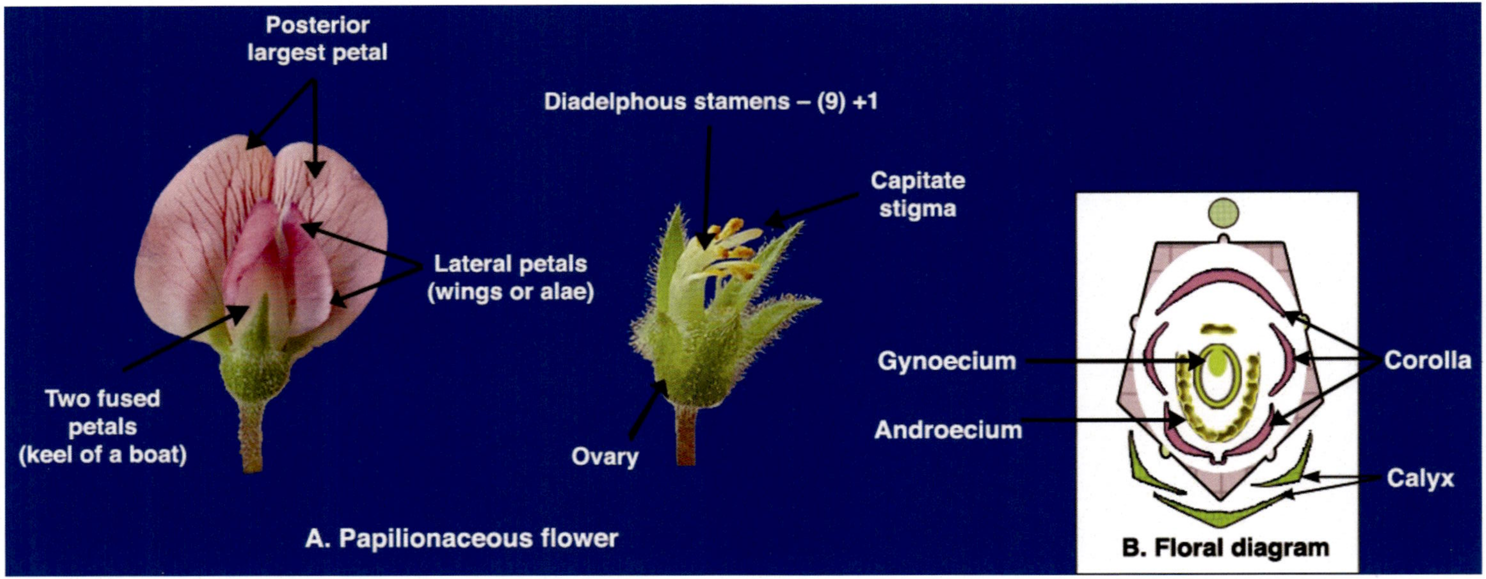

Figure 11.1: Floral Biology and Diagram of Chickpea (*Cicer arietinum* L.) (pg. 128)

Chapter 12

Breeding Pigeonpea (*Cajanus cajan* L.)

Introduction

Pigeonpea is also known as red gram, it is a protein-rich perennial grain legume. It is considered a unique legume crop due to unique traits, including wide variation in maturity period, natural outcrossing, deep root system and drought tolerance. Because of its resilient capacity against various non-favourable environmental conditions, it is usually cultivated as an annual crop in tropical and subtropical regions worldwide. In India, it is the important grain legume next to chickpea. It is mainly grown as *kharif* season. India is the largest producer in the world. It has diverse uses like human food (*i.e.*, green peas; de-hulled split), feed (crushed dry seeds), animal fodder (*i.e.*, green leaves, silage), and fuelwood (dry stems). In addition, being a pulse crop, it improves soil fertility by fixing atmospheric nitrogen.

Background Information

1.	*Common name*	Arhar, tur, red gram
2.	*Botanical Name*	*Cajanus cajan* L.
3.	*Family*	*Fabaceae* (*Leguminosae*)
4.	*Chromosome number*	2n = 2x = 22
5.	*Centre of origin*	India

6.	*Nature of pollination*	Often cross-pollination (cross-pollination range 3-40 per cent)
7.	*Pollination mechanisms*	Bisexuality, Cleistogamy, Entomophily- Melittophily
8.	*List of wild relatives*	*Cajanus cajanifolius, C. scarabaeoides, C. albicans, C. acutifolius, C. conferiflorus, etc.*

List of Wild Species with Key Traits

Sl. No.	*Wild Species*	*Key Traits*			
		Agronomic/ Quality	*Biotic Stress Resistance*	*Abiotic Stress Resistance*	*Other*
1.	*C. acutifolius*	Bold seed	Pod borer and Pod fly resistance	-	Cytoplasmic male sterility
2.	*C. albicans*	High seed protein, More seeds per pod	Pod borer, Pod fly, and sterility mosaic disease resistance	Salt tolerance	
3.	*C. cajanifolius*	High seed protein	-		Nuclear and Cytoplasmic male sterility
4.	*C. crassus*	High seed protein, good forage source	Sterility mosaic disease resistance	-	
5.	*C. lineatus*	Cleistogamy	Pod fly and sterility mosaic disease resistance	-	Cytoplasmic male sterility
6.	*C. platycarpus*	Photoperiod insensitivity, High flower and pod setting	Phytophthora blight, Sterility mosaic diseases, Pod borer and Cyst nematode resistance	Salt tolerance and Aluminum toxicity resistance	-

Sl. No.	*Wild Species*	*Key Traits*			
		Agronomic/ Quality	*Biotic Stress Resistance*	*Abiotic Stress Resistance*	*Other*
7.	*C. scara-baeoides*	Dwarfism and earliness, high seed protein	Pod borer, Pod fly, Sterility mosaic disease	Drought and salt tolerance Aluminum toxicity resistance	Cytoplasmic male sterility
8.	*C. sericeus*	High seed protein	Pod borer, Pod fly resistance Sterility mosaic virus, Phytophthora blight resistance	Salt tolerance	Cytoplasmic male sterility

Classification of Pigeonpea

A. Based on Flowering Pattern

1. **Determinate type (DT):** Flowering duration is short and flowers are borne more or less at the same level at the top end. Inflorescence arrangements are slightly corymb shape bunch. *e.g.*, GT-100, Prabhat, Pragati, Durga, Vambhan 1, AL 15, *etc.*
2. **Non-determinate type (NDT):** Flowering duration is longer and flowers borne at several points along the branches. Inflorescence arrangements terminal panicle, *e.g.*, most of the cultivars are NDT types, *e.g.*, BDN-2, GT 101, GT103, *etc.*

B. Based on Maturity Duration

A tremendous variation in maturity period (*i.e.*, 90 to 260 days) has been reported in pigeonpea. It is widely adapted to a wide range of environments and cropping system. The different maturity groups with key varieties are mentioned in Table 12.1.

Table 12.1. The Different Maturity Groups with Key Varieties in Pigeonpea

Sl.No.	*Group*	*Flowering Days (Range)*	*Maturity Days (Range)*	*Name of Variety*
1	Extra early	<60	<130	ICPL 88039, PAU 881, VLA 1, PA 16, GT 102
2	Early	61-90	131-150	Pusa 2018-4, UPAS 120, Prabhat, GT-100, GTH-1 (Hybrid), AL 882, CORG 9701, IPA 15-6, PA 6, PUSA 992, GT 105, IPH-15-03 (Hybrid)
3	Mid-early	91-130	151-165	TJT 501, PT 0012, BDN 711, GT 101, GT 103, TS 3R, PRG 176, WRGE 93, GT 107
4	Medium	131-160	166-185	ICPL 87119, BDN 716, JKM 189, CO 8, GRG 811, GT 104, GT 106, GT 108
5	Late	>160	>200	Gwalior 3, Rajendra Arhar 1, Bahar, MA 6, IPA 203

Gene Pool

A gene pool of pigeonpea is grouped into three categories

Group	*Species*
GP-1 (Gene pool 1)	Comprises of all cultivars and landraces of *C. cajan*.
GP-2 (Gene pool 2)	Wild species with many valuable traits and may produce partially fertile hybrid in interspecific crosses; *e.g. C. scarabaeoides, C. cajanifolus, C. reticulum, C. lanceolatus, C. lineatus, C. trinervirus, C. latisepalus, C. albicans, C. sericeus, C. acutifolius, etc.*
GP-3 (Gene pool 3)	The species includes which are not directly crossable with cultivated species, *e.g., C. cinereus, C. crassus, C. goensis, C. rugosus, C. platycarpus, C. volubilis, C. mollis,* and *C. confertiflorus.*

Distribution

Pigeonpea is one of the oldest cultivated crops in the world. In India, it has been cultivated since the Vedic period (2000 – 600 BC). Extensively distributed crop species and is found in Asia, Africa, Latin America, and Caribbean regions. India is the largest producer in the world, account of total 90 per cent of the world's production, it is mainly grown in the states of Maharashtra, Uttar Pradesh, Madhya

Pradesh, Karnataka, Gujarat, Andhra Pradesh and under the irrigated belt of Punjab, Haryana and Rajasthan.

Progenitor: *Cajanus cajanifolius and Atylosia lineate* → Genus *Cajanus* and *Atylosia* have many similarities to each other. The view is that *Cajanus* ascended from *Atylosia*. In Western Ghats (West Bengal and Odisha) *Atylosia* species are known as wild tur.

Floral Biology, Selfing and Crossing Techniques

Inflorescences	:	Racemose type; terminal or auxiliary with long peduncle; terminal raceme found in non-determinate types and auxiliary in determinate types.
Flower	:	Small bract, Pedicellate (hairy pedicel), zygomorphic, irregular, complete, perfect, pentamerous except pistil, slightly perigynous, papilionaceous.
Calyx	:	Five sepals, persistent, gamosepalous, calyx tube campanulate (bell-shaped) and sepaloid.
Corolla	:	Five petals [1+2+(2)], polypetalous, unequal size, vexillary aestivation, **Papilionaceous** [one posterior largest petal (standard vexillum) that encloses two lateral petals (wings or alae) and two fused forming a unique structure like a keel of a boat].
Androecium	:	Diadelphous stamens – (9) +1; one posterior stamen is free, but the remaining 9 are fused to form an incomplete stamina tube that encloses the ovary, six filaments are long, while the remaining four stamens, including the free posterior, have short filaments. Anther-dithecous, basifixed and dorsifixed. The burst anther longitudinally.
Gynoecium	:	Monocarpellary, ovary- superior, unilocular, 2-9 ovules, Placentation- marginal, style – long, filiform and glabrous with curved-thick and capitate stigma.
Fruits	:	Legume or pod (dehiscence formed by marginal placentation) Pod is pubescent, covered with glandular and non-glandular hairs bearing 2–7 seeds.
Seed	:	Ex-albuminous with two cotyledons.
Floral formula	:	⊕ ⚥ $K_{(5)}\ C_{1+2+(2)}\ A_{(9)+1}\ \underline{G}_{1}$

Anthesis: Anthesis usually occurs in the bud stage (1 day before the opening of the flower). Anthesis occurs during the day, beginning from 06.00 A.M. and lasting till 4.00 P.M. The highest anthesis period recorded is between 09.00 and 10.00 A.M.

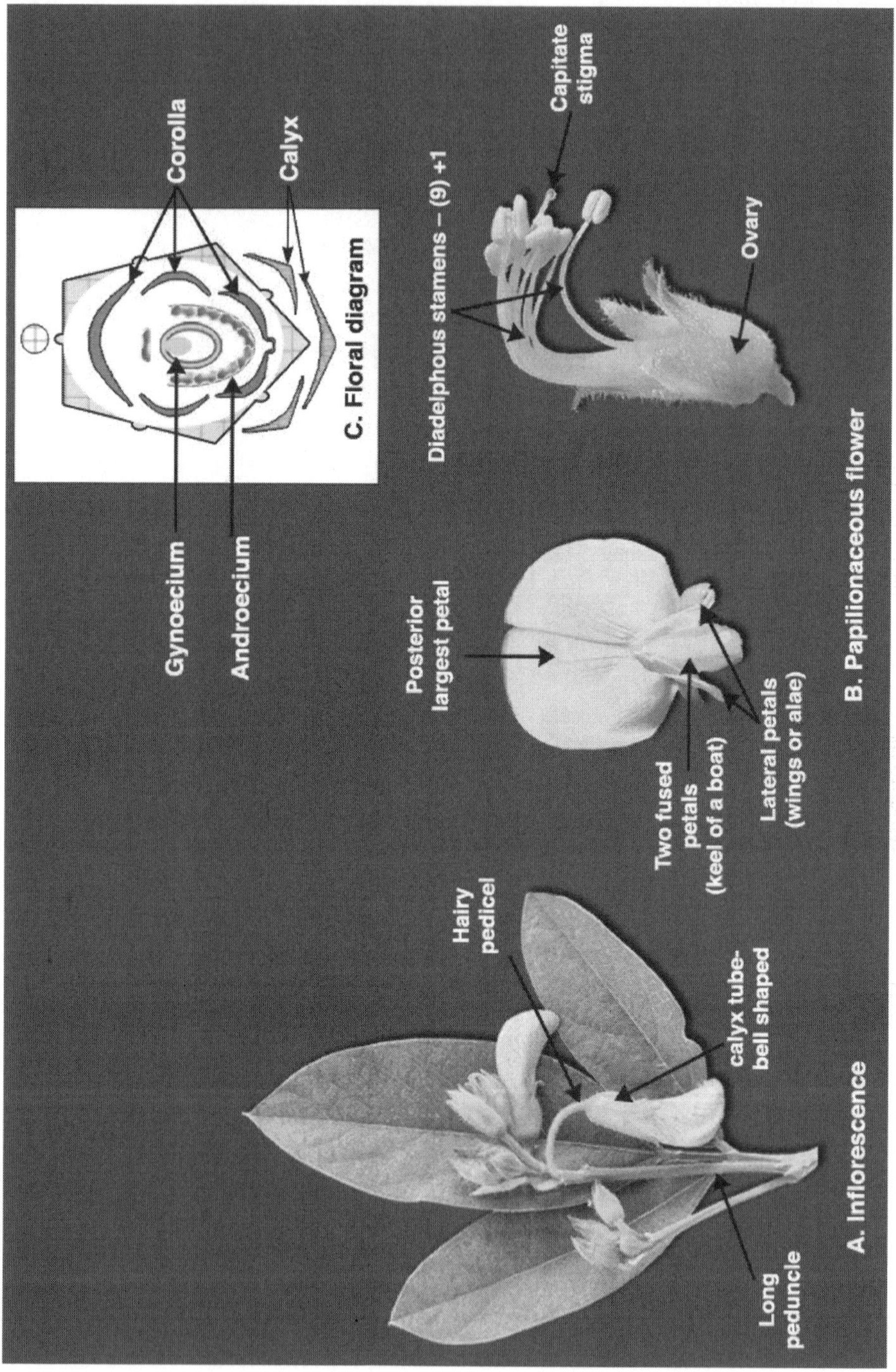

Figure 12.1. Floral Biology and Diagram of Pigeonpea (*Cajanus cajan* L.)

Selfing: Pigeonpea is a often cross-pollinated species, 3-40 per cent cross-pollination is reported due to visit of honeybees and other insects. To ensure self-pollination, bagging should be done at bud stage. Label it after bagging,

Artificial Crossing and Pollination: In the case of artificial crossing, one or two buds/clusters likely to open the next day or two days later should be selected for emasculation; the rest should be removed. The bud is held at the base between the thumb and forefinger. The anterior sepal is exposed with the help of forceps, and the keel is pushed down or a small cut at the base of the keel petal and cut the central axis down to the distal end. The exposed anthers are then softly removed from the filaments. Pollination is done immediately after emasculation and after bagging labelled it. Care should be taken that is no any mechanical damage to the other floral parts (stigma, ovary) at the time of emasculation and pollination.

Breeding Objectives

1. **Breeding for higher yield:** The yield improvement in pigeonpea is the ultimate objective. Due to photosensitivity, and indeterminate growth habit, the resultant low harvest index. It improves by improving different yield attributing traits *viz.*, number of primary branches per plant, number of pods per plant, pod length, number of seeds per pod, seed size, *etc.*
2. **Breeding for wider adaptability:** Essential for yield stability over the environment *i.e.*, diverse environments
3. **Breeding for shattering resistance:** In the case of non-determinate varieties a shattering resistance can save yield losses.
4. **Breeding for early and uniform maturity and determinant type:** The majority of cultivars of pigeonpea are non-determinant, long duration, and yield losses occur due to shattering. It is essential to breed determinant-type cultivars that fit into the cropping sequence without yield compromise.
5. **Breeding for biotic stress resistance**: The yield losses occur due to various diseases and pest. Developing resistant cultivars against a particular disease or pest is the most convenient option to improve yield in pigeonpea. The important diseases *viz.*,
   - ☆ *Fusarium* wilt: It is controlled by single dominant gene. Some identified resistance sources/genotypes are BDN 1, BDN 2, ICP 8863, ICPL 87119, ICPL 87051, ICPL 87ICPL 20109, ICPL 20096, ICPL 20115, ICPL 20116, ICPL 20102, and ICPL 20094.
   - ☆ Sterility mosaic: It is also known as green plague, is governed by four autonomous loci, two duplicate dominants (*i.e.*, *Sv1* and *Sv2*), and two recessive (*i.e.*, *sv3* and *sv4*) genes.
   - ☆ Phytophthora stem blight: It is governed by a dominant gene (*Pd1*). A resistant source can be identified by sick plot- screening.

- Major pests are:
  - Pod borer (*Helicoverpa armigera*): The resistant breeding with conventional mean is rather difficult. Gene pyramiding for two or more insect-resistant genes (cry1Ab and cry1F domain) is one of the best options for developing resistance genotypes. Besides, a wild relative *Cajanus acutifolius* also reported a resistance source against *Helicoverpa*. A cultivar ICPL 332 (Abhay) is the first resistant cultivar developed by ICRISAT and ICAR. Some other resistance sources are ICP 903, ICPL 84060, ICPL 84660.
  - Pod fly: Reported resistance sources are PDA 89-2E, PDA-88-IE, ICP 7946, *etc.*

6. **Breeding for abiotic stress resistance:** Drought, water lodging, and salinity are important abiotic stresses.
7. **Breeding for quality:** The quality is determined through various parameters such as
   - Milling quality: High-quality split (dal)
   - Physical appearance includes Grain size, grain shape and grain colour; White colour is mainly preferred in Gujarat.
   - Chemical composition: Includes protein, starch content. The wild relatives, *C. scarabaeoides*, *C. albicans* and *C. sericeus*, were described to have high protein content (28.5–30.5 per cent) and used as donors in breeding high protein lines. Line HPL 40 was reported >30 per cent protein. The average protein content 23 per cent. Red seed coat contains more polyphenols (Tannin) than the white seed coat. So, preference is towards a white seed coat. Red grain contains a lesser amount of Sulphur containing amino acid.
8. **Breeding for vegetable purposes:** Pigeonpea as a green vegetable (immature seeds) is very common in Gujarat, Maharashtra and Karnataka. Also, in the tribal zones of various states. The popular vegetable purpose varieties are T 15-15, HY-3C, TTB-6, ICP 7035, ICP 7979, BRG 2, *etc.* The essential characteristics associated with vegetable purpose varieties are
   - Early podding,
   - Green long pods (7-8 cm) with attractive large seeds
   - Round-the-year green pods production,
   - Manyfold picking of pods,
   - Non-sticky green pod surfaces,
   - Ovules are completely developed,
   - Easy shelling,
   - Long shelf life of green pods and shelled grains.

9. **Exploitation of heterosis with diverse CMS systems:** Attempts have been made to exploit hybrid vigour in pigeonpea by utilizing CMS. So far nine CMS system (A_1-A_9) have been recognized (Table 12.2). Though, A_2 and A_4 CMS systems are well characterized and exploited commercially in pigeonpea hybrids development in India.

Table 12.2: Different CMS Systems and their Sources

Sl.No.	*CMS Systems*	*Sources*
1.	A_1	*Cajanus sericeus*
2.	A_2	*Cajanus scarabaeoides*
3.	A_3	*Cajanus volubilis*
4.	A_4	*Cajanus cajanifolius*
5.	A_5	*Cajanus cajan*
6.	A_6	*Cajanus lineatus*
7.	A_7	*Cajanus platycarpus*
8.	A_8	*Cajanus reticulatus*
9.	A_9	*Cajanus lanceolatus*

10. **Breeding converts pigeonpea from an often cross-pollinated to a self-pollinated crop:** In pigeonpea breeding, natural cross-pollination has both advantage (for hybrid development) and disadvantage (for genetic deterioration). Purity maintenance is a major problem due to natural cross-pollination. The stable genetic solution is only a self-pollination with cleistogamous flowers. The unique line reported from a cross between *Cajanus cajan* and *Cajanus lineatus* having easily distinguished floral traits allows only 1-2 per cent outcrossing. A single recessive gene governs cleistogamous trait, and it easy to introduce in other lines. The line ICPL 87154 has partial cleistogamous was developed earlier in ICRISAT with low (<1 per cent) natural outcrossing.
11. **Breeding cultivars with speedy seed filling rates:** A speedy pod filling is a unique and essential trait for pigeonpea. It helps develop varieties with uniform flowering and pod maturity, enabling easy insect control and yield advantages.

Breeding Approaches (Methods)

1. **Plant Introduction:** PI 3704 (Hy-1), PI 4628 (Hy-2), PI 2817-1-A (Hy-3A), PI 4839 (Hy-4), *etc.* introduced from USA; Prabhat short duration variety from IARI, ICPL 87 from ICRISAT.
2. **Pure line selection:** *e.g.*, UPAS 120, BDN-2, ICPL 8863 (Maruthi), BDN 711, *etc.*

3. **Hybridization and selection:** Inter-varietal; Pedigree methods of selection, back cross. Interspecific: *C. cajanus* × *C. scaraboides*; *C. cajanus* × *C. cajanifolius.* Transfer pod borer resistance from the wild relatives (*i.e., C. scarabaeoides*) into the cultivated background–transfer of cytoplasmic male sterility system.
4. **Mutation breeding:** Generally used to improve specific traits and adaptation. Mainly gamma rays, EMS and fast neutrons have been used in pigeonpea. Several varieties with improved attributes have been released for commercial cultivation are given in Table 12.3

Table 12.3. List of Mutant Cultivars of Pigeonpea Released in India

Sl.No.	*Cultivars*	*Year*	*Source of Mutagen*	*Characters Improved*
1.	Vishakha-1 (TT 6)	1976	Fast neutrons	High yield, bold seeded
2.	Co 3	1977	EMS (0.6 per cent)	High yield, bold seeded, higher shelling, field dormancy for 15-20 days
3.	Trombay Vishakha-1	1983	Fast neutrons	Increased seed size with all desirable traits of parent variety T-21
4.	Co 5	1984	Gamma rays (160 Gy)	Early maturity, photoperiod insensitivity and drought tolerance
5.	TAT 5	1984	Fast neutrons	Increased seed size (50 per cent), high TGW and early maturity (140 days)
6.	TAT 10	1984	Fast neutrons (25 Gy)	Medium-large size grain and extra-early maturity (115-120)
7.	TT-401	2007	-	High yield, tolerance to pod borer and pod fly damage
8.	TJT-501	2009	-	High yielding, early maturity, tolerance to Phytophthora blight

5. **Heterosis breeding:** Exploitation of hybrid vigour through different hybrids. In pigeonpea, two-line and three-line breeding systems are used to develop hybrids.

Two-Line hybrid breeding systems:

☆ **GMS system:** Three genes, namely, *ms1*, *ms2* and *ms3*, have been reported in pigeonpea for imparting genic male sterility. ICPH-8, a world's first GMS-based hybrid, was developed at ICRISAT in 1991 from a cross between male-sterile Prabhat (determinate) and ICPL161 (indeterminate).

- ☆ **TGMS system:** Like rice two-line hybrid breeding, a temperature-sensitive genetic male sterility system in pigeonpea has been identified. The line was developed by cross between a wild species *Cajanus sericeus* with *Cajanus cajan* (cultivar ICPA 85010). In some genotypes of pigeonpea the fertility status is affected by temperature, for example, when a temperature ≥ 25°C plant remains fully fertile, but it becomes sterile when the temperature ≤ 24°C.

Three-line hybrid breeding (CGMS) system: The genetics of fertility restoration of A_4 cytoplasm was studied and two dominant genes (*Rf1* and *Rf2*) were found controlling the fertility of the hybrids. Two cytoplasm A_2 and A_4 are being used in commercial hybrid development. A Medium maturing hybrid GTH 1 (GT 288A × GTR 11), a world's first A_2 cytoplasm (*Cajanus scarabaeoides*) based CGMS hybrid developed at Sardarkrushinagar Dantiwada Agricultural University, Sardarkrushinagar (Former it was GAU), Gujarat in year 2004. Later on, A_4 cytoplasm (*Cajanus cajanifolius*) utilized by ICRISAT to develop commercially viable hybrids *viz.*, ICPH 2671 (Pushkal), ICPH 2740 and ICPH 3762 (Parbati). The list of hybrids is described in Table 12.4.

6. **Genomic-assisted Breeding:** With the help of genomics tools, various markers genes and QTLs associated with Fusarium wilt resistance, sterility mosaic disease resistance, fertility restoration, hybrid purity assessment, plant type, and earliness have been identified. The introgression of such genes can be possible in cultivated lines from wild sources by various genomics tools.
7. **Speed breeding:** The pigeonpea life period is photoperiod dependent. These techniques allow Rapid generation advancement by establishing optimum photoperiod and day length. Saxena and his co-worker at ICRISAT developed photo-insensitive genotypes *viz.*, ICPL 85024, ICPL 87093, ICPL 00004, and ICPL 00151. This technique helps to conserve genetic variability and helps in genomic or marker-assisted selection.

Breeding Institutes

☆ International Level

i International Crop Research Institute for Semi-Arid Tropics (ICRISAT), Hyderabad, India.

ii. International Centre for Agricultural Research in the Dry Areas (ICARDA), Beirut, Lebanon, Syria.

☆ **National level:** Indian Institute of Pulses Research (IIPR), Kanpur, (Uttar Pradesh).

☆ **State level:** Regional Research Stations-State Agricultural Universities.

Breeding Achievements

A. **Varieties:** T-15-15, GT-1(Vegetable), BDN-2, GT-100, GT-101, GT-102, GT-103, GT-104, GT-105, GNP-2 (Dual purpose), GJP-1, AGT-2, AVPP-1, BSMR-856, BSMR-736, ICPL-87119, ICPL-87, BRG 5, Tirupati Kandi 59, Chhattisgarh Arhar-1, LRG 133-33, Gujarat Tur 107, Gujarat Tur 108, Renuka, Phule Kaveri, Gujarat Tur 109, Phule Pallavi, Naam-88, Gujarat Tur 110, Gujarat Tur 111, *etc.*

B. **Hybrids:** As per Table 12.4

Table 12.4. List of Pigeonpea Hybrids Based on GMS and CMS System Released in India

System	*Name of Hybrids*	*Cytoplasm*	*Released Year*	*Institute*
GMS based	ICPH-8	A_2	1991	ICRISAT
	PPH-4	A_2	1994	PAU, Ludhiana
	COPH 1		1994	Coimbatore, TNAU
	COPH 2		1997	
	AKPH 4104		1997	PDKV-Akola
	AKPH 2022		1998	
CGMS based	GTH 1	A_2	2004	SDAU, Gujarat
	ICPH 2671 (Pushkal)	A_4	2010	ICRISAT
	ICPH 2740		2012	
	ICPH 3762 (Parbati)		2014	
	IPH 15-03	A_2	2020	IIPR, Kanpur

Chapter 13

Breeding Green Gram (*Vigna radiata* L.)

Introduction

Green gram is also known as mungbean, moong, or golden gram. In India, it is the third most important pulse crop after chickpea and pigeonpea. India is the largest producer, consumer, and importer of moong. It is short duration, requires minimum inputs, wider adaptability. Due to its short duration, its best suited for crop rotation, intercropping, catch crop, relay cropping. It is consumed in the form of full dry seeds and in the form of dal prepared by splitting the seeds in a mill. Used for making roti or along with rice as moong dal khichdi. The sprouted moong is a highly nutritious food since the increase in amino acids like ascorbic acid, thiamin, and niacin compared to the dried form. It is a good source of protein and comprises higher folic acid (folate) and iron content than other legumes. Due to the presence of the Rhizobium and Bradyrhizobium bacteria in the root nodules, it fixes the atmospheric nitrogen and thus improves soil fertility.

Background Information

1.	*Common name*	Moong, Mungbean, Greenbean
2.	*Botanical Name*	*Vigna radiata* (L.) R. Wilczek var. *radiata*
3.	*Family*	*Fabaceae* (*Leguminosae*)
4.	*Chromosome number*	2n = 2x = 22

5.	*Centre of origin*	India
6.	*Nature of pollination*	Self-pollination
7.	*Pollination mechanisms*	Bisexuality, Cleistogamy
8.	*List of wild relatives*	*Vigna radiata* var: *sublobata*, *Vigna hainiana*, *V. angularis*, *V. aconitifolia* var. *sylvestris*, *V. trilobata*, *etc.*

The Gene Pool of *Vigna* Species

A gene pool of *Vigna* species is grouped into three categories

Gene Pool	*Species*
GP-1 (Gene pool 1)	Comprises of *V. radiata* var. *radiata; V. radiata* var. *setulose; and V. radiata* var. *sublobata*
GP-2 (Gene pool 2)	Wild species with many valuable traits and may produce partially fertile hybrid in an interspecific cross; *e.g. V. grandiflora; V. mungo* var. *mungo; V. mungo* var. *silvestris; V. stipulacea; V. subramaniana; V. tenuicaulis; V. trilobata; V. trinervia; V. umbellata*
GP-3 (Gene pool 3)	The species consist that are not directly crossable with cultivated species *e.g., V. aconitifolia; V. angularis; V. dalzelliana; V. glabrescens; V. grandis; V. vexillate;* except species *V. glabrescens* which produced fertile offspring when cross with *V. radiata*

Distribution

Green gram is believed to have been domesticated in India ~3500 years ago. The wide diversity found in West Asia. It spread to Southeast Asia, China, and Taiwan. It is cultivated in various countries *viz.*, India, Pakistan, Bangladesh, Sri Lanka, Philippines, Taiwan, Thailand, Nepal and Southern Asian countries. In India, Maharashtra, Uttar Pradesh, Madhya Pradesh, Karnataka, Gujarat, Andhra Pradesh, Tamil Nadu and Rajasthan are major growing states.

Floral Biology, Selfing and Crossing Techniques

Inflorescences	:	Racemose type; terminal or auxiliary with long peduncle clusters of 10–20 cleistogamous flowers.
Flower	:	Small bract, Pedicellate (hairy pedicel), zygomorphic, irregular, complete, perfect, pentamerous except pistil, slightly perigynous, papilionaceous.

Calyx	:	Five sepals, persistent, gamosepalous, calyx tube campanulate (bell-shaped) with linear teeth.
Corolla	:	Yellow colour, slightly curved, five petals [1+2+(2)], polypetalous, of very unequal size, vexillary aestivation, **Papilionaceous** (one posterior largest petal (standard vexillum) that encloses two lateral petals (wings or alae) and two fused forming a unique structure like a keel of a boat, commonly referred to as a keel that protects the stigma and style.
Androecium	:	Diadelphous stamens – (9) +1; one posterior stamen is free, but the remaining 9 are fused to form an incomplete stamina tube that encloses the ovary. Anther-dithecous, basifixed and dorsifixed. The bursting of anther longitudinally.
Gynoecium	:	Monocarpellary, ovary- superior, unilocular, 2-9 ovules, Placentation- marginal, style – long, filiform and glabrous with curved-thick stigma.
Fruits	:	Legume or pod (dehiscence formed by marginal placentation) Pod is linear glandular and non-glandular hairs bearing 10–14 seeds.
Seed	:	Ex-albuminous with two cotyledons.
Floral formula	:	⊕ ⚥ $K_{(5)}\ C_{1+2+(2)}\ A_{(9)+1}\ \underline{G}_{1}$

Anthesis: Anthesis occurs in bud stage (1 days before the opening of flowers). Anthesis take place during 9:00 AM to 3:00 AM.

Artificial Crossing and Pollination

In the case of artificial crossing, one or two light green colours with the appropriate size of buds per cluster which is likely to open the next day later, should be selected for emasculation; the rest should be removed. Emasculation is carried out in the evening time. The bud is held at the base between the thumb and forefinger. The upper half of the standard, wing, and keel petals are opened with the help of dissecting needle or fine-tip forceps to expose the stigma and anthers. The exposed anthers are then softly removed from the filaments by fine forceps. Tagging of emasculated bud by tying around the internodes below the emasculated bud with coloured cotton thread or a tag. Pollination should be done by gently rubbing the feathery stigma with dehisced anther on the next day early in the morning.

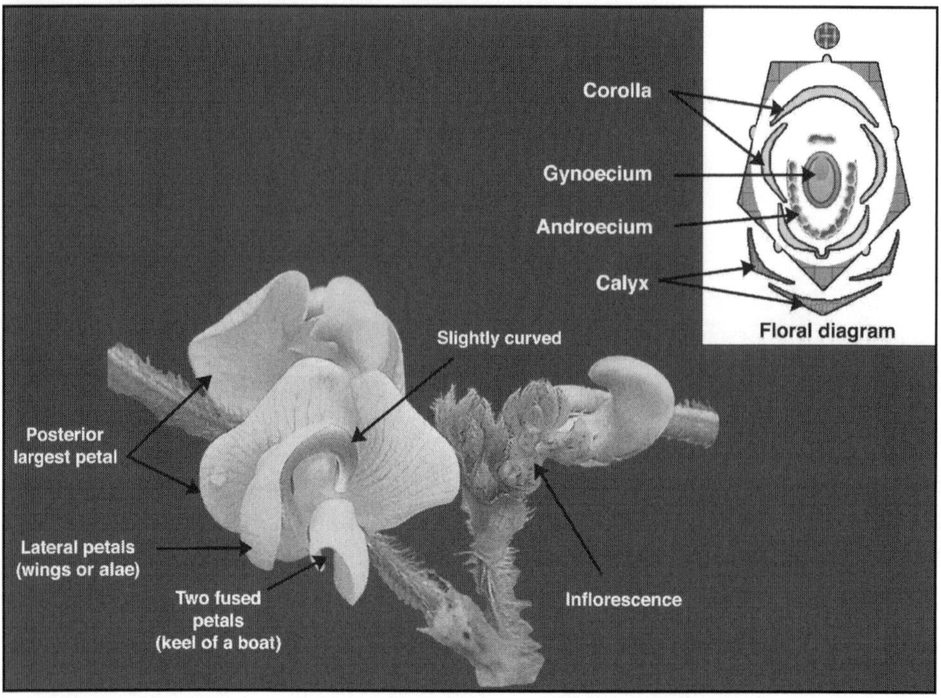

Figure 13.1. Floral Biology and Diagram of Green Gram (*Vigna radiata* L.).

Breeding Objectives

1. **Breeding enhances yield:** Through improving various target yield attributes like the number of seeds per pod, number of pods per plant, number of branches, *etc.*
2. **Synchronous maturity:** *e.g.*, GM-4, Pusa bold-1, SML 668, PDM 139 (Samrat).
3. **Breeding for summer/spring season:** The ideal plant type should have a short duration, high harvest index, minimum photoperiod sensitivity, maximum initial growth and determinate growth habit, maximum pods set in first reproductive flush, long pod, non-shattering, ideally vegetative growth should finish with flowering.
4. **Breeding for *Kharif* season:** The ideal plant type should have a medium late duration (90-120 days), maximum clusters per plant, long pod, more number of seeds per pod, medium plant height, shattering tolerance, medium seed dormancy, *etc.*
5. **Breeding variety suitable for multi-season:** *i.e.*, *kharif* and *summer*: The early and photo and thermo insensitive variety can fit, *e.g.*, Gujarat Mungbean 4.
6. **Breeding medium duration cultivar:** suitable for dry land conditions.

7. **Breeding for shattering resistance:** This feature is important for non-determinant cultivars to avoid yield losses.
8. **Breeding resistant to biotic stresses:** The yield weakened in Mungbean due to various diseases (*i.e.*, mungbean yellow mosaic virus (MYMV), powdery mildew, and *Cercospora* leaf spot) and insects (*i.e.*, pod borer, bean flies and bruchids).

   MYMV: *Vigna mungo*, a wild relative species has been used as a resistant donor. A yellow mosaic disease-resistant allele was transferred into cultivated mungbean and used to develop yellow mosaic virus-resistant mungbean cultivars/resistant varieties. So far number of cultivars have been identified. Some examples are Pusa 9531, Pusa Bold, IPM 99-125 (Meha), PAU 911, Pant moong -6, HUM 12, Sweta, MH 125 (Basanti), TMB-37, TJM 3, MH 318, Pusa 1431, SGC 16 and GAM 5, and IPM 410-3 (Shikha),

   Powdery mildew: VBN 2, KKM-3, AKM 9904 (PKV AKM 4), AKM 9911, TM 2000-2, TM-96-2, IPM 410-3 (Shikha) and SGC 16.

   Bruchids: Resistant source introduced from wild species; a completely resistant cultivar is TC1996.
9. **Breeding for quality and Mineral Biofortification**

   Cooking quality: Uniform cooking in a short time, highest digestibility among grain legumes from 83 to 90 per cent.

   Nutritional value: High protein, increase of methionine content

   Physical: Seed size, colour *etc.*

   Milling quality: High *dal* recovery- 80 per cent and more, Seed is less hard

   Minerals: High Fe and Zn content

   Nutrient bioavailability: Reduced phytic acid content and trypsin inhibitors; sprouted Mungbean declined up to 76 per cent phytic acid, enhancing Fe and Zn bioavailability up to 2.4 and 3.0 times, respectively.

Breeding Approaches (Methods)

1. **Introduction –** A large number of exotic lines were introduced from AVRDC Taiwan and other countries. The varieties *viz.*, Shining Ming No. 1 and Jalgaon- 781 are selected from China moong germplasm; while the varieties like Pant Moong 5 and 2, Pusa101, Pusa Vishal, WGG2 Pusa 9531, *etc.* are examples of primary introduction from AVRDC, Taiwan germplasm.
2. **Pure line selection:** Because of the self-pollination nature pure line selection approach is used in varietal development. Examples GAM 5, Utkarsh, Pant Mung 5, Suketi, *etc.*

3. **Mutation breeding:** Mutagenesis has been used to improve specific traits and adaptation. Various mutagens have been used in mungbean improvement. Among gamma rays and EMS most successful. Several varieties of mungbean with improved attributes have been released for commercial cultivation are given in Table 13.1.

Table 13.1. The List of Mutant Cultivars of Mungbean Released in India

Sl.No.	*Cultivars*	*Year*	*Source of Mutagen*	*Characters Improved*
1.	Pant Moong 2	1982	Gamma rays (100 Gy)	Resistance to virus diseases (MYMV), more pods and high yield
2.	Co 4	1982	Gamma rays (200 Gy)	High yield, early maturity, and resistance to drought
3.	ML 26-10-3	1983	Gamma rays	Resistance to a yellow mosaic virus (YMV) and high yield
4.	TAP-7	1983	Gamma rays	Early maturity (5-7 days), resistance to mildew and leaf spot, higher yield (23 per cent)
5.	MUM-2	1992	EMS (0.1 per cent)	High yield and resistance to diseases
6.	BM 4	1992	-	High yield, early maturity with tolerance or resistance to YMV
7.	LGG 450	1993	-	High yield, early maturity with tolerance or resistance to YMV
8.	LGG-407	1993	-	High yield, early maturity with tolerance or resistance to YMV
9.	TARM-2	1994	Gamma rays	High yield, medium late maturity and resistance to powdery mildew disease
10.	TARM-18	1996	-	High yield and resistance to powdery mildew disease
11.	TARM-1	1997	Gamma rays	High yield, resistance to powdery mildew disease and medium maturity
12.	TMB-37	2005	-	Yellow mosaic virus-resistant, high yielding and large seed size
13.	TM-96-2	2007	-	Powdery mildew, Cercospora leaf spot disease resistance, early maturity and suitable for fallow rice cultivation

Sl.No.	*Cultivars*	*Year*	*Source of Mutagen*	*Characters Improved*
14.	TJM-3	2007	Gamma rays	Resistant to yellow mosaic virus, powdery mildew, Rhizoctonia root-rot diseases, early maturity and large seeds
15.	TM 2000-2	2010	-	Resistance to powdery mildew and suitable for rice fallows

4. **Hybridization and selection**: Both intra and inter-specific hybridization are used in mungbean improvement.

   Intraspecific Hybridization: Most important method is the pedigree selection, various varieties have been developed, *e.g.*, IPM512–1 (Soorya), GM 7, GM 6, GM 4, GM 3, Pusa 0672, DGGV-2, Meha, Pant Mung-6, *etc.*

   Distant hybridization: Transfer to resistance for biotic stresses, improve methionine content from *V. mungo*: developed varieties with improved plant types. A wide hybridization with three wild species *viz. V. mungo* (urdbean), *V. umbellate* (rice bean), and *V. angularis* (adzuki bean) produced viable hybrid when mungbean used as a female parent. A wide cross between *V. radiata* × *V. mungo* and *V. radiata* × *V. umbellate* have been attempted successfully to transfer to mungbean yellow mosaic virus (MYMV) resistance in Mungbean. The cultivars, *viz.*, HUM 1, Pant mung 4, and IPM 99–125 are developed by wide hybridization.

5. **Biotechnological approach**

   Plant tissue culture: Micropropagation and double haploid breeding to develop homozygous line; Embryo culture is used to overcome post fertilization barrier, *i.e., V. radiata* × *V. mungo, V. radiata* × *V. umbellate.*

   Genomic-assisted Breeding: With the help of genomics tools, various markers genes and QTLs associated with agronomic traits, powdery mildew, *Cercospora* lcaf spot, mungbean yellow mosaic Indian virus (MYMIV) have been identified.

Breeding Institutes

- ☆ **International level:** World Vegetable Center (WVC), Taiwan.
- ☆ **National level:** Indian Institute of Pulses Research (IIPR), Kanpur, (Uttar Pradesh).
- ☆ **State level:** Regional Research Stations-State Agricultural Universities.

Breeding Achievements

GM 1, GM 2, GM 3, GM 4, GAM 5, GM 6, GM 7, Utkarsh, Pusa Ratna, Pusa 1371, Pusa 1431, Pant Mung 8, DGGV 2, CO 8, PAU 911, MH1142, IPM512–1 (Soorya), IPM 205–7 (Virat), GAM 8: Hara Moti, Shalimar Moong -3, MH 1762, MH 1772, GM 9, GM 10, Lam Pesara 610, VBN 7, *etc.*

Chapter 14

Breeding Black Gram (*Vigna mungo* L.)

Introduction

Black gram warm season food legume also known as urdbean. It is an ancient crop of India. India is the largest producer, consumer, and importer of urdbean. It is short-duration, requires minimum inputs, and has wider adaptability. It is consumed as dal prepared by splitting the seeds in a mill. Dal converted into flour that is used to prepare several food items, for making roti or along with rice as moong dal khichdi. It is very nutritious and recommended for people with diabetes. The sprouted urd is rich in vitamin C.

Background Information

1.	*Common name*	Urd, Urdbean, Uradbean, Urad dal
2.	*Botanical Name*	*Vigna mungo* (L.) Hepper
3.	*Family*	*Fabaceae* (*Leguminosae*)
4.	*Chromosome number*	2n = 2x = 22
5.	*Centre of origin*	India
6.	*Nature of pollination*	Self-pollination
7.	*Pollination mechanisms*	Bisexuality, Cleistogamy
8.	*List of wild relatives*	*Vigna mungo* var. *silvestris*, *Vigna hainiana*, *V. angularis*, *V. aconitifolia*, *V. sublobata*, *etc.*

Varietal Classification of *Vigna mungo* (L.) Hepper

Based on agronomic traits, it classified into three groups

Variety	*Important traits*
Mungo	Early maturing and Bold seeded
Viridis Bose	Late maturing and seed shinning or greenish dull type
Silvestris Lukoki, Marechal, and Otoul,	Wild type

Distribution

The urdbean is widely distributed in South and East Asia, mainly in India. India is the primary centre of diversity, while Southeast Asia is the secondary centre of diversity.

Floral Biology, Selfing and Crossing Techniques

The floral biology more or less similar to the Mungbean flower

Inflorescences	:	Racemose type; terminal or auxiliary
Flower	:	Pedicellate (hairy pedicel), zygomorphic, irregular, complete, perfect, pentamerous
Calyx	:	Five sepals, persistent, gamosepalous
Corolla	:	Yellow colour, slightly curved, five petals [1+2+(2)], polypetalous, of very unequal size, vexillary aestivation.
Androecium	:	Diadelphous stamens – (9) +1; Anther-dithecous, basifixed and dorsifixed.
Gynoecium	:	Monocarpellary, ovary- superior, unilocular, 2-8 ovules, Placentation- marginal, style – long, filiform and glabrous with curved-thick stigma.
Fruits	:	Legume or pod (dehiscence formed by marginal placentation) Pod is linear glandular and non-glandular hairs.
Seed	:	Ex-albuminous with two cotyledons.
Floral formula	:	⊕ ⚥ $K_{(5)}\ C_{1+2+(2)}\ A_{(9)+1}\ \underline{G}_{1}$

Anthesis: Anthesis occurs in bud stage (before the opening of the flower). Anthesis take place in night during 9:00 PM to 3:00 AM.

Artificial Crossing and Pollination

In the case of artificial crossing, one or two light green colours with the appropriate size of buds per cluster likely to open the next day should be selected

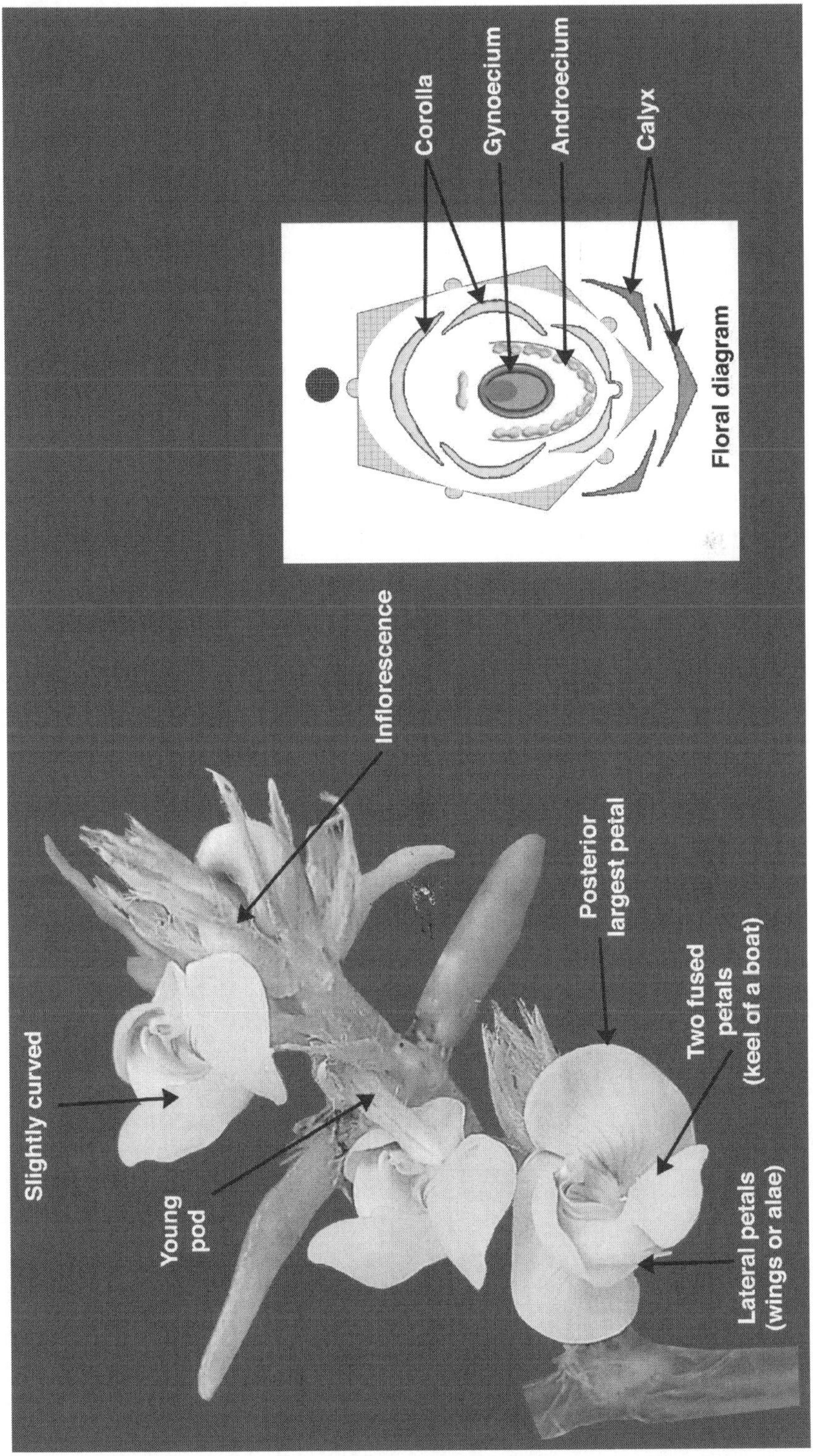

Figure 14.1. Floral Biology and Diagram of Black Gram (*Vigna mungo* L.)

for emasculation; the rest should be removed. The rest procedure of *emasculation and pollination are similar to mungbean.*

Breeding Objectives

1. **Breeding for enhancing yield:** Through improving target yield attributes like the number of seeds per pod, number of pods per plant, number of branches, *etc.* The longer duration varieties have more yield advantages compared to short duration.
2. **Breeding for the summer/spring season:** The ideal short-duration plant type is important.
3. **Breeding variety suitable for multi-season** *i.e., kharif* and summer: The early and photo and thermo insensitive variety is suitable. The variety LBG 752 released from Guntur (AP) in 2009 is photo insensitive.
4. **Breeding short-duration cultivars without yield reduction:** The yield and earliness correlate negatively. So, the breed variety with earliness and yield stability.
5. **Breeding cultivars suitable for mechanical harvesting:** This can possible by altering plant type. For example, main stem bearing plant. Recently scientists from IIPR developed a main stem-bearing variety IPU 13–1.
6. **Breeding resistant to biotic stresses:** The yield weakened in urdbean due to various disease (*i.e.,* yellow mosaic disease (YMD)- mungbean yellow mosaic virus (MYMV) and Mungbean yellow mosaic India virus (MYMIV), powdery mildew, and *Cercospora* leaf spot) and insects (*i.e.,* pod borer, stem flies, jassid and thrips).

   YMVD: Through conventional breeding, many resistant varieties have been developed. However, due to a narrow genetic base and rapid emergence of new begomovirus species the resistance breaks after some time. The examples of resistant cultivars are Vamban 2, Sarla, Pant U 35, KU 300, TU 94-2, KU 96-3, LU 391, NUL 7, LBG 752, VBG 04-008, Pant Urd 10, VBN 10, GAU 4 (Shayamal), *etc.*

   Powdery mildew: The resistance is governed by a single recessive gene. The reported resistance sources are Pant U 30, LBG 642, P 115 Line 6203. Important resistant cultivars are TU 40, Co 5, LPG 787, LBG 17, WBU 26, WBU 108, LBG 402, NUL 7, VBG 04-008.

   Cercospora Leaf Spot: The reported resistant sources are IC 11008, HPBU 98, HPBU 51, Pant U 26, UPU 95–1, and UG 407. Resistant cultivars are Jawahar Urd 2, and 3, Pant U 19, Mash 48, Mash 21, RBU 38, KB 512 and SBC 40.

   Leaf Crinkle Virus: NDU 94–6 is a resistant source reported in India. Resistant varieties are Himachal Mash 1 and ADT 3.

Stem flies: Resistant genotypes are UG 218, PDU 1, PDU 5, AKU 7, Co 305, UP 95–1, and LBG 707.

Pod borer: UG 737, PLU 557, and TAU 1 are reported to be highly resistant genotypes.

Bruchids: it is a severe store pest. Genotypes Mash 59, VM 2011 and VM 2166, VM 2164 showed resistance.

7. **Breeding resistant to abiotic stresses:** Important stresses are heat and salinity.

8. **Breeding for quality and Mineral Biofortification**

   Nutritional value: High protein, increase methionine content: high protein *e.g.*, K-1 (24.2%) and ADT-2 (21.5%).

   Physical: Seed size, attractive colour, lustre *etc.*

   Milling quality: high *dal* recovery- 80 per cent and more, Seed is less hard.

   Minerals: High Fe and Zn content.

Breeding Approaches (Methods)

1. **Selection methods:** Selection methods pureline, and mass selection are predominant in urdbean improvement. Several cultivars have been developed through selection methods. *e.g.* T9, T27, Naveen, Mash 2, Gwalior 2, Azad urd 1, Co1, Co2, Co3, Co5, GU 1, GAU 4 (Shayamal), *etc.*
2. **Mutation breeding:** Mutagenesis has been used to improve specific traits and adaptation. Among the different mutagens, gamma rays and EMS are most successful. Several varieties of urdbean with improved attributes have been released for commercial cultivation are given in Table 14.1;

Table 14.1. List of Mutant Cultivars of Urdbean Released in India

Sl.No.	*Cultivars*	*Year*	*Source of Mutagen*	*Characters Improved*
1.	Co 4	1978	EMS (0.02 per cent)	Early maturity, erect, photosensitive, and determinate type
2.	TAU 1	1985	Gamma rays	High yield, resistance to powdery mildew and larger seed size
3.	Manikya	1988	-	High yield, early maturity, large seed size
4.	TPU-4	1992	Gamma rays	High yield and high seed weight
5.	Vamban 2	1997	-	High yield, early maturity with tolerance or resistance to YMV
6.	TU-94-2	1999	-	High yield and resistance to yellow mosaic virus disease
7.	DU-1	2007	Gamma rays (200 Gy)	High grain yield and resistance to insects

3. **Hybridization and selection**: Like mungbean, intra- and inter-specific hybridization is also used in urdbean improvement.

   Intraspecific Hybridization: Most important method is the pedigree selection method, various varieties have been developed, *e.g.*, Birsa Urd 1, LBG 17, Pant U 19, Pant U 35, KU 300, KU 301, NDU 99–2, Pant U-40, WBU 108, *etc.*

   Distant hybridization: A wide hybridization with three species *viz. V. radiata*, *V. umbellate*, and *V. silvestris* are successfully used. The important cultivars are developed *viz.*, VBN 7, Vamban 5, Vamban 6, Mash 114, Mash 1008, Mash 1137, VBG04-008, VBN 8, TU 40.

4. **Biotechnological approach**

   Plant tissue culture: Micropropagation and double haploid breeding to develop homozygous line; Embryo culture is used to overcome post fertilizer barrier *in wide hybridization.*

   Genomic-assisted Breeding: Mapping or tagging of important traits *viz.*, agronomic traits, mungbean yellow mosaic Indian virus (MYMIV) and mungbean yellow mosaic virus (MYMV) are possible.

Breeding Institutes

International level: World Vegetable Center (WVC), Taiwan.

National level: Indian Institute of Pulses Research (IIPR), Kanpur, (Uttar Pradesh).

State level: Regional Research Stations-State Agricultural Universities.

Breeding Achievements

T9, GU 1, GAU 4 (Shayamal), ADT 6, DU-1, CO 6, GBG 1, IPU 11-02, KPU 524-65 (Kota Urd 3), KU 301, LBG 17, OBG 33 (Shashi), Pant U 31, Pant U 35, Pant U 9, Pant Urd 10, Pratap Urd-1, VBN 9, VBN 10, VBN 11 (VBG 12-062), Vallabh Urd 1, Mash 1190, BDU-12, DBG-16, Pant Urd 11, TJU 130, Narmada, KPU 18-1, Mash 878, Phule Rajan, *etc.*

Chapter 15

Breeding Cowpea (*Vigna unguiculata* L.)

Introduction

Cowpea is an important grain legume of dry regions of the tropics and subtropics in the world. Due to a robust tap root system, the crop can be grown under severe conditions such as drought and sandy soils. Nutritional point of view, grains are rich in protein, carbohydrates, calcium and folic acid. It has multipurpose uses like food, feed, fodder, forage and vegetable. The cowpea hulls are fed to livestock as a nutritious supplement to cereal fodder. It improves soil fertility by the biological fixation of atmospheric nitrogen. The yard-long bean, *V. unguiculata* ssp. *unguiculata* cultivar group *Sesquipedalis* evolved and is mainly grown in Asian countries, *i.e.*, India and China, where it is consumed as a fresh pod with seed as a vegetable.

Background Information

1.	*Common name*	Black-eyed pea, horse bean
2.	*Botanical Name*	*Vigna unguiculata* L. Walp
3.	*Family*	*Fabaceae* (*Leguminosae*)
4.	*Chromosome number*	2n = 2x = 22
5.	*Centre of origin*	West and Central Africa

6.	*Nature of pollination*	Self-pollination
7.	*Pollination mechanisms*	Bisexuality, Cleistogamy
8.	*List of wild relatives*	*V. unguiculata* var. *dekindtiana, V. unguiculata* var. *cylindrical, V. unguiculata* var. *spontanea*

Distribution

The cowpea originated and was domesticated in West or Central Africa. The distribution of cowpea to other parts of the world, such as Asia, Europe and the Americas, through different ways. The distribution of cowpea in India was from West Africa over North Eastern Africa.

Floral Biology, Selfing and Crossing Techniques

The floral biology is more or less similar to other Fabaceae crops, *i.e.*, pigeonpea flower.

Inflorescence	:	Racemose type; unbranched axillary- produced several flowers at terminal end
Flower	:	Found in alternate pair, pedicellate, zygomorphic, irregular, complete, perfect, pentamerous
Calyx	:	Five sepals, persistent, gamosepalous-two are large, and three are small (2 +3) arrangements
Corolla	:	White, purple or pale violet colour, straight, five petals [1+2+(2)], polypetalous, unequal size, vexillary aestivation.
Androecium	:	Diadelphous stamens – (9) +1; Anther-dithecous, basifixed and dorsifixed
Gynoecium	:	Monocarpellary, ovary- superior, unilocular, 2-8 ovules, Placentation- marginal, style – long, filiform and glabrous with curved-thick stigma
Fruits	:	Pod (dehiscence formed by marginal placentation) is long and cylindrical and constricted between the seeds
Seed	:	Ex-albuminous with two cotyledons
Floral formula	:	⊕ ⚥ $K_{(5)}\ C_{1+2+(2)}\ A_{(9)+1}\ \underline{G}_{1}$

Anthesis: Flowers open in the morning from 6:00-10:00 AM. Anthesis occurs in the bud stage (before opening of flower). Anthesis takes place from 10:00 PM to 1:00 AM.

Selfing: Individual flower bagging is tedious and has a poor seed-setting response. Cover whole plants with insect-proof cages/nylon bags to protect the flower from visiting insects.

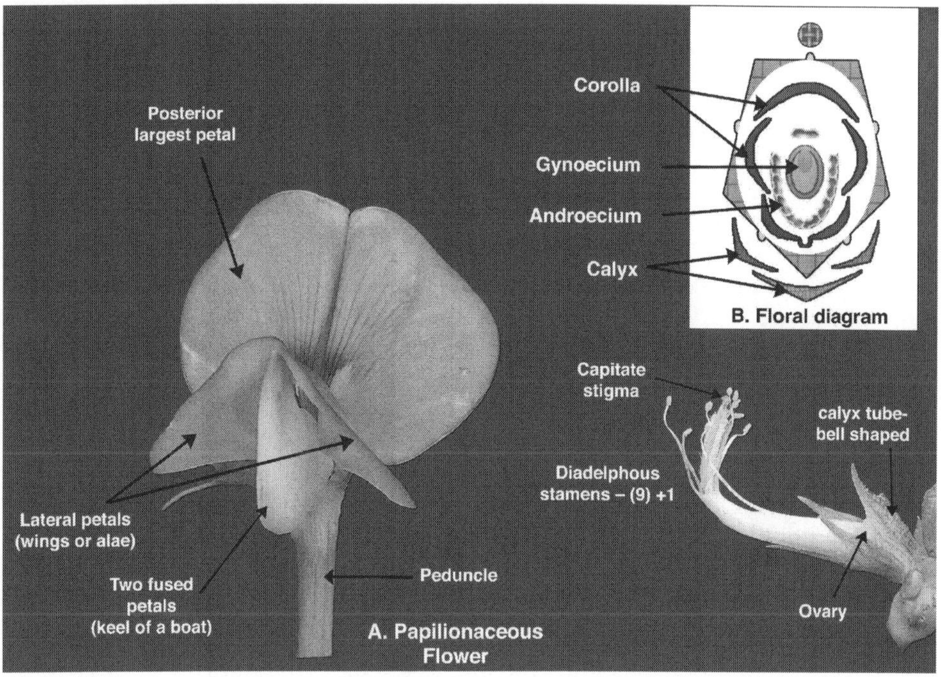

Figure 15.1. Floral Biology and Diagram of Cowpea (*Vigna ungiculata* L.).

Artificial Crossing and Pollination

In the case of artificial crossing, one or two appropriate sizes of buds per cluster likely to open the next day should be selected for emasculation; the rest of the immature and over-mature buds should be removed. *The emasculation technique is similar to pigeonpea.* The pollination is done just after emasculation or the following day through freshly opened flowers and tag it. To protect the emasculated flower before pollination and after pollination by a visit of insects by covering whole plants with insect-proof cages/nylon bags.

Breeding Objectives

1. **Breeding for enhancing yield:** The yield defined based on the utility of cowpea

   Seed purpose: High seed yield

   Green pod yield: Vegetable cowpea have more green pod yield

   Green fodder purpose: High green biomass
2. **Breeding variety suitable for multi-season:** The early and photo and thermo insensitive variety is suitable, *e.g.*, Vamban 2
3. **Breeding for short-duration cultivar**

4. **Breeding resistant to biotic stresses:** Disease *viz.*, Ascochyta blight, yellow mosaic disease (YMD), bacterial blight, powdery mildew, and insects, *viz.*, pod borer, aphid, jassids.
5. **Breeding resistant to abiotic stresses:** Drought, heat and salinity are important.
6. **Breeding for quality**
   - ✰ Nutritional value: High protein and other minerals.
   - ✰ Physical: Seed size, attractive colour, luster, *etc.*; Pod size, colour, shape.

Breeding Approaches (Methods)

1. **Selection methods:** It include pure line and mass selection, which are predominant methods in cowpea improvement; *e.g.*, Pusa Komal developed through pure line selection.
2. **Mutation breeding:** Mutagenesis has been used to improve specific traits and adaptation. Among the different mutagens, gamma rays and DMS most successful. Several varieties of cowpea with improved attributes have been released for commercial cultivation and are given in Table 15.1.

Table 15.1. List of Mutant Cultivars of Cowpea Released in India

Sl.No.	*Cultivars*	*Year*	*Source of mutagen*	*Characters improved*
1.	V16 (Amba)	1981	DMS	High yield, resistance to fungal and bacterial diseases
2.	V37 (Shreshtha)	1981	DMS	High yield, high vegetative growth, suitable for fodder also
3.	V38 (Swarna)	1981	DMS	High yield, early maturity, synchronous flowering, better quality pods and grains, resistance to diseases
4.	V240	1984	DMS	High yield, resistance to fungal, viral and bacterial diseases
5.	Co 5	1984	Gamma rays (300 Gy)	More nutritive forage cowpea, high yield (16 per cent), comparable for intercropping with fodder cereals
6.	Cowpea-88	1990	-	High grain yield, high green fodder yield, resistance to yellow mosaic virus
7.	COCP 702 [=CoVu 702 and CO(CP) 7]	2002	Gamma rays (200 Gy)	High yield and good quality

3. **Hybridization and selection**
   - ☆ Intra-specific hybridization: Pedigree method of selection, Backcross methods of selection. Pusa Dofasli cross derivatives of Pusa Phalguni × Philippine selections; IIHR 16: Arka Garima x Pusa Komal.
   - ☆ Inter-specific hybridization: Arka Garima: selection from cross derivatives of TUV 762 × *V. unguiculata* var. *sesquipedalis.*
4. **Biotechnological approach**

   Genomic-assisted Breeding: Mapping or tagging of important traits *viz.*, agronomic traits, biotic and abiotic traits.

Breeding Institutes

☆ **International Level**

i. International Center for Agricultural Research in the Dry Areas (ICARDA), Beirut, Lebanon, Syria.
ii. International Institute of Tropical Agriculture (IITA), Ibadan, Nigeria.
iii. World Vegetable Center (WVC), Taiwan.

☆ **National Level**

i. Indian Institute of Pulses Research (IIPR), Kanpur (Uttar Pradesh).
ii. Indian Institute of Vegetable Research (IIVR), Varanasi, (Uttar Pradesh).
iii. Indian Institute of Horticultural Research (IIHR), Bengaluru, (Karnataka).

☆ **State level:** Regional Research Stations-State Agricultural Universities.

Breeding Achievements

A. **Seed cowpea:** Co7, Pusa sampada, Pusa phalguni, Pusa rituraj, Pusa komal, GC3, GC4, RC-101, GC 5, GC 6, GC 7, V-38 (Swarna), Vamhan – 1, Phule Sonali, VBN 4, GC 8, Pant Grain Cowpea -67, *etc.*

B. **Vegetable purpose:** Vamban 2, Anand vegetable cowpea 1, Arka samrudhi, Arka mangala, Arka suman, Arka garima, Kashi kanchan, Kashi vishan, *etc.*

C. **Fodder purpose:** GFC 1, GFC 2, GFC 3 (*Kharif* season), GFC-4 (*Summer* season), C-88, IGFRI- 5450 (Kohinoor), Bundel Lobia-1, UPC-287, UPC-5286, IGFRI - DC – 215, *etc.*

Chapter 16

Breeding Guar [*Cyamopsis tetragonoloba* (L.) Taub]

Introduction

Guar is also known as cluster bean, an important grain legume of arid and semi-arid regions of India. The crop can withstand drought and high-temperature conditions due deep root system. In India, cluster bean is grown multipurpose, *viz.*, green pods are for vegetables, green/dry plants for cattle feed/fodder and green manure; besides, it is a cheap source of vegetable protein. The guar seed is a source of *guar gum,* also known as guaran. It is a non-toxic and eco-friendly agrochemical (a galactomannan) naturally-found hydrocolloid in seed endosperm. In the north-western part of India, it is chiefly grown as a source of gum (guar gum/ galactomannan) due to its unique qualities and wide applicability.

Background Information

1.	*Common name*	Guar, Cluster bean
2.	*Botanical Name*	*Cyamopsis tetragonoloba* (L.) Taub
3.	*Family*	*Fabaceae* (*Leguminosae*)
4.	*Chromosome number*	2n = 2x = 14
5.	*Centre of origin*	Africa
6.	*Nature of pollination*	Self-pollination
7.	*Pollination mechanisms*	Bisexuality, Cleistogamy
8.	*List of wild relatives*	*C. senegalensis, C. serrata* and *C. dentata*

Distribution

A significant diversity is found in India through cluster bean originated in Africa. The crop is mainly distributed on marginal and sub-marginal lands of arid and semi-arid regions. In India, crop is primarily distributed in northwestern states, *viz.*, Rajasthan Haryana, Gujarat, Uttar Pradesh, Punjab and Madhya Pradesh.

Floral Biology, Selfing and Crossing Techniques

The floral biology is more or less similar to other legumes crops,

Inflorescence	:	Racemose type; 40–60 flowers in the branch type and 50–70 in the erect type with sparsely-branched inflorescences
Flower	:	Pedicellate, zygomorphic, irregular, complete, perfect, pentamerous
Calyx	:	Linear teeth like unequal five sepals, persistent, gamosepalous
Corolla	:	Purplish to pink colour, straight, five petals [1+2+(2)], polypetalous, unequal in size, vexillary aestivation
Androecium	:	Diadelphous stamens – (9) +1; Anther-dithecous, basifixed and dorsifixed
Gynoecium	:	Monocarpellary, ovary- superior, unilocular, 5-12 ovules, Placentation- marginal, style – short, head-shaped stigma
Fruits	:	Pod (dehiscence formed by marginal placentation). The pods are flattened and borne in clusters
Seed	:	Ex-albuminous with two cotyledons
Floral formula	:	⊕ ⚥ $K_{(5)}\, C_{1+2+(2)}\, A_{(9)+1}\, \underline{G}_{1}$

Anthesis: Anthesis occurs in the bud stage (throughout the day). Anthesis takes place during 5.00 AM. and maximum around sunrise 7.00 to 9.00 AM. Dehiscence of anther before the opening of flower around 10.00 PM to 1.00 AM.

Artificial Crossing and Pollination

In the case of artificial crossing, one or two appropriate sizes of mature lower buds per cluster likely to open the next day should be selected for emasculation; the rest immature and over-mature buds should be removed. *The emasculation technique is similar to chickpea.* The pollination is done just after emasculation or the next day morning through freshly open flowers and tagging it.

Breeding Objectives

1. **Breeding for enhancing yield:** The yield defined based on the utility of guar
   - ☆ Seed purpose: High seed yield

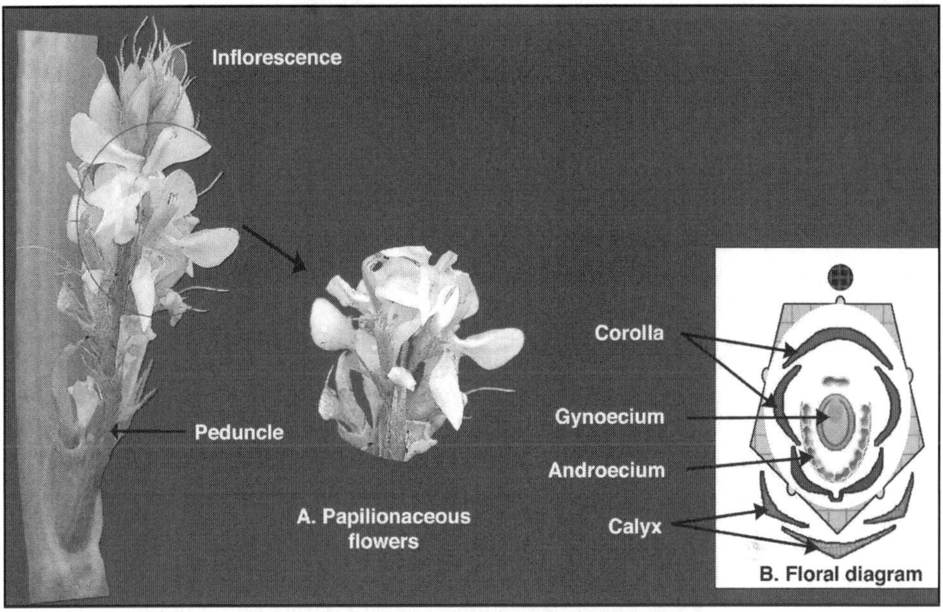

Figure 16.1. Floral Biology and Diagram of Cluster Bean (Guar) (*Cyamopsis tetragonoloba* L.)

- ✰ Green pod yield: Vegetable cluster bean more green pod yield
- ✰ Green/dry fodder purpose: High green and dry biomass

2. **Breeding for a short duration; determinate growth habit cultivar**
3. **Breeding variety suitable for multi-season:** The early and photo and thermo insensitive variety is suitable.
4. **Breeding resistant to biotic stresses:**
   - ✰ Disease: Alternaria leaf spot: resistant genotypes are GAUG-9003, GAUG-9005, GAUG-9406, GG-1, *etc.*
   - ✰ Bacterial blight: Promising resistant lines are GG-1, GAUG-9406, RGC-1027; powdery mildew, and insects, *viz.,* pod borer, Aphid, jassid.
5. **Breeding resistant to abiotic stresses:** Drought, heat and salinity are important.
6. **Breeding for quality**
   - ✰ Nutritional value: High guar gum and protein content; there are positive associations between seed yield and per cent guar content. However negative association is reported between seed weight and guar gum content.

Breeding Approaches (Methods)

1. **Selection methods:** Selection methods, including pureline and mass selection, are predominant methods that are effective due to the self-pollinated nature of the crop.
2. **Mutation breeding:** Mutagenesis has been used to generate variability for specific traits, *viz.*, early maturing, high seed gum content, disease tolerance and yield components. Among the different mutagens, gamma rays and X-rays are the most successful.
3. **Hybridization and selection**:
   - ☆ **Intra-specific hybridization:** Hybridization is difficult and painstaking due to tiny flowers. Various approaches like pedigree and backcross method of selection are used in cluster bean varietal improvement.
   - ☆ **Inter-specific hybridization:** It is used limitedly due to pre-fertilization barriers like stigmatic incompatibility and poor pollen germination. Interspecific hybridization attempted among *C. tetragonoloba* × *C. senegalensis* and *C. tetragonoloba* × *C. serrata,* however, between *Cymopsis tetragonoloba* × *C. serrata* has been found successful.
4. **Biotechnological approaches:**
   - ☆ **Tissue culture techniques:** Micropropagation and genetic transformation are used to support guar varietal improvement by escaping post-fertilization barriers.
   - ☆ **Genetic transformation:** Gene transformation (*i.e., rsu3*) is essential to improve the galactomannan content in cluster bean seeds.
   - ☆ **Genomics-assisted breeding:** Used to identify QTLs/genes associated with photo-insensitive, drought tolerant, disease/pest resistant cultivars with higher yield.

Breeding Institutes

☆ **International Level**

i. International Center for Agricultural Research in the Dry Areas (ICARDA), Beirut, Lebanon, Syria.

ii. World Vegetable Center (WVC), Taiwan.

☆ **National Level**

i. Central Arid Zone Research Institute (CAZRI), Jodhpur, Rajathan

ii. Indian Agricultural Research Institute (IARI), New Delhi

iii. Indian Institute of Vegetable Research (IIVR), Varanasi, Uttar Pradesh

iv. Indian Institute of Horticultural Research (IIHR), Bengaluru, Karnataka

☆ **State level:** Regional Research Stations-State Agricultural Universities.

Breeding Achievements

A. **Seed purpose**: RCG 1038, GC1, GC 2, GC 3, Sona, Suvidha, Naveen, PLG-85, RGC-471, Karan Guar-14, Karan Guar-15, *etc.*

B. **Vegetable purpose:** Pusa Mausmi, Pusa Sadabahar, Pusa Navbahar, PLG-850, Sharad Bahar, Pusa Domausami, Goma manjari, VRCB-95, *etc.*

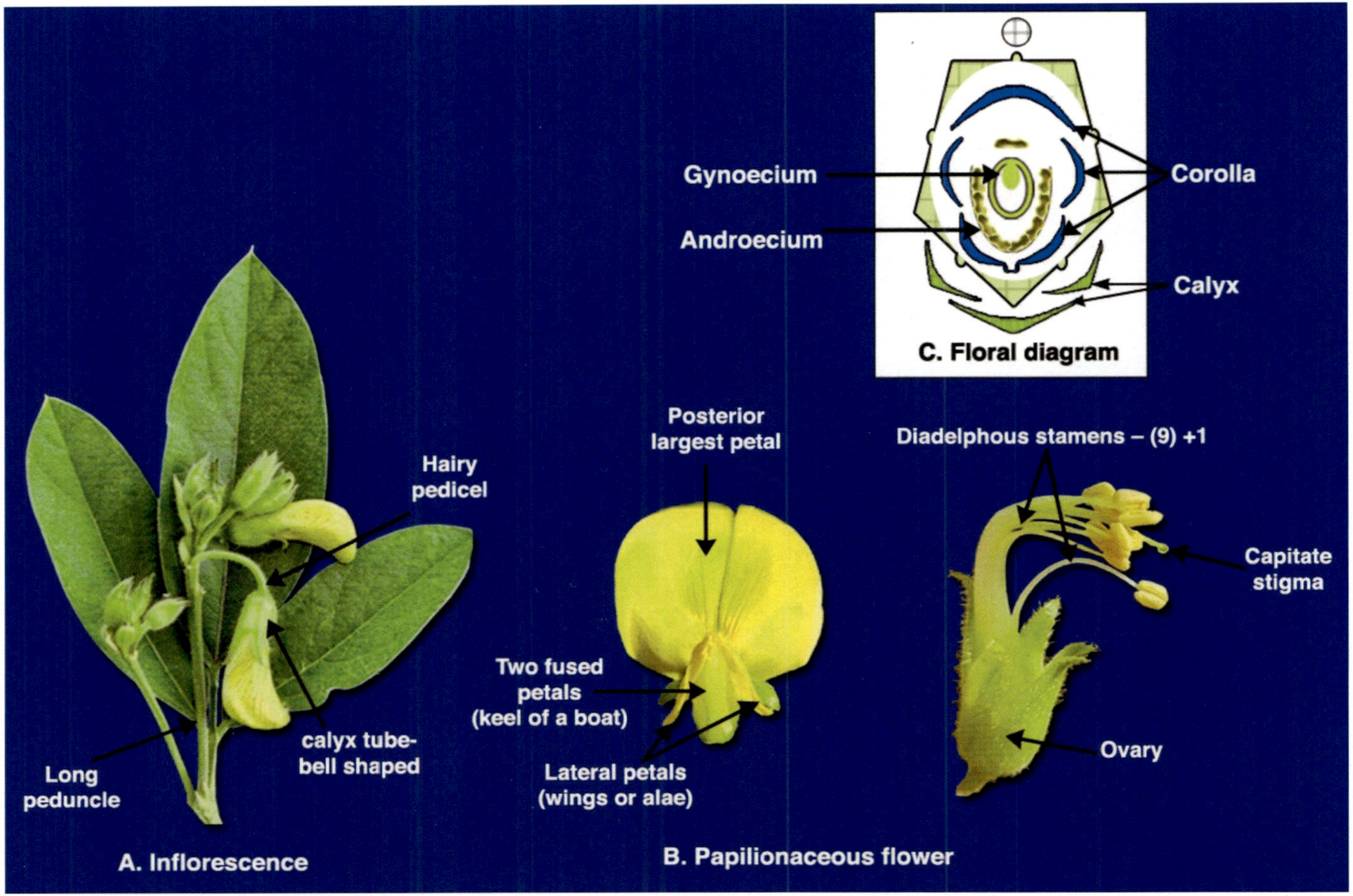

Figure 12.1. Floral Biology and Diagram of Pigeonpea (*Cajanus cajan* L.) (pg. 140)

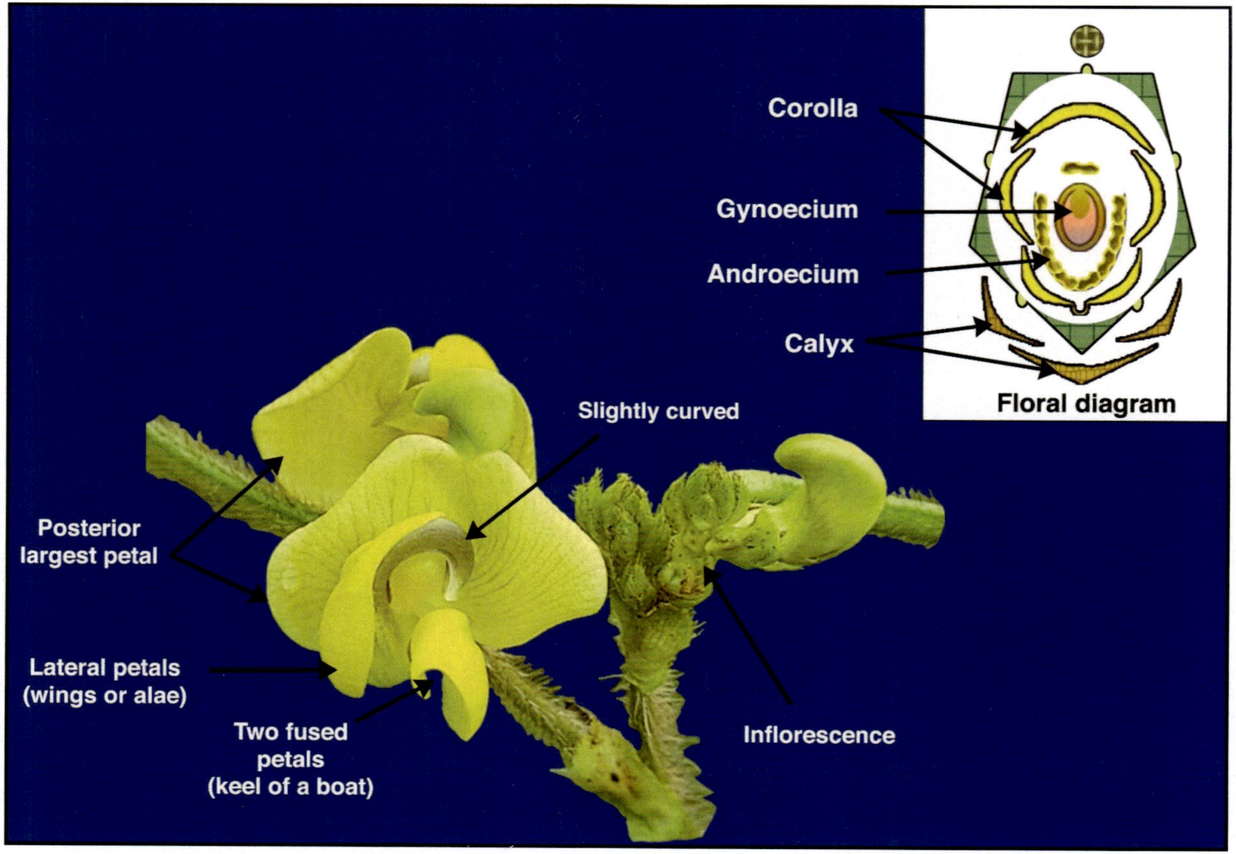

Figure 13.1. Floral Biology and Diagram of Green Gram (*Vigna radiata* L.). (pg. 150)

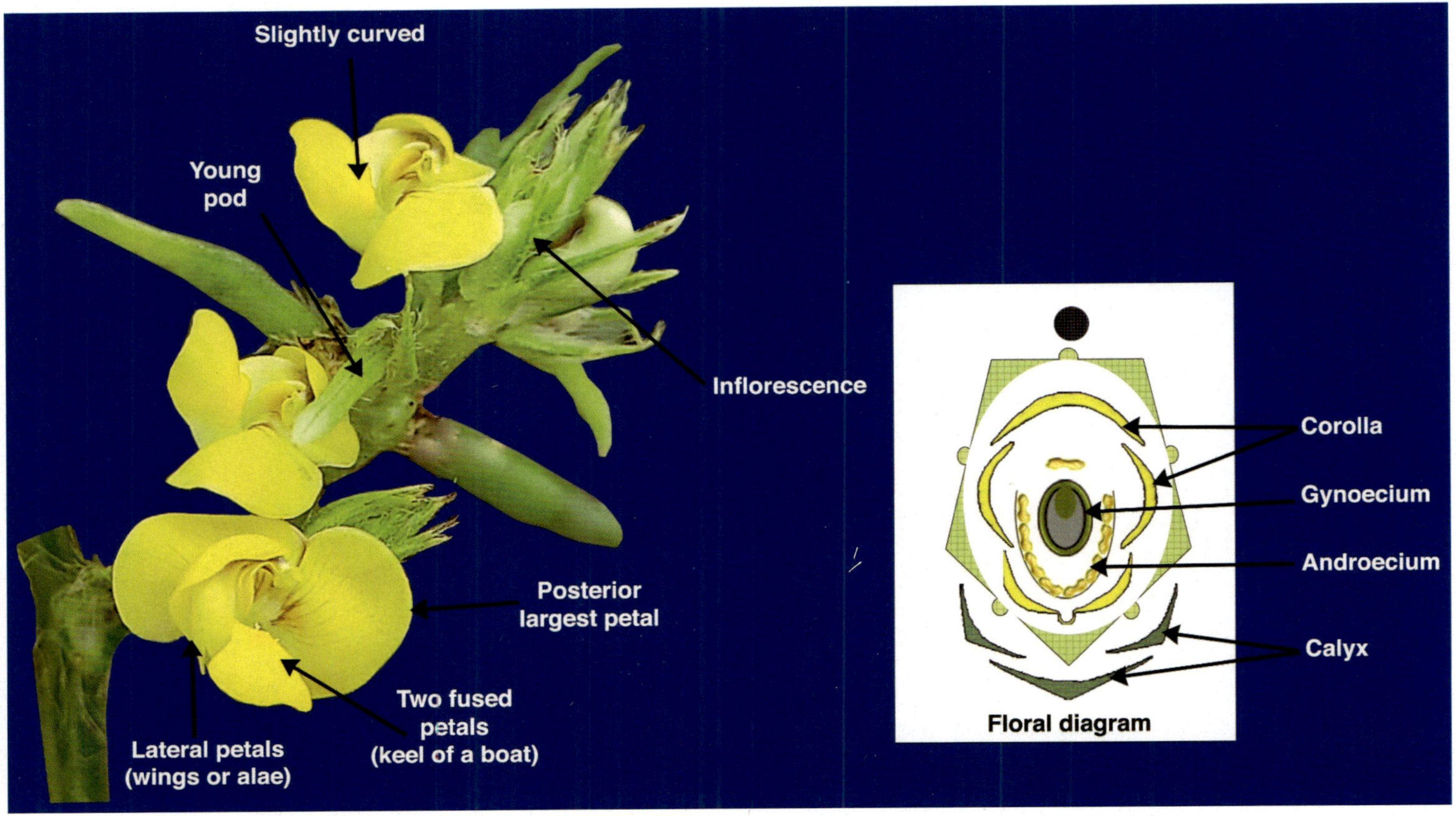

Figure 14.1. Floral Biology and Diagram of Black Gram (*Vigna mungo* L.) (pg. 157)

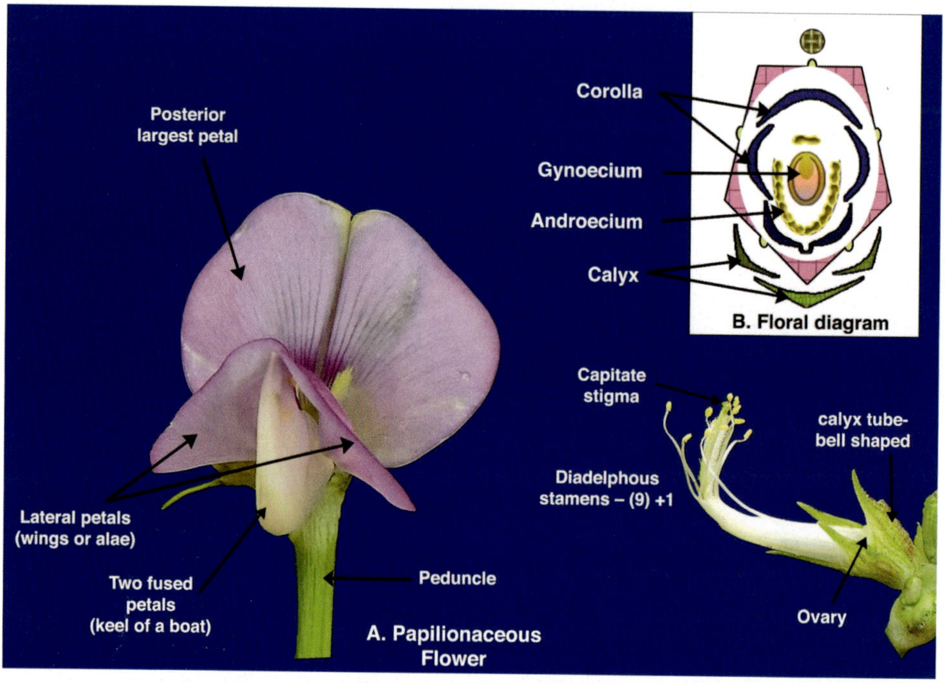

Figure 15.1. Floral Biology and Diagram of Cowpea (*Vigna ungiculata* L.). (pg. 163)

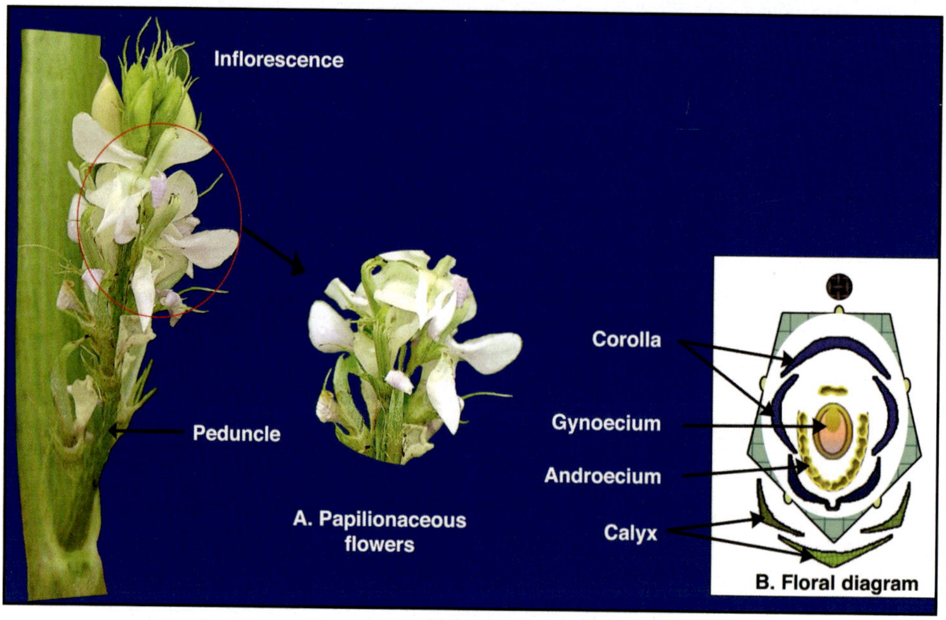

Figure 16.1. Floral Biology and Diagram of Cluster Bean (Guar) (*Cyamopsis tetragonoloba* L.) (pg. 169)

Section-IV

Breeding Oilseeds, Fibre and Sugar Crops (Groundnut, Mustard, Soybean, Sunflower, Sesame, Castor, Cotton and Sugarcane)

Chapter 17

Breeding Groundnut (*Arachis hypogaea* L.)

Introduction

Groundnut is an economically important annual oil seed legume in the world. The word groundnut is derived from the Greek words "*Arachis*" meaning legume and "*hypogaea*" meaning below the ground. India is the second largest producer after China. It mostly cultivated in subtropical and warm temperate zone. Groundnut has fundamentally indeterminate growth behaviours. It is principally grown for vegetable oil as the kernels contain 45-55 per cent oil. Nutritional point of view, its rich source of protein (around 25 per cent), carbohydrates (10–20 per cent), vitamins, dietary fibre, minerals and antioxidants. The de-oiled cake of groundnut is a very rich source of protein. It feeds for cattle or used as a manure, whereas the plant haulm is used as fodder for livestock. It helps improve soil health and fertility by harbouring symbiotic nitrogen-fixing bacteria in root nodules. Groundnut crop is considered as unpredictable legumes because of

- ✰ Unique pod development mechanism, the pods are borne below ground positively geotropic. Thus, the performance is unpredictable before harvest.
- ✰ Overall, crop performance and oil percentage are highly influenced by the environment.

- ☆ Less boron leads to low shelling per cent, and moisture stress at the immature seed stage leads to lower yield and reduced well-developed kernels.
- ☆ The availability of excess moisture leads to more vegetative growth and a reduction in yield.

Background Information

1.	*Common name*	Groundnut, peanut, monkey nut, goose nut, manila nut, jack nut, earth nut
2.	*Botanical Name*	*Arachis hypogaea* L.
3.	*Family*	*Fabaceae* (*Leguminosae*)
4.	*Chromosome number*	2n=4x=40
5.	*Centre of origin*	Brazil and South America (Bolivia)
6.	*Nature of pollination*	Self-pollination
7.	*Pollination mechanisms*	Bisexuality, Cleistogamy, Keel petal
8.	*List of wild relatives*	*Arachis monticola, A. prostrate, A. silvestres, A. lignose, A.villosa, A. pusilla, A. cardenasii*

Subspecies of Groundnut and Varietal Association

According to the patterns of reproductive and vegetative branching and on the pod morphology, the cultivated species is grouped into two cultivated subspecies *viz., A. hypogaea* subsp. *hypogaea* and *A. hypogaea* subsp. *Fastigiate* [Krapovickas (1968)]. The important features of two species are mentioned in Table 17.1. Krapovickas and Gregory (1994) further classified these two species into six botanical varieties.

Table 17.1. The Important Features of *A. hypogaea* Species

A. hypogaea subsp. hypogaea	*A. hypogaea subsp fastigiate*
It comprises *Virginia* bunch (semi-spreading) and *Virginia* runner (spreading) types; the other species is Var. *hirsute* (Peruvian humpback and Chinese dragon)	It comprises var. *fastigiata* (Valencia type) and var. *vulgaris* (Spanish type), generally known as bunch-types; other species are Var. *peruviana and* Var. *aequatoriana*
The first two nodes bear vegetative branches, next two branches bear inflorescences	Inflorescences are borne on the second and succeeding nodes of primary branches
Foliage dark green; inflorescence simple; seed dormancy usually presence	Foliage light green; inflorescence either simple or compound; seed dormancy usually absence

Group of Cultivated Types

In India, the cultivated groundnut types are grouped into three categories

i) **Spreading - *Virginia* runner:** The central axis is short and upright, while the main branches are spread flat along the ground.

ii) **Semi-spreading - *Virginia* bunch:** Intermediate bunch and spreading types are classified as semi-spreading.

iii) **Bunch type-Valencia and Spanish bunch:** The central axis is elongated and straight, and the main branches are sloping to the central axis.

Table 17.2. Morphological Comparison of Spreading and Bunch Types of Groundnuts

Sl.No.	*Traits/Features*	*Bunch*	*Spreading*
1	Botanical Group	*Spanish* and *Valencia*	*Virginia*
2.	Plant growth habit	Erect	Spreading
3.	Flowers on the main stem	Present	Absent
4.	Lateral branches	Mostly reproductive	Always vegetative
5.	Branching pattern	Sequential	Alternate
6.	Foliage colour	Light green	Dark green
7.	Leaflet size	Large and oblong	Smaller and pointed
8.	Pod size	Small to medium	Medium to bold
9.	Pod arrangement	Cluster at base of the plant	Scattered along with branches
10.	Seed size	Smaller	bolder
11.	Seed dormancy	Generally absent	Present
12.	Oil content Range	42-46 per cent	48-55 per cent
13.	Maturity group	Early	Late
14.	Growing season	Both *Kharif* and Rabi	*Kharif*

Genetic Origin of Cultivated Groundnut

The cultivated groundnut is an allotetraploid (2n=4x=40) with two different genomes 'AA' and 'BB'. It is believed to be the result of hybridization between two wild diploid species and following chromosome doubling. The species *A. duranensis* is believed AA genome donor, while *A. ipaensis* is donor of BB genome.

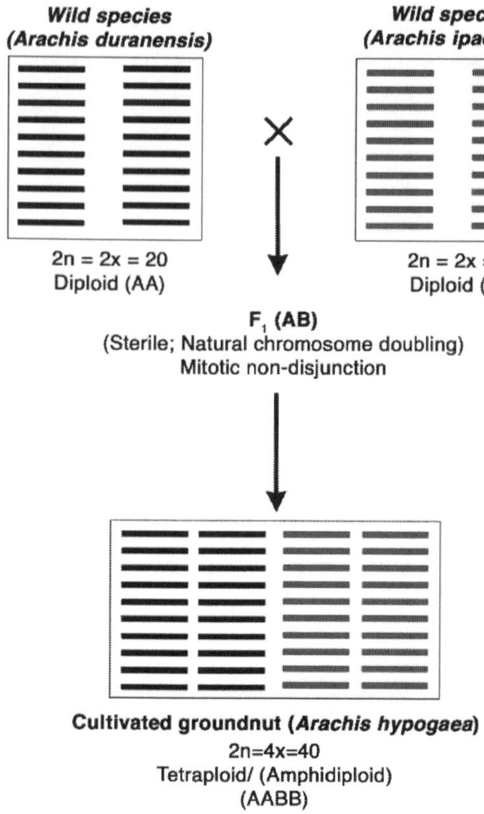

Figure 17.1. Genetic Origin of Cultivated Groundnut.

Gene Pool

A gene pool of peanut is grouped into three categories

Sl.No.	*Gene pool*	*Species*
1.	Gene pool 1 (Primary)	Includes tetraploid species, *viz. Arachis hypogaea* and *A. monticola*
2.	Gene pool 2 (Secondary)	Includes all the diploid species of section *Arachis viz., A. cardenasii, A. chacoense, A. correntina, A. duranensis nom.nud., A. batizocoi, A. ipaensis,* and *A. villosa*
3.	Gene pool 3 (Tertiary)	Includes the species outside the genus *Arachis*

Distribution

Peanut is cultivated throughout the world's warm temperate and tropical territories. Commercial cultivation in more than 100 countries, including Asia, Southeast Asia, African countries, the Caribbean; USA; Brazil, Argentina and Mexico in Latin America. In India major areas are confined to the five states *viz.*, Gujarat, Andhra Pradesh, Tamil Nadu, Karnataka and Maharashtra.

Floral Biology, Selfing and Crossing Techniques

Inflorescence	:	Simple (solitary flower) or compound (racemose type includes two to five flowers)
Flower	:	Papilionaceous, sessile, zygomorphic, irregular, complete and perfect
Calyx	:	Gamosepalous with five sepals are in two groups, *i.e.*, one awn-like opposite the keel and the rest four remain opposite to the back of the standard petal; Pedicel is actually an extended hypanthium (calyx tube)
Corolla	:	Five petals [1+2+(2)], polypetalous, unequal in size, vexillary aestivation, papilionaceous (one posterior largest petal (standard vexillum) that encloses two lateral petals (wings or alae) wrap around the keel
Androecium	:	Ten monadelphous stamens, of which two are sterile, remaining eight are fertile. Four are with globose, dorsifixed and monothecous anthers, while four are adnate, introrse, oblong anthers. The filaments are fused upto two-thirds of their length
Gynoecium	:	Ovary monocarpellary, superior and unilocular, 1 to 5 ovules, Placentation- marginal, style is glabrous throughout its length enclosed within the calyx tube (hypanthium); club-shaped stigma with hairy end. A stalk of the ovary elongates to form a peg and curve towards the soil.
Fruits	:	Legume or pod bearing 1–5 seeds, depending on the variety
Seed	:	Ex-albuminous with two cotyledons
Floral formula	:	⊕ ⚥ $K_{(5)} C_{1+2+(2)} A_{(10)} \underline{G}_{1}$

Anthesis: Anthesis of flowers occurs in the morning between 6.00 and 8.00 hours. The anthers dehiscence takes about 1-2 hrs before the opening of the flowers, and fertilization finishes before mid-day (6 hours after pollination).

Selfing technique: To ensure the cent per cent self-pollination, covers the whole plant with a muslin cloth bag.

Artificial Crossing and Pollination

Select mature flower buds (generally pink or light purple) of the female

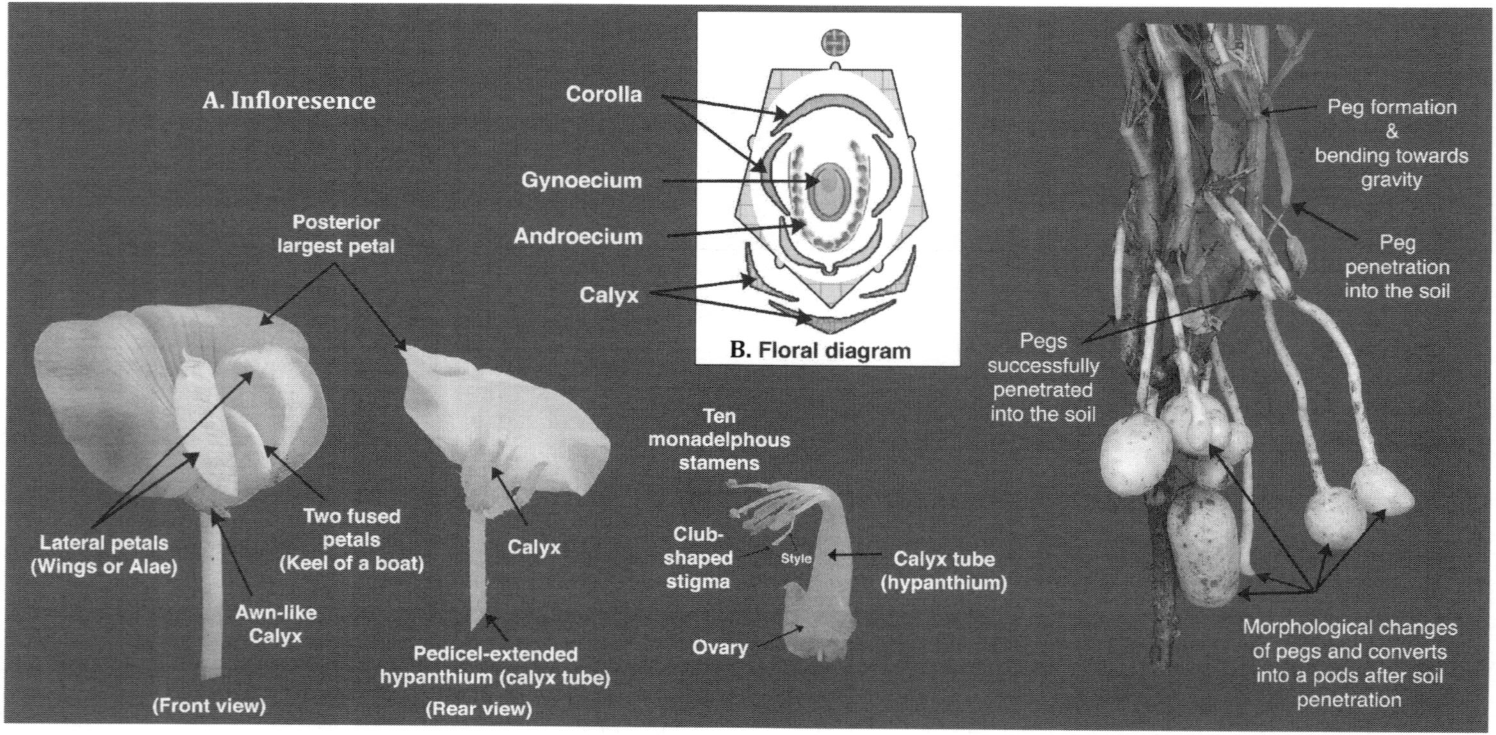

Figure 17.2. Floral Biology and Diagram of Groundnut (*Arachis hypogaea* L.).

parent, which are ready to open the next day and very close to the soil surface, are selected and emasculated in the evening during 1.30 - 6.00 PM. Rest over imatured and premature buds in the inflorescence and developing peg resulted from self-pollination if any in axils of the leaf should be removed with forceps. For emasculation, hold the bud gently between the thumb and the index finger. Cut the standard and a portion of the wing petals with the help of a razor blade by cutting on the depressed side at two-thirds length below the tip. Anther gradually pulls the calyx and corolla by holding it at the tip of the flower bud. Then pull down the keel by forceps and carefully remove all anthers with filaments in a downward movement. Do not touch the stigma by forceps during the whole procedure. Pollination is performed on the following day morning from 6.00 to 10.00 AM. After 5-7 days of pollination, successfully crossed flowers will be produced gynophores (pegs). Tagging a respective cross by introducing a small wire ring of 4 mm diameter with the pegs.

Post Fertilization Events: A Peg Formation Mechanism in Groundnut

After fertilization, the flower drops, but the hypanthium with style remain persists at the base of the ovary for a week (4-6 days) during this time, the embryo development pause, but in response to auxin formation, a stalk-like structure or peg (carpophore or gynophore) develops due to the elongation of intercalary meristematic cells at the basal region of the ovary. A peg carries the ovary with the fertilized ovules at its tip. It has positive gravitropism, so it moves toward and enters the soil. Once the peg pierces the ground and reaches a depth of 2-4 cm, the peg growth stops, and subsequently, the development of the embryo resumes resultant the ovary develops into a pod due to favourable environmental factors like darkness, mechanical stimulus, and Ca and nutrients uptake by the developing pod. *Hence, in groundnut the flowers are above the ground level, but peg and pod formation occur below the ground.*

Breeding Objectives

1. **Breeding for higher yield:** In terms of yield, increasing the number of pods and seeds are the main targeting attributes. It can be improved by various components traits *viz.*, Seed size (Virginia types have larger kernels than Spanish type), number of pods per plant, number of seeds per pod, pod yield per plant, shelling outturn, growth habit and duration are also impact on yield. Pod yield traits inheritance by additive and non-additive gene action, while shelling outturn is also controlled by additive and non-additive gene effects with a pair of genes without dominance. *Various factors are responsible for low productivity in groundnut are*
   - ☆ Non-synchronous flowering effect in a number of pod filling.
   - ☆ Success in hybridization is significantly less due to tiny and complex flower structure, low flower-to-pod ratio (7:1), few seeds per pod *etc.*

- ☆ Maximum pod losses occur during harvesting due to weak peg attachment.
- ☆ Seed dormancy: Prolonged in spreading types, while absence in bunch type
- ☆ Low harvest index and very less seed multiplication ratio, *i.e.* 1:5.

2. **Breeding for earliness:** Important to escape drought, and disease, and prevent germination due to early monsoon.
3. **Breeding for fresh seed dormancy:** The problem persists in Spanish cultivars. The Spanish types cultivars grown during *kharif* may face early N.E. monsoon rains that result in the sprouting of seeds. So, it is essential to breed varieties having dormancy. Semi-spreading varieties are dormant, *e.g.*, TMV 7 and slightly dormant varieties, *viz.*, BSR1, ALR 2 reported dormancy for 15 days.
4. **Breeding varieties for quality and mineral biofortification**
   a. **Breeding for high oil content and quality:** Groundnut oil is considered a stable and nutritive as it contains oleic (40-50 per cent) and linoleic acids (25-35 per cent). The oil content in groundnut germplasm has been reported from 46.5 to 63.1 per cent. The yield and oil content in groundnut are not linked traits. Inheritance studies reported that oil is governed by a non-additive genetic component and two alleles.
   b. **Oil quality:** It comprises the fatty acid composition, iodine value, oleic-to-linoleic acid ratio (O/L), and shelf life. A high oleic in oil is vital for good shelf life, nutritional quality and essential for processing purposes (confectionaries). Spanish-Valencias have higher linoleic acid, while the Virginia types usually have higher oleic acid content.
   c. **Breeding for table purposes (confectionery) groundnut:** Includes unique physical and chemical attributes acceptable to consumers, traders and manufacturers. Among essential physical parameters include attractive seed size, shape, pink or tan seed colour, flavour, and high sound mature kernel (SMK), free from aflatoxin contamination are essential. In the case of the chemical composition includes high protein (>30 per cent), sugars (>6 per cent), low oil (<45 per cent) and free fatty acid (FFA) (<1 per cent), and high O/L, *etc.*, are essential parameters.
   d. **Biofortification:** Cost-effective and sustainable solution to improve Fe and Zn concentration.
   e. **High shelling percentage:** More than 75 per cent is ideal. The variety having thin shelled shows high shelling percentage.

f. **Aflatoxin resistance:** Aflatoxin is a toxic, carcinogenic secondary metabolic substance produced by fungi *Aspergillus flavus* and *A. parasiticus*. So far, very little success has been achieved in resistance toward aflatoxin contamination due to

- ❖ Complex resistance nature
- ❖ No precise selection method is available for segregating generations
- ❖ No any genotypes which having all the seeds remain free from infection

5. **Breeding resistance to biotic stresses**

- ☆ **Disease:** Rust and early and late leaf spot are major foliar diseases varieties having multiple disease resistance are Girnar-1, ALR-2, GPBD-4, ICGS (FDRS) 10, AK 265, Phule Bharti (JL 776), GJG 32 (ICGV 03043) Phule Morna (KDG 123), Phule Warna (KDG 128), Phule Unnati (RHRG-6083).
- ☆ Soil-borne disease- stem rot and collar rot are major. The important resistant genotypes available for stem rot are PBS 18037, NRCGCS 19 and NRCGCS 319. While in the case of collar rot, a lack of stable resistant source is available. However, some moderate tolerance genotypes are reported *viz.*, J11, JCG 88, TG-37A, O.G. 52-1, ICGV 00350, ICGV 93261, ICGV 92195, ICGV 92035, ICR 48, *etc.*
- ☆ Bud necrosis: Important cultivars reported field level resistant are G.G. 7, Kadiri 3, ICGS 11, ICGS 37, ICGS 44, ICGV 86325, R 8808, K 134, DRG 12, JCG 88, CSMG 884, Chandra, *etc.*
- ☆ Nematodes: Root-knot nematodes and lesion nematodes are important. The resistance sources are reported in wild species, *viz., A. cardenasii, A. diogoi, A. batizocoi, etc.*
- ☆ Insect-Pests: *Spodoptera, Helicoverpa*, red hairy caterpillars, leaf miner are major pests. Various sources of resistance for pest like *Spodoptera*, thrips and jassids have been identified in groundnut germplasm and are used in breeding programs.

6. **Breeding resistance to abiotic stresses:** Drought, salinity, high temperature, and aluminium toxicity affect groundnut productivity. The genotypes K 134, ICGS-1, ICGS-5, ICGS-37, *etc.* are found to be tolerant to drought.

Breeding Approaches (Methods)

1. **Introduction and selection:** Introduction and selection of exotic cultivars; M-13 (high oil content) is selected from NC 13 (U.S. cultivars). TMV 10 (bold-seeded Virginia type) is a reselection from Argentinean cultivars R.S. 87.

2. **Pure line and mass selection:** The various cultivars have been developed from selection are TMV-2, TMV-3, TMV-7, TMV 8, Pollachi-1, S-206, S-230, TMV-12, Dh-8, ICGS-11, RS-138, Konkan Gaurav, SG-84, Tirupati-3, ALR-2, VL-Moongphali-1, JL-50.
3. **Mutation breeding:** In India, the mutation-breeding project was initiated in BARC, Trombay, Mumbai (1957-58). The varieties released from this centre are commonly known as Trombay groundnut. The mutation is used to improve specific traits and adaptation. The different varieties that have been developed are given in Table 17.3.

Table 17.3. List of Mutant Cultivars of Groundnut Released in India

Sl.No.	*Cultivars*	*Year*	*Source of Mutagen*	*Characters Improved*
1.	T.G. 1 (Vikram)	1973	X-Rays (750 Gy)	Large seed size, maturity (135 days), oil content (47-48 per cent) and TMV resistance
2.	T.G. 3	1973	X-Rays (150 Gy)	Increased pod number, high yield and good adaptability
3.	T.G. 4 (T.G. 14)	1976	X-Rays (150 Gy)	Uniform maturity, high yield and good adaptability
4.	MH-2	1978	Spontaneous mutant	Higher yield (4060 kg/ha), very dwarf (10-12 cm), early maturity (110 days), resistance to *Cercospora personata* and easy digging
5.	BP-1	1979	Gamma rays (450 Gy)	Improved seed size, semi-erect habit and high yield
6.	BP-2	1979	Gamma rays (450 Gy)	Improved seed size, semi-erect gigas type and high yield
7.	B.G. 1	1979	X-Rays	Higher yield, semi-erect habit, gigas type and large seed
8.	B.G. 2	1979	X-Rays	Higher yield, large seed and semi-errect habit
9.	T.G. 17	1982	X-Rays (150 Gy)	High yield, plant type (no secondary branches) and high harvest index
10.	Co 2	1984	EMS (0.2 per cent)	High yield and higher shelling percentage
11.	VRI 2	1989	EMS (0.2 per cent)	Large seed size, suitable for both rainy and post-rainy season
12.	TGS 1 (Somnath)	1989	Gamma rays (200 Gy)	Early maturity, large seed size, high yield, sequential flowering and high oil content

Sl.No.	*Cultivars*	*Year*	*Source of Mutagen*	*Characters Improved*
13.	TAG-24	1991	Gamma rays (200 Gy) and X-Rays (750 Gy)	Earliness, semidwarf habit, yield stability, high harvest index, shorter internodes, dark green, small leaves and high-water use efficiency
14.	TG-22	1992	X-rays (150 Gy)	High yield, medium-large seed size, seed dormancy, tolerance to acidic soils
15.	TKG-19A	1993	X-Rays (750 Gy) and X-Rays (150 Gy)	High yield, large seed size, maturity (120 days), slightly reduced oil content with increased protein percentage, tolerant to acidic soils
16.	TG-26	1996	Gamma rays (300 Gy)	High yield, semi-dwarfness, tolerance to peanut bud necrosis disease, fresh seed dormancy, compact pod setting, response to inputs
17.	R9251	1997	Gamma rays	Early maturity, tolerant to peanut bud necrosis disease
18.	TPG-41	2002	X-Rays (150 Gy)	High yield, large seed, maturity 120 days, 25 days seed dormancy, released for all India post-rainy situation under irrigation
19.	Mutant 28-2	2003	EMS (0.5 per cent)	Large seed, resistance to late leaf spot, *Spodoptera litura* and thrips
20.	TG-37A	2004	Gamma rays (300 Gy)	Semi-dwarfness, compact pod setting, high yield and smooth pod surface
21.	TG-38	2006	Gamma rays (300 Gy)	Erect growth habit with sequential branching, semidwarf height, medium-size dark green leaflets, compact pod setting with smooth pod surface, tolerance to stem rot and dry root rot
22.	TLG-45	2007	X-Rays	High pod yield, semidwarf plant and good quality
23.	T.G. 39	2008	Gamma rays	Semi-erect with medium height and alternate flowering, Seed contains 50 per cent oil, 26.5 per cent protein, 12.6 per cent carbohydrate and 4.5 per cent sucrose. Its oil contains 59 per cent oleic acid and 23 per cent linoleic acid.

Sl.No.	Cultivars	Year	Source of Mutagen	Characters Improved
24.	T.G. 51	2008	Gamma rays	Tolerance to stem rot and root rot diseases erect with semidwarf height and sequential flowering. Seed contains 49 per cent oil with 43 per cent oleic acid and 37 per cent linoleic acid.
25.	TDG 39	2009	Gamma rays	Large seeds, 120 days maturity and high content of oleic acid
26.	GPBD 5	2010	Crossing with T.G. 49	Large seed

4. **Hybridization and Selection:** Both inter-varietal and wide hybridization methods have been used in peanut improvement.
   - ☆ **Inter-varietal hybridization:** Pedigree selection method is extensively used; various varieties have been developed are mentioned in Table 17.4.

Table 17.4: List of Groundnut Cultivars Developed through Pedigree Selection

Sl.No.	Name of Varieties	Group	Pedigree
1.	Gujarat Groundnut 8 (GG 8)	Spanish Bunch	27-5-1 × JL-24
2.	GG 16	Spreading	JSP 14 × JSSP 4
3.	Gujarat Junagadh Groundnut-HPS-1	Virginia runner (Spreading)	JSP 21 × VG 5
4.	GJG 9	Spanish bunch	G.G. 5 × ICGV-90116
5.	GJG 31	Spanish	G.G. 2 × PBS-21065
6.	GJG 17	Virginia runner	JSSP 11 × GG 6
7.	GJG 22	Virginia bunch	JSSP 17 × GG 20
8.	GJG 19	Virginia runner (Spreading)	JSSP 12 × LGN 2
9.	GJG 32	Erect	(F 334 A-B-14 × NCAc 2214) × ICG 2241) × (ICGMS 42 × Kadiri 3) × ICGMS 28 × (F 334A-B-14 × NCAc 2214)] × LI × (Robut 33-1-1-5)
10.	GJG 33	Erect (Spanish bunch)	[(ICGV 92069 × ICGV 93184) SIL 4 × (ICGS 44 × ICGS 76)]
11.	GG HPS 2	Virginia Bunch	JVBHPS 2 × GG 20
12.	GG 41	Virginia runner (Spreading)	GG-21 × TPG 41
13.	GG 23 (Sorath Kiran)	Virginia bunch (semi-spreading)	JVR 503 × ALG 234

Sl.No.	Name of Varieties	Group	Pedigree
14.	GG 35 (Sorath Gold)	Spanish bunch	TG 37A × JVB 2113
15.	Raj Mungfali 3 (RG 559-3)	Virginia	(TKG 19A × Kadiri 3) × TKG 19A
16.	Girnar 1	Spanish	14-4-B-19-B × NCAc 17090
17.	Girnar 2	Virginia bunch	M 13 × R 33-1
18.	Girnar 3	Spanish bunch	Girnar 1 × ICGS 11
19.	Girnar 4	Virginia bunch	ICGV 06420 × Sun Oleic 95R
20.	GG 101 (Sorath Keshav)	Spanish bunch	TG-37A × KDG 123
21.	GG 24 (Sorath Urja)	Virginia bunch (semi-spreading)	JSSP 35 × KDG-128
22.	GG 42 (Sorath Shaan)	Virginia runner (Spreading)	KDG 128 × VG 9375

Source: www.jau.in/ and www.dgr.org.in/

- ✩ **Distant hybridization:** Wide hybridization has been used groundnut for the transfer of specific traits. The variety GPBD 4 was developed by cross between [KRG 1 × *A. cardinadii* (ICGV 86855)]. The nematode resistance genotype TxAG-7 was developed by wide hybridization between *A. hypogaea* and A. *cardenasii.* While genotype TxAG-6 was developed from a complex cross of *A. batizocoi* × (*A. cardenasii* × *A. diogoi*)

5. **Biotechnological approaches:** Various biotechnological techniques are using in peanut improvement
   a. Tissue culture techniques- Embryo rescue is used to get wide cross successfully, *e.g. A. puscilla* × *A. hypogaea.*
   b. Marker-assisted backcross breeding (MABC): Various markers and QTLs have been identified linked to yield, disease and pest resistance.
      - ✩ The late leaf spot and rust-resistance cultivars TAG 24, ICGV 91114 and JL 24 were developed through MABC.
      - ✩ Introgression of rust-resistant QTL from GPBD 4 in popular farmers' cultivars, *viz.*, JL 24, ICGV 91114, and TAG 24.
      - ✩ High oleic genotypes ICGV 06110, ICGV 06142 and ICGV06420 were developed through MABC using high oleic parent Sun Oleic 95R. Recently at ICAR–DGR, two new varieties, Girnar 4 and Girnar 5, with nearly 80 per cent of oleic acid content developed using marker-assisted introgression of high oleic content.
   c. Transgenic breeding for disease resistance: *Agrobacterium*-mediated insertion of *CryIA, cry1EC and cry1AcF* genes for insect resistance.

d. It helps to identify drought tolerance genotypes,
e. Genotype resistance to aflatoxin contamination, and
f. Quality improvement, *i.e.*, Beta-carotene enrichment.

Breeding Institutes

- **International Level:** International Crop Research Institute for Semi-Arid Tropics (ICRISAT), Hyderabad, India.
- **National level:** Indian Institute of Groundnut Research (IIGR), Junagadh, Gujarat, India.
- **State level:** Regional Research Stations-State Agricultural Universities.

Breeding Achievements

A. **Spreading types:** GAUG-10, GG-11, GG-12, GG-13, GG-14, GJG-HPS-1 and GJG-16, GJG-17, GG 41, GG 42 (Sorath Shaan)

B. **Semi-spreading types:** GG-20 and GG-22, Girnar 2, Girnar 4, GG 23 (Sorath Kiran), GG 24 (Sorath Urja)

C. **Bunch type:** GG-2, GG-4, TG-26, GG-5, GG-7, TPG-41, GJG 8, GJG-9, Kadiri-5, Kadiri-6, GJG 32, GJG 33, GG HPS 2, GJ 35 (Sorath Gold), Girnar 3, TG 37A, Gujarat Groundnut 37 (Sorath Gaurav), Gujarat Groundnut 38, Gujarat Groundnut 39, Girnar 5, Gujarat Groundnut 40, Girnar 6, GG 101 (Sorath Keshav), *etc.*

Chapter 18

Breeding Mustard (*Brassica juncea* L.)

Introduction

Brassica is the third most important oil seed crop in the world, whereas second in India, next to groundnut. It comprises different species: Rai (Indian mustard), yellow sarson, brown sarson, gobhi sarson, toria, karan rai, black mustard and taramira. In India, major breeding works are focused on Indian mustard, comprising 90 per cent areas of rapeseed-mustard. Mustard is an important source of edible oil, besides it also used industrial lubricant and a base for polymer synthesis; oil cakes are used as animal feeds. Mustard oil is the best for edible since it has the lowest amount of saturated fatty acids compared to other vegetable oils.

Background Information

1.	*Common name*	Rai, Sarson, rayda, brown mustard, Indian mustard, Chinese mustard, *etc.*
2.	*Botanical Name*	*Brassica Juncea* L.
3.	*Family*	*Brassicaceae* (*Cruciferae*)
4.	*Chromosome number*	2n=4x=36
5.	*Centre of origin*	China
6.	*Nature of pollination*	Self-pollination but cross-pollination occurs up to 22 per cent

7.	*Pollination mechanisms*	Bisexuality, anemophily and entomophily-melittophily
8.	*List of wild relatives*	*B. japonica, B. nigra* (2n=16), *B. oleracea* (2n=18), *B. carinata* (2n=34), *B. carica, B. napus* (2n=38),*B. compestris* (2n=20)

Classification of *Brassica* Species

Common Name	*Botanical Name*	*Genome Symbol*	*Chromosome Number*	*Remarks*
Indian rape/ Rapeseed/Toria	*Brassica rapa* L. (Syn. *B. compestris*) Spp. toria; yellow sarson and brown sarson	AA	2n = 2x = 20	Diploid
Black mustard	*B. nigra* L.	BB	2n = 2x = 16	
Cauliflower/ Cabbage	*B. oleracea* L.	CC	2n = 2x = 18	
Taramira/Salad	*Eruca sativa* L.	EE	2n = 2x = 22	
Radish	*Raphanus sativus* L.	RR	2n = 2x = 18	
White mustard	*Sinapis alba/B. hirta*	SS	2n = 2x = 24	
African mustard	*B. tournefortii*	TT	2n = 2x = 20	
Indian mustard/ Brown mustard	*Brassica juncea* L.	AABB	2n = 4x = 36(AABB)	Allotetra-ploid (Amphidi-ploid)
Gobhi sarson	*Brassica napus* L.	AACC	2n = 4x = 38(AACC)	
Ethiopian mustard	*Brassica carinata* L.	BBCC	2n = 4x = 34(BBCC)	

Genetic Relationship and Evolution of *Brassica* Species

The genetic relationship and evolution of different *Brassica* species were well explained by a Korean botanist **Nagaharu U** in 1935, commonly known as the 'U triangle'. It explains the genetic association between the six *Brassica* species. Out of six three common diploid species result due to ancestral genome *viz., B. oleracea* (2n =2x= 18, CC), *B. rapa* (2n=2x=20, AA) and *B. nigra* (2n= 2x= 16, BB). While three allotetraploid **(Amphidiploid)** species *viz., B. napus* (2n= 4x= 38, AACC), *B. juncea* (2n= 4x= 36, AABB) and *B. carinata* (2n= 4x= 34, BBCC) produced by combination of two parental diploid *Brassica* species (Figure 18.1).

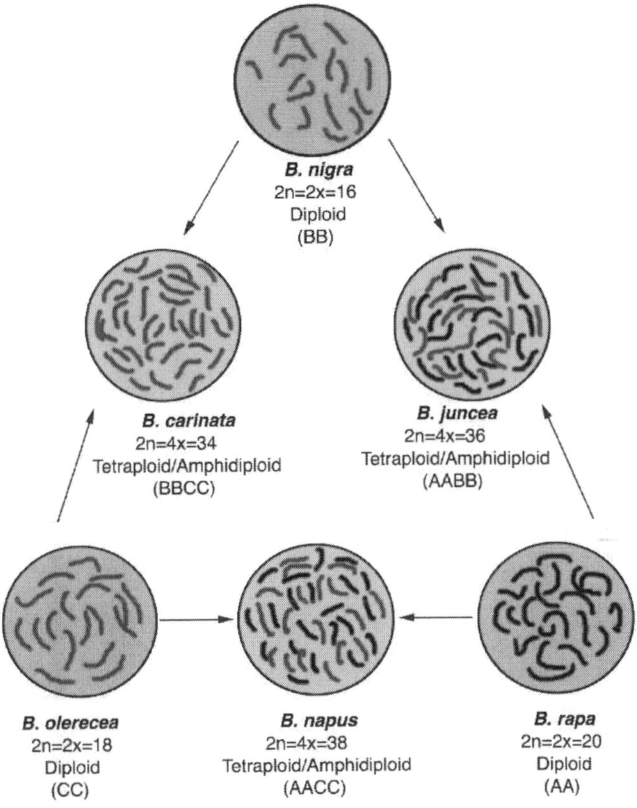

Figure 18.1. Genetic Relationship-Evolution of different *Brassica* Species.

A Typical Intergenic Amphidiploid Hybrid- *Raphanobrassica*

A Russian geneticist **G. D. Karpechenko** in 1927, reported a typical intergenic hybrid called *Raphanobrassica.* A cross between radish (*Raphanus sativus;* 2n= 18; RR) and cabbage (*Brassica oleracea;* 2n= 18: CC). The F_1 hybrid (2n=18) is found to be sterile due to synaptic failure; through the chromosome doubling, this hybrid showed a fertile plant having 2n=36 chromosomes (18 bivalents) (Figure 18.2). The hybrid had reversed trait behaviour of parental species, hence remains only in theoretical importance.

Distribution

Indian mustard is believed to be one of the earliest domesticated plants. It originated in China, and from there, it was introduced to India and then spread to Afghanistan and other countries. It is grown across Europe, Africa, Asia, America, and Australia. India is one of the major mustard-growing countries in the world.

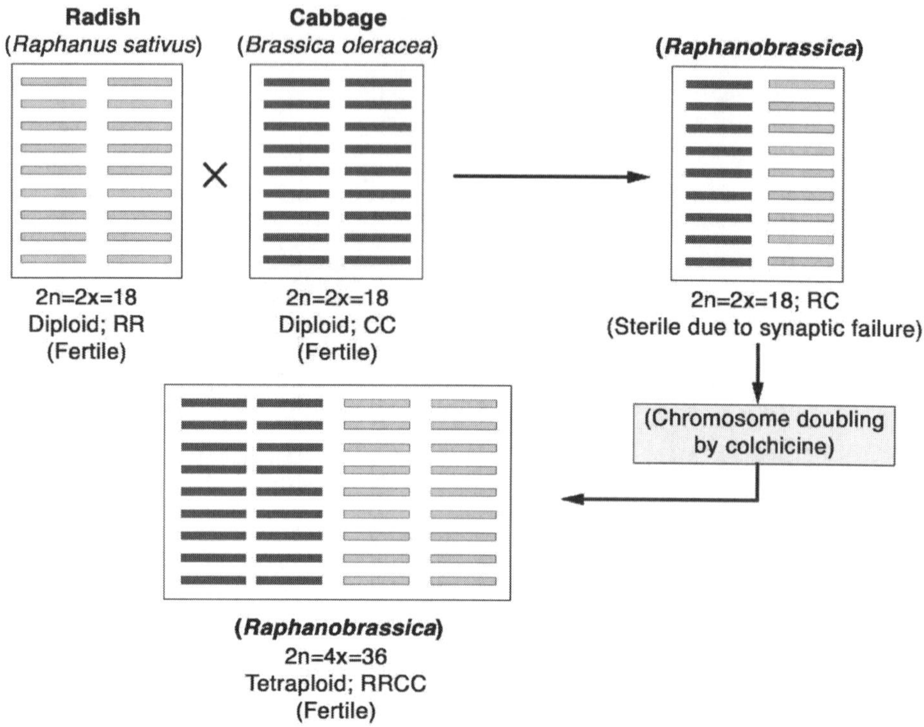

Figure 18.2. Illustration about Synthesis of *Raphanobrassica*.

Floral Biology, Selfing and Crossing Techniques

Inflorescence	:	Racemose; Corymb raceme
Flower	:	Ebracteate, pedicellate, actinomorphic (Regular) or rarely zygomorphic, complete, perfect, tetramerous, hypogynous, cruciform and yellow in colour
Calyx	:	Sepals are four in number, pale yellow to cream in colour, polysepalous, arranged in two whorls; each whorl contains two sepals, aestivation-imbricate
Corolla	:	Petals are four in number, bright yellow in colour, polypetalous, alternate with sepals, cruciform clawed, valvate aestivation,
Androecium	:	Stamens – 6, polyandrous, tetradynamous (stamens arranged in two whorls, two outer short and four inner long), in the bud stage, the immature stamens are always below the stigma, anther- dithecous, basifixed and introrse or extrorse

Gynoecium	:	Two syncarpous carpels, ovary- superior, unilocular, but later becomes bilocular due to development of false septum called replum. Placentation- parietal, style – short and thick, stigma – bilobed or capitate
Fruits	:	Siliqua (long and narrow)
Seed	:	Numerous, minute, non-endospermic (Exalbuminous)
Floral formula	:	Ebr. ⊕ ⚥ $K_{2+2}\ C_4\ A_{2+4}\ \underline{G}_{(2)}$

Anthesis: The flowers begin to open from 8.00 AM to 12.00 PM. The flowers continue to open for 3-4 days, and on the 4th to 5th day, the petals and sepals are shedding. As the flowers open, the anthers begin to dehisce from the apex downwards. At the time of dehiscence, the pollen liberating sides of the anthers, remaining towards the stigma and slightly shaking of the flowers by wind, insects *etc.* is sufficient to transfer pollen. Large numbers of bees visit the flowers soon after they open, resulting in a certain amount of cross-pollination take place.

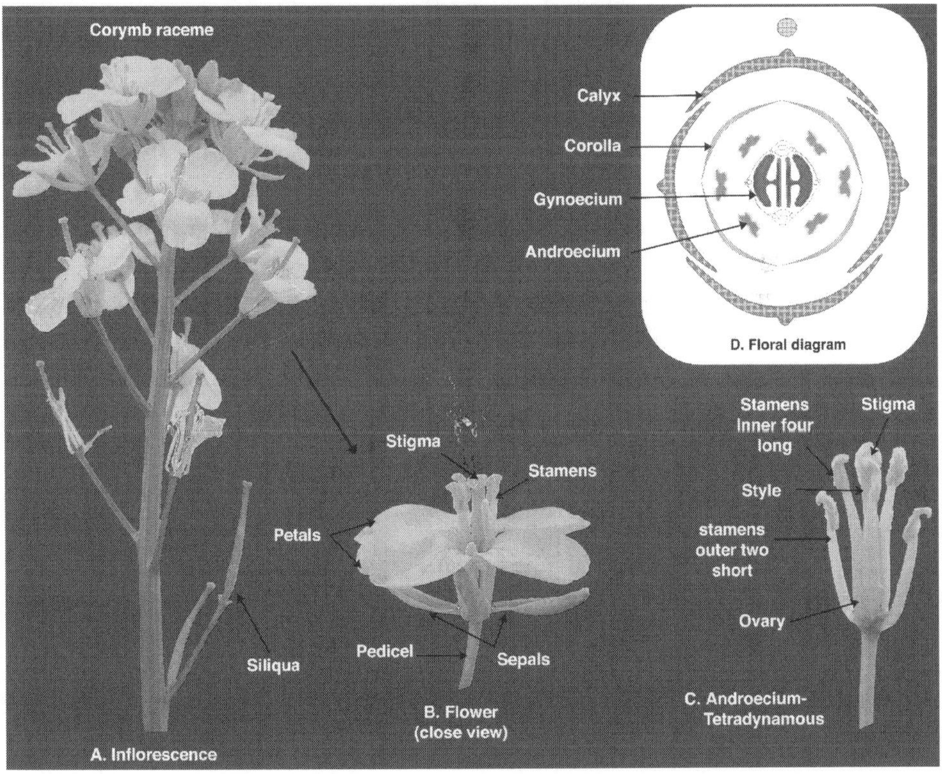

Figure 18.3. Floral Biology and Diagram of Mustard (*Brassica juncea* L.).

Selfing technique: To ensure the cent per cent self-pollination, cover either the whole plant with a muslin cloth bag or the bagging of an individual branches.

Artificial Crossing and Pollination: Flower buds that will open the next day are selected and the remaining buds and flowers on the flowering branch are removed. The petals as well as the stamens of selected buds, are removed with the help of forceps, and the emasculated flowers are bagged. Ripen anthers are collected in the morning of the next day and pollination is made by dusting pollens on stigma. After pollination, the flowers are again bagged. The pollen grains are reported to remain viable for seven days, while the stigma is receptive from three days prior to the opening of flowers to three days after opening.

Breeding Objectives

1. **Breeding for higher seed and oil yield:** The value of oil yielding crop can be determined by seed and oil yield. Seed yield is under the control of complex polygenic systems, it can be improved by different component trait *viz*, number of branches, number of siliquae, number of seeds per silique, harvest index, test weight *etc.* Similarly, oil content is also under polygenic control, oil content varying from 35 per cent (*E. sativa/B. carinata*) to 46 per cent (*B. rapa* spp. yellow sarson).
2. **Breeding short duration variety:** It is essential to fit in different cropping systems, developed short duration (mature in 100-120 days) varieties, *e.g.*, Pusa Agrani, Pusa Mahak, Pusa Tarak, Pusa Mustard 25, Pusa Mustard 27, Pusa Mustard-28, Kranti, Pant Rai 19, *etc.*
3. **Ideal plant type:** It includes dwarf plants with basal branching suitable for mechanical harvesting and determinant growth habit.
4. Breeding for desirable agronomic traits like lodging and shattering resistance: *e.g.* BIO-902.
5. Breeding for resistance to *orobanche e.g.*, Durgamani.
6. **Breeding for stress resistance:**

   i. Biotic stress: Different diseases and pests like

Disease-Pest	*Resistant or Tolerant Varieties/Lines/Genotypes/Source*
Powdery mildew	TM-4, GM-2
White rust	NDUH-YJ-6, Jwahar Mustard-1, RK-01-03, ELM-079, Bio-YSR, BEC-144, BEC-286, EC-399299, EC-399301, RC-781, JM-1, Heera, *etc.*
Alternaria blight	TM-4, RH 8113, ELM-079
Aphid	T 6342, RL 1359

   ii. Abiotic stress: Drought, heat, salt, and temperature are major abiotic stress limiting the seed yield in mustard.

- ☆ Salt tolerant genotypes are CS-52, CS-54, CS-56, CS-60, *etc.*
- ☆ Heat tolerant genotypes: Pusa Vijay, Pusa Mustard 25, Pusa Mustard 27, Pusa Mustard 28, *etc.*
- ☆ Drought resistant/tolerant genotypes: RH 819 and RH 781.

7. **Breeding for nutritional quality**

- ☆ High oil content (>45 per cent)
- ☆ Oil quality: Test, colour (it correlated with seed coat colour), high protein content and low saturated fatty acid content (around 7 per cent)
- ☆ **Double zero and triple zero type variety:** The high erucic acid in mustard is having industrial value, while for edible point of view, it should be low (< 2 per cent in oil). The variety with < 2 per cent erucic acid is termed as zero type ('0'). Most Indian varieties possess more than 50 per cent erucic acid content in oil, which may cause diseases like Myocardial fibrosis in adults and Lipidosis in children. In year 2016, IARI, released variety PM-30 (Pusa Mustard-30) having < 2 per cent erucic acid. Other examples are Heera, NUDHYJ- 5, *etc.* The variety having, <2 per cent erucic acid in oil and < 30 micro -moles of glucosinolate per gram of its de-oiled meal, it is known as double zero type ('00'). Best example of such type is canola cultivars (trade name).
- ☆ **Glucosinolate:** A group of compounds that give the characteristics of unpleasant flavour to vegetables and condiments. In meal, it is relatively high, therefore, restricted to use in cattle feed. A remarkable achievement was made by improving oil quality by reducing toxic components like erucic acid and glucosinolates in *Brassica*. The best canola quality mustard also refers to double zero mustard variety Pusa Double Zero Mustard 31 (PDZ-1) and PDZM 33 content very low glucosinolates (<30ppm) and Erucic acid (<2 per cent) has released in 2016-17 by ICAR–IARI. In addition to the '00' type, the variety has less fibre content is called triple type ('000').
- ☆ Linolenic acid should be reduced (<3 per cent) as it very unstable and is easily oxidised to give unpleasant smelling substance in oil.

What is Canola?

It is a registered trademark of Canadian Oil Association. It refers to cultivars of oilseed brassicas in seed oil having < 2 per cent erucic acid and < 30 μmoles/g of glucosinolate of its deoiled meal. Varieties with canola quality are also termed as 'double low' or double zero (00). In India Canola type variety is TERI R 9903. 'Canola' quality varieties are commonly developed from either or two species of *Brassica napus* and *Brassica campestris*. Efforts to develop Canola quality *Brassica juncea* varieties have succeeded in India.

The Characteristics of Canola

- Higher yield with < 2 per cent erucic acid
- Short duration
- Suitable for livestock feed as oil meal contains less than 30 micromoles glucosinolates per gram of defatted meal,
- Used as salad oil for its light colour and texture
- Suitable for the baking industry

8. **Development of hybrid through diverse CMS-FR system:** *B. juncea* lacks natural cytoplasmic male sterility and fertility restoration (CMS-FR) system. The origin of CMS system in *Brassica* by two ways, *viz.*, spontaneous and alloplasmic in *Brassica* species
   1. **Spontaneous CMS system:** It is arising without any intentional intervention, *e.g.*, 'Polima' CMS in *B. napus*.
   2. **Alloplasmic CMS system:** Alloplasmic lines of *B. juncea* containing cytoplasm of *B. oxyrrhina* (*oxy*), *B. tournefortii* (*tour*), *Diplotaxis siifolia* (*sif*), *Trachystoma ballii* (*trachy*), *Raphanus sativus* (*ogu*), *Moricandia arvensis* (*mori*) and *Erucastrum canariense* (*can*), *D. catholica* (*cath*), *D. berthautii* (*bar*), *etc.* have been identified and are used for the development of CMS lines and restorers. Initially, hybrids were developed using a mori-based CMS system, but not being popular due to poor stability. Presently public and private sectors are using *ogu* (*Raphanus sativus*) based CMS in the hybrid development of mustard.

Breeding Approaches (Methods)

1. **Selection approach:** It comprises methods like mass selection and pureline selection; various cultivars were developed, *viz.*, Kranti, Varuna, Krishna, Yellow sarson–Benoy, Jhumka, Ragini, *etc.*
2. **Hybridization and selection:** Includes bulk method, pedigree method, backcross, population improvement approach, *i.e.*, recurrent selection, synthetic composite, *etc.*

   Pedigree selection: Various cultivars developed by this method are yellow sarson: Gujrat Sarson-1, Indian mustard: GDM 4, GDM 5, Pusa Swarnim, Pusa Mustard 25, Pusa Mustard 26, Pusa Mustard 27, Pusa Mustard 28, Pusa Mustard 29, Pusa Mustard 30, PDZM 31, Pusa Mustard 32, PDZM 33 *etc.*

   Backcrossing and its modification: Various cultivars *viz.*, JM-1, JM-2, JM-3, Pusa Mustard 21, Pusa Mustard 22, Pusa Karishma, *etc.*, are developed.
3. **Mutation Breeding:** Method used to create new variability, improve specific trait. Various cultivars have been released for commercial cultivation are given in Table 18.1.

Table 18.1. List of Mutant Cultivars of Rapeseed-Mustard Released in India

Sl.No.	Cultivars	Year	Source of Mutagen	Characters Improved
1.	RLM 198	1975	Gamma rays	Improved oil content, early maturity (5-6 days) and high yield
2.	TPM-1	2007	-	Yellow seed coat and tolerance to powdery mildew
3.	NUDB-38(rapeseed)	2004	Gamma rays and EMS	Early maturity and high-yield potential
4.	Mutant NUDB-26-11 (rapeseed)	2007	Gamma rays and EMS	High yield potential

4. **Heterosis breeding:** Commercial hybrid has been developed in a mustard crop based on a three-line breeding system (CGMS). The different hybrids has been developed commercially in mustard by *Raphanus sativus* (*ogu*), *Moricandia arvensis* (*mori*) based CMS-FR system. The list of hybrids are listed in Table 18.2.

Table 18.2: List of Mustard (*B. juncea*) Hybrids Released Commercially in India

Sl.No.	Name of Hybrid	CMS-FR System	Released Year	Name of Institutes
1.	NRCHB 506	*Moricandia*	2008	DRMR, Bharatpur
2.	Coral 432 (PAC 437)	*Ogura*	2010	Advanta Seeds India, Pvt.
3.	DMH-1 (Dhara Mustard Hybrid-1)	CMS,"126-1"	2010	NDDB and Delhi University
4.	Coral 437 (PAC 437)	*Ogura*	2012	Advanta Seeds India, Pvt.
5.	PHR 126	*Ogura*	2021	PAU, Ludhiana
6.	RCH 1	*Ogura*	2021	
7.	SVJH 108	*Ogura*	2021	Shakti Vardhak Hybrid Seeds Pvt., Ltd., Hisar

5. **Biotechnological approaches:**

   Tissue culture techniques:

   Embryo rescue: Employed for wide crossing - resynthesise of alloploid Brassica species: Thrived short duration cultivars *viz.*, Pusa Agarni. *Diplotaxis erucoides*-induced cytoplasmic male sterility in *Brassica juncea* is rescued by the *Moricandia arvensis* restorer.

Somaclonal variation: The successful example is the Pusa Jai Kishan (BIO-902) variety developed from the somaclone of the Varuna variety.

Protoplast fusion: Development of somatic hybrids, *e.g. M. arvensis + B. juncea; D. catholica + B. juncea*; Alloplasmic CMS, *e.g.* Synthesized a stable CMS system *in B. juncea* by placing its nucleus in the cytoplasm of a wild species *Moricandia arvensis.* A CMS *B. juncea* line was developed by placing its nucleus in the cytoplasm of an alien wild species, *D. catholica,* through sexual hybridization followed by backcrossing.

Transgenic breeding: In India, the first transgenic hybrid Dhara Mustard Hybrid-11 (DMH 11) based on *barnase-barstar* was developed by Delhi University.

Marker-assisted/Genomic assisted breeding: Various QTLs and markers identified and are used in various disease resistance and oil quality improvement.

Breeding Institutes

- **National Level:**
  - i. Indian Institute of Rapeseed-Mustard Research (IIRMR), Bharatpur (Rajasthan).
  - ii. Indian Institute of Agriculture Research, New Delhi.
- **State level:** Regional Research Stations-State Agricultural Universities.

Breeding Achievements

- **Varieties:** Kranti, Krishna, Varun, Pusa Jaikishan (BIO 902), Pusa Bold, GM-1, GM-2, GM-3, GDM-4, GDM-5, RLC 2, RLC 3, Pusa Mustard 32, Radhika, Pusa Double Zero Mustard 31, Pusa Double Zero Mustard 33, KMR 16-2 (Surekha), Pusa Bold- (Pusa Bold-WRR2), Rohini, Pusa Mustard 34, CS 64, Gujarat Mustard 6, Gujarat Mustard 7, Gujarat Mustard 8, Pusa Double Zero Mustard 35, Pusa Double Zero Mustard 36, *etc.*
- **Hybrids:** DMH-1, NRCHB 506, Coral 432 (PAC 437), Coral 437 (PAC 437), PHR 126, SVJH 108, *etc.*

Chapter 19

Breeding Soybean (*Glycine max* L. Merrill)

Introduction

Soybean, a 'Golden Nugget of the Orient', is known worldwide as a miracle crop. It is a major oilseed in the world and also in India due to the exclusive convergence of oilseed, leguminous and nutraceutical properties. It is typically used for vegetable oil production and protein processing. The seed contains an average of 20 per cent vegetable oil, 40 per cent protein, 22 per cent carbohydrate, and 5 per cent crude fibre. Besides, it is also an excellent source of good quality unsaturated oil. This versatile plant is a natural soil fertilizer by fixing 50 kg of nitrogen per hectare. The important soybean-based foods are soy paste (miso), soy sauce (shoyu), soy curd (tofu), soy milk and tempeh (a fermented, cake-like product).

Background Information

1.	*Common name*	Soja bean, soya bean
2.	*Botanical Name*	*Glycine max* L. Merrill
3.	*Family*	*Fabaceae* (*Leguminosae*)
4.	*Chromosome number*	2n=4x=40
5.	*Centre of origin*	China

6.	*Nature of pollination*	Self-pollination
7.	*Pollination mechanisms*	Bisexuality, Cleistogamous, Keel petal
8.	*List of wild relatives*	*Glycine soja, G. gracilis, G. tomentolla, G. usuriensis*

The genus *Glycine* has two subgenera *viz.*, Soja (annuals) and *Glycine* (perennials). The subgenus *soja* has two species: *Glycine max* (2n=4x=40) cultivated and *Glycine soja* (2n=4x=40) wild annual type. Cultivated soybean *Glycine max* is considered to be domesticated from its wild annual soybean *Glycine soja*.

Distribution

The soybean crop was domesticated in China about 1500–1100 BC. In India, the earliest soybean reference described the proof of soybean variety grown in Calcutta Botanical Garden. However, the commercial cultivation of soybean in the country commenced in the late 1960s. Now it is one of the became major oilseeds in the country. In India, major areas are confined to Madhya Pradesh, Maharashtra and Rajasthan.

Floral Biology, Selfing and Crossing Techniques

Inflorescence	:	Axillary racemose
Flower	:	Papilionaceous, zygomorphic, irregular, complete and perfect
Calyx	:	Tabular, gamosepalous with five sepals
Corolla	:	Five petals [1+2+(2)], polypetalous, of very unequal size, vexillary aestivation, papilionaceous
Androecium	:	Ten diadelphous (9+1) stamens
Gynoecium	:	Monocarpellary, ovary- superior and unilocular, 1 to 5 ovules, Placentation- marginal; capitate stigma
Fruits	:	Legume or pod bearing 1–5 seeds, pod borne in cluster (3-5 pods) on short stalks
Seed	:	Ex-albuminous with two cotyledons
Floral formula	:	⊕ ⚥ $K_{(5)}\ C_{1+2+(2)}\ A_{(9)+1}\ \underline{G}_{1}$

Anthesis: Anthesis takes place in the bud stage, anthesis begins in the morning and continues till before noon hours. Stigma becomes receptive about 24 hours before anthesis and remains receptive 48 hours after anthesis.

Selfing technique: Although soybean is largely self-pollinated, inflorescences is covered under paper bag before commencement of anthesis to ensure the cent per cent self-pollination.

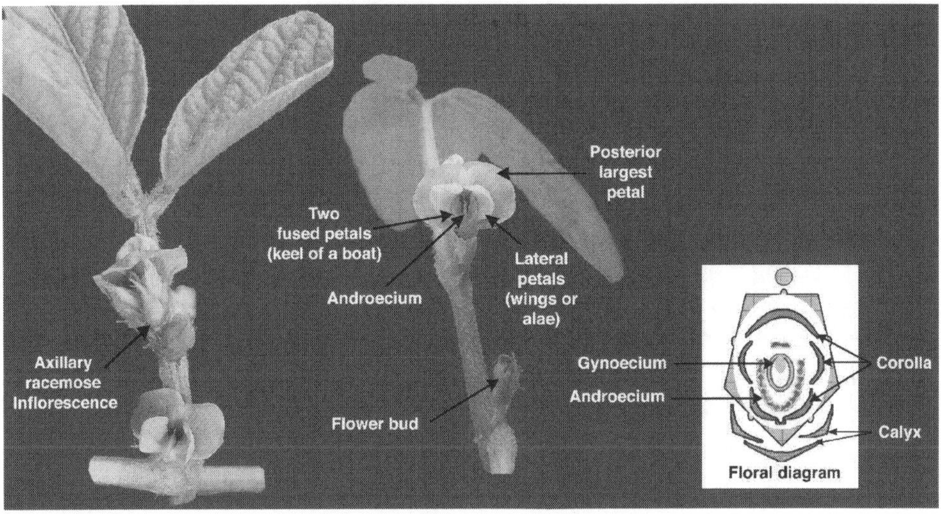

Figure 19.1. Floral Biology and Diagram of Soybean (*Glycine max* L. Merrill).

Artificial Crossing and Pollination

Select the inflorescences comprise of young buds, select mature flower buds, and remove immature ones of a female parent. For emasculation, hold the bud very gently and remove the calyx with the help of forceps. After gradually pulling the corolla, holding it at the tip of the flower bud with forceps and removing all anthers with filaments. Do not touch or disturb the pistil, otherwise, the flower bud will be dropped. Pollination is performed on the following day morning. Collect the matured pollen from the male parent by removing a floral organ with forceps. Collect pollen and gently brush it on a stigma. Tagging and bagging of pollinated inflorescences.

Breeding Objectives

1. **Breeding for high-yielding cultivars:** The yield is polygenic in nature, and it can be determined through various polygenic components like plant height, number of branches, number of pods, number of seeds per pod, and bold seed size.
2. **Breeding for early type and unique growth habit:** Earliness and dwarf plant type correlate positively. Three types of growth habits are determinate, semi-determinate and nondeterminate. The determinants growth pattern governs by two genes *Dt1* and *Dt2*. Determinate type is governed by recessive allele dt1, semi-determinate type growth by *Dt2*, while allele *Dt1* responsible for nondeterminate type. Due to the short growing season, determinate varieties are desirable in India. Examples of determinate varieties are NRC 7, NRC 12, JS 71-05, *etc.*

3. **Breeding for non-pod shattering type:** Pod shattering is a major problem in high temperature and low humid/drought conditions. A Locus NST1A and two major genes, Pdh1 and SHAT1–5 (NAC gene) has identified that are associated with pod dehiscence in soybean.
4. **Breeding for thermo-photo insensitive:** Essential for yield stability in varying thermo and photo periods, for example, cultivar JS 20-34.
5. **Breeding cultivar suitable for mechanical harvesting:** Ideal trait is having a high insertion point of the lowermost pod. Various cultivars, *viz.*, MACS 58, NRC 37, Type 49, Durga, Punjab 1 have been developed for mechanical harvesting.
6. **Breeding for quality and biofortification-elimination of anti-nutritional factors**
   i. **Oil content and quality:** Improvement of oil content in soybean is essential for value addition. Governed by multiple genes and shows genotype × environment interaction: Oil content and yield is a positive correlation but a negative association with protein content. Oil concentration affects seed coat, *i.e.*, yellow seed coat varieties are richer in oil than other seed coat colours, *viz.*, black, brown and green.
   ii. **Protein content and quality:** The protein content negatively correlates with oil content. During breeding for quality parameters, both things should be kept in mind.
   iii. **Seed coat colour:** In soybean, various coat colours, *viz.*, yellow, black, brown, green, or a combination of all the above also found, it also influences the oil concentration and consumer preference.
   iv. **Improve oil protein digestibility oil flavour and fragrance:** Poor protein digestibility and off-flavour are undesirable traits limiting bioavailability. The genetics of responsible biomolecules are now well-known. For example, genetic Elimination of Kunitz Trypsin Inhibitor (KTI) improves protein digestibility. Conversely, the catalytic oxidation of unsaturated fatty acids by lipoxygenases Lox1, Lox2 and Lox3 is responsible for off-flavour. The environment also influences the concentration of these biomolecules. The removal of these undesirable biomolecules by conventional and genomic-assisted breeding. In India NRC127 is the first Kunitz Trypsin Inhibitor (KTI)-free soybean variety developed through marker-assisted backcrossing (MABC). KTI was also eliminated from the genotypes JS97–52, JS93–05, NRC7, MACS450, S97–12, *etc.* through marker-assisted breeding. The cultivar NRC142 developed through marker-assisted breeding possesses dual traits advantage, *viz.*, zero KTI and lipoxygenase-2.
   v. **Improved minerals bioavailability:** A heat-stable anti-nutritional factor like phytic acid, 1,2,3,4,5,6-inositol hexaphosphoric acid, are present in soybean seeds. It binds with nutritionally essential metals,

especially Zn, Ca and Mg, forming phytic acid-metal complexes (phytin). A novel soybean line M 766 with low phytic acid developed through molecular markers technique.

vi. **Enhanced oil shelf life:** Fast oxidation and development of a fishy smell in stored soybean oil is associated with high Linoleic and α-linolenic acid (PUFA). Soybean oil with high oleic acid and low α-linolenic acid possesses enhanced oxidative stability, flavour and storability. The self-life of oil can be improved by increasing monounsaturated fatty acid (oleic acid) and reducing α-Linolenic acid.

7. **Breeding for disease resistance:** *i.e.*, Soybean rust (SBR); Soybean Mosaic Virus; Yellow Mosaic Disease are major diseases.
8. **Breeding for abiotic stress tolerance:** *i.e.*, Drought, salinity, high and low temperature, floods, acidity, mineral toxicity and nutrient deficiency

Breeding Approaches (Methods)

1. **Introduction and selection:** Various cultivars introduced or improved by selection *viz.*, Monetta, Palmetto, Bragg, Seminole, Lee, Improved Pelican, *etc.* from the USA, variety Gujarat Junagadh Soybean 3 (GJ Soy 3), *etc.*
2. **Pure line selection:** Variety Co 1, JS 2 was the pure line selection of Tehri Garhwal material.
3. **Mutation breeding:** Generally used to improve specific traits and adaptation. The varieties developed are given in Table 19.1.

Table 19.1: List of Mutant Cultivars Released in India

Sl.No.	*Cultivars*	*Year*	*Source of Mutagen*	*Characters Improved*
1.	VL Soya-1	1985	Spontaneous mutation	Black seeds and high yield
2.	MACS-450	2000		Semi-determinate growth habit, medium maturity, and high yield
3.	TAMS 98-21	2007	Gamma rays (250 Gy)	Multiple diseases and pest resistance, resistance to bacterial pustules, myrothecium leaf spot and soybean mosaic virus diseases, high yield

4. **Hybridization and selection:** Both normal and wide hybridization methods have been used in soybean improvement.
   a. **Intervarietal hybridization:** Pedigree selection and its modifications, *i.e.*, single seed descent (SSD). C. A. Brim 1966 described the detailed procedure of SSD in soybean crop as the 'modified pedigree method'.

   b. **Distant hybridization:** Different species, *viz., G. argyrea, G. canescens,* and *G. tomentella* have been successfully hybridized with soybean. The perennial *Glycine tomentella* (Hayata) and the annual *Glycine soja* (Sieb. and Zucc.) were hybridized successfully with the domesticated soybean to produce various breeding lines.

5. **Biotechnological approaches:**
   a. **Embryo rescue:** Important for wide hybridization
   b. **Genomic/Marker-assisted breeding:** Selectable gene markers for yield, oil and protein quality, Pyramid the desirable quality traits, various disease and abiotic stress resistant gene are used in the development of various cultivars. V.C. Concibido and associates in 1996 first-time effective use of Marker Assisted Selection (MAS) in plant breeding demonstrated in soybean cyst nematode resistance.
   c. **Transgenic breeding:** *Agrobacterium*-mediated insertion of *Bar* gene for resistance against herbicide; *CryIAc,* for lepidopteran insect resistance. The most widely adapted transgenic variety is Roundup-ready.
   d. **Genome editing:** Promising and powerful tool for accurate genetic modification and gene function identification. However, the ultimate success merely depends on transformation efficiency in soybean.
   e. **Speed breeding:** Emerging techniques for shortening the breeding cycle by generation advancement, thereby accelerating soybean improvement.

Breeding Institutes

- ✫ **International level:** International Center for Agricultural Research in the Dry Areas (ICARDA), Beirut, Lebanon, Syria.
- ✫ **National level:** Indian Institute of Soybean Research (IISR), Indore, Madhya Pradesh, India.
- ✫ **State level:** Regional Research Stations-State Agricultural Universities.

Breeding Achievements

Gujarat Soybean 1, Gujarat Soybean 2, Gujarat Soybean-3, JS 93-05, JS 335, JS 80-21, JS 75-46, MACS 58, Monetta, Parbhani Sona (MAUS 47), Pratishta (MAUS 61-2), Shakti (MAUS 81), PK 472, NRC 2 (Ahilya 1), NRC-12 (Ahilya 2), NRC-7 (Ahilya 3), NRC-37 (Ahilya 4), VL Soya 99, JS 21-72, NRC 152, Indore Soy 136, JS 22-12, NRC 165, G.Soy 4: Sorath Sonali, NRC-149, NRC- 197, Pusa Soybean 21, Gujarat Soybeen 5, *etc.*

Chapter 20

Breeding Sunflower (*Helianthus annuus* L.)

Introduction

Sunflower is the fourth most important edible oil seed crop after soybean, mustard and groundnut. It is cultivated worldwide for its oil and confectionary purposes. It belongs to genus *helianthus* name is derived from the Greek words *helio* means 'sun' and *anthos* meaning 'flower'. Due to photo-insensitive, day-neutral and short-duration features, it can be grown in any season, *i.e., kharif*, rabi and spring/summer and also fit into different cropping sequences. Oil content ranges from 46-52 per cent with high-quality non-cholesterol properties. The main uses of oil are for human consumption, salad oil, making soaps, paints, biofuel, candles, and the cosmetic industry.

Background Information

1.	*Common name*	Suryamukhi, Surajmukhi
2.	*Botanical Name*	*Helianthus annuus* L.
3.	*Family*	*Asteraceae* (*Compositae*)
4.	*Chromosome number*	2n=2x=34
5.	*Centre of origin*	North America

6.	*Nature of pollination*	Cross-pollination
7.	*Pollination mechanisms*	Protandry, Entomophily- Melittophily, Anemophily
8.	*List of wild relatives*	*H. petiolaris, H. argophyllus, H. giganteus, H. rigidus, H. argophyllus*

Sunflower belongs to the subdivision *Tubiflorae* and tribe *Heliantheae*. The genus *Helianthus* contains of 49 species and 19 subspecies. Two domesticated species, *Viz.*, A cultivated diploid *H. annuus* originated from diploid (2n = 34) annual wild *H. annuus* species used as an edible oil crop, whereas, hexaploid *H. tuberosus* (2n=6x=102) is perennial cultivated used as edible tuber and fodder.

Distribution

Sunflower is native to North America; Native Americans cultivated it 4000 years ago but commercialization of the plant took place in Russia. The American Indians were the first to domesticate the multi-headed ornamental plant into a single-headed plant with various seed colours. In India, sunflower is mainly grown in the Southern and Central Peninsular regions covering Karnataka, Andhra Pradesh, Telangana, Maharashtra and Tamil Nadu.

Floral Biology, Selfing and Crossing Techniques

Inflorescence	:	The capitulum or head consisted of few to many flowers enclosed by an involucre of bract. Flowers are of two types. **Ray florets:** Arranged in single tabular border rows, generally pistillate or neuter **Disc florets:** Arranged in spirals from the centre of the head.,generally bisexual
Flower	:	**Ray florets:** Bracteate, zygomorphic, sessile, incomplete, pistillate, sometimes neuter, ligulate and epigynous **Disc florets:** Actinomorphic, complete, bisexual, tubular, heteromerous
Calyx	:	Gamosepalous, sepals reduced to numerous hairy structures called pappous, persistent, act like a parachute that help in the dispersal of fruit
Corolla	:	**Ray florets:** Petals five, gamopetalous, ligulate, superior, aestivation- valvate, ligulate and irregular **Disc florets:** Same as ray florets but tubular and regular

Androecium	:	**Ray florets:** Absent **Disc florets:** Stamens five, syngenesious (anthers are fused into a tube around the style), epipetalous, alternate the petals, anther-dithecous, basifixed, introrse, filament simple and free
Gynoecium	:	**Ray florets:** Absent or if present bicarpellary syncarpous, ovary- inferior, unilocular, single ovule in each locule, placentation- basal, style–long, simple and slender, stigma –bifid **Disc florets:** Same as ray florets
Fruits	:	Cypsela (dry, indehiscence fruit)
Seed	:	Ex-albuminous, with two cotyledons
Floral formula	:	**Ray florets:** Pistillate: Br. % ♀ $K\infty\ C_{(5)}\ A_0\ \overline{G}_{(2)}$ Neuter: Br. % N $K\infty\ C_{(5)}\ A_0\ \overline{G}_0$ **Disc florets:** Br. ⊕ ⚥ $K\infty\ C_{(5)}\ \overset{\frown}{A_{(5)}\ \overline{G}_{(2)}}$

Anthesis: Protandrous condition: anthesis occurs between 6:00 and 10:00 AM during warm sunny days. It may delay under cool, cloudy or wet weather conditions. Anthesis begins with ray florets and then switches to the disc florets. Pollen remains viable for up to 12 hours, while stigma remains receptive for 2-3 days.

Selfing technique: Sunflower is a highly cross-pollinated crop; therefore, the entire inflorescence (head) is covered under a paper bag or cotton cloth bag to achieve selfing pollination

Artificial Crossing and Pollination

Select the inflorescence comprised of open flowers in which anthesis begins. Hand emasculation is performed in the early morning by removing the anther tubes with the help of forceps. Emasculation can also be done by spraying Gibberellic acid (100ppm) after bud initiation for three days in the morning is effective. Pollination is performed just after emasculation on the same day morning by collecting the matured pollen from the male parent that are already bagged before flowering. Tagging and bagging of pollinated inflorescences.

Classification of Sunflower Cultivars According to Plant Height

Sl.No.	*Types*	*Plant Height (cm)*	*Flowering Habit*
1.	Giant tall	180 to 430	Late
2.	Semi-dwarf	130 to 180	Medium late to early
3.	Dwarf types	60 to 130	Early

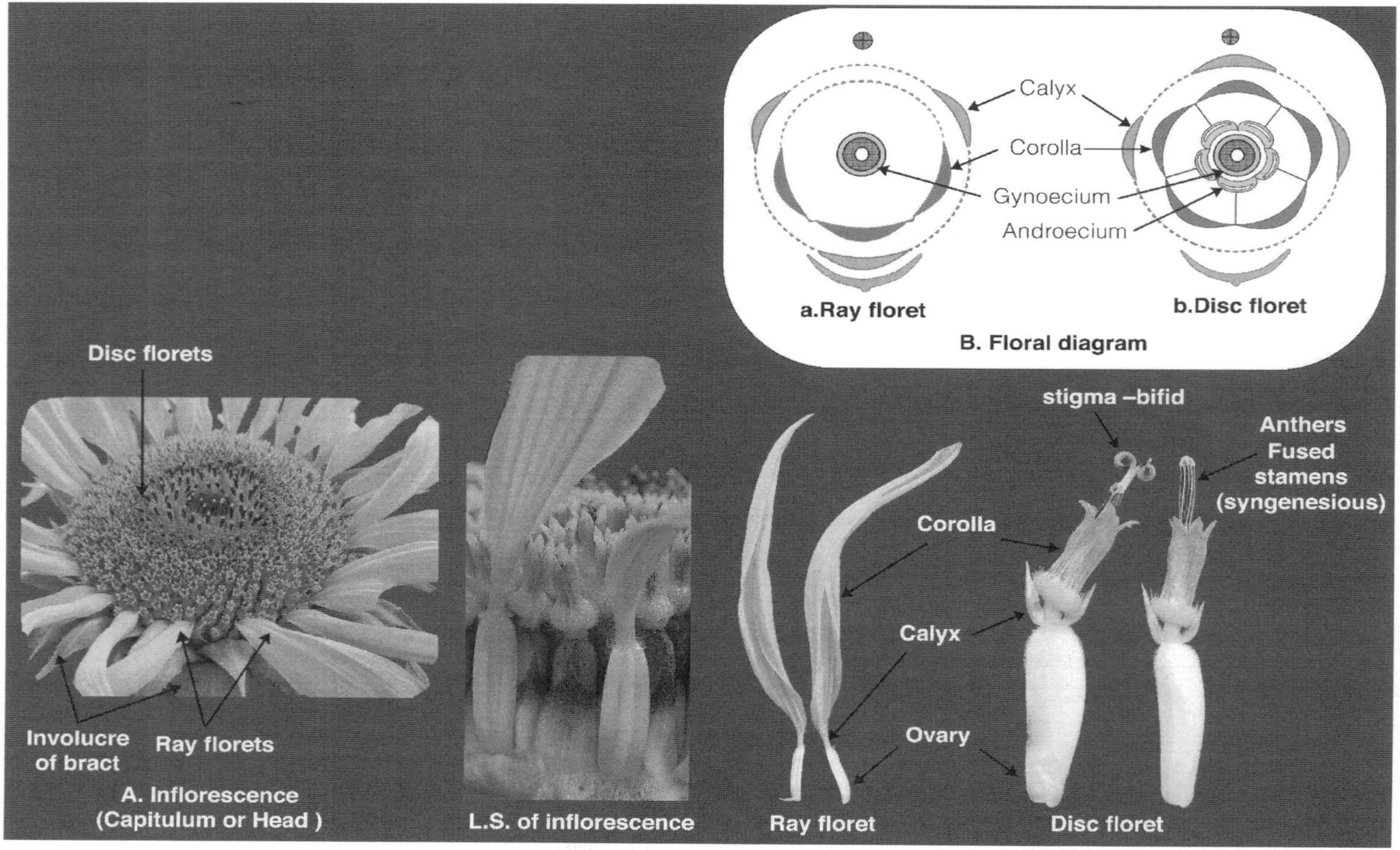

Figure 20.1. Floral Biology and Diagram of Sunflower (*Helianthus annuus* L.).

Essential Features of Sunflowers that Favour Cultivation in India

1. Photoperiod insensitive, day neutral, so it can be cultivated in any season
2. Short duration (75 to 110 days)
3. Determinant growth habit
4. High seed multiplication ratio (more than1:100)
5. Suitable for mechanical farming
6. Drought tolerance capacity
7. Suitability of processing like oil extraction

Breeding Objectives

1. **Breeding for high seed yield:** High seed yield is the ultimate goal of sunflower breeding. Seed yield is complex and polygenic and exhibits G × E interaction. Reported both additive and non-additive genetic controlled. Seed yield can significantly increase by breeding concentrating on yield components like seed size, head shape and size, and number of seeds per head.
2. **Breeding for early maturity:** Earliness is essential for dry land and irrigated conditions.
3. **Breeding for lodging resistance and dwarf type:** Dwarf type and lodging resistance is a positive association. Lodging resistance essential for saving yield losses
4. **Breeding varieties with high oil content and quality oil:** Negative correlation reported between oil content and seed yield. For example, seeds with thin shells yield more oil content. Oil content is a polygenic character under the control of additive genes; hence it exhibits high to medium heritability. Breeding with wild species like *H. niveus* (Benth.) and *H. salicifolius* may be necessary for increasing oil's quality and protein content. A negative correlation between oil and protein contents has been reported. The substitute approach to increase the protein content in seed meal is to hull seed. Using molecular breeding tools, several QTLs have been identified as associated with increasing seed oil content and altered fatty acid composition of oil.
5. **Breeding for self-fertile lines:** Protandry and self-incompatibility mechanisms force hand pollination for effective seed setting. Hand pollination in a large scale is cumbersome and laborious. To overcome such problems by use of self-fertile lines.
6. **Breeding for diversification of cytoplasmic source.** More than 70 diverse CMS systems have been identified in the sunflower hybrid breeding program. However, only a single source of CMS (PET-1) has been used due to a lack of restoration and maintainers in the rest CMS sources. Continuous use of a single source fears the possible vulnerability

of hybrids to diseases and insect pests associated with deploying a single cytoplasmic. It leads to a narrow genetic base causing a yield plateau. Modification of the cytoplasmic source is the best option to overcome this problem in sunflowers.

Table 20.1. List of Diverse Cytoplasm that Induce Male Sterility and their Source in Sunflower

Sl.No.	*Name of Cytoplasm*	*Source*
1.	CMS-PET1	*Helianthus petiolaris*
2.	CMS PEF1	*H. petiolaris* sp fallax
3.	CMS GIG1 (CMG2);CMS GIG2	*H. giganteus*
4.	CMS 514A	*H. tuberosus*

7. **Breeding resistance to biotic stress:** The major biotic stress includes diseases and insect pest

   Disease: Important diseases are powdery mildew, rust, downy mildew, charcoal rot, *Alternaria,* sunflower leaf curl virus, *etc.*

   Pest: *Helicoverpa, spodoptera,* Grass hopper, Jassid, *etc.*

   Various conventional and molecular breeding tools have been used to identify resistant genes from wild species, *i.e.,* Wild *H. annuus, H. tuberosus, H. argophyllus, H. petiolaris, etc.* They are used to breed resistant cultivars against various biotic stress.

8. **Breeding resistance to abiotic stress:** Several abiotic stresses include drought, heat stress, low soil fertility, *etc.,* among which drought and salinity are the most distressing environmental stress. The best results in enhancing drought resistance of cultivated sunflowers have been achieved using wild *Helianthus* species. In the case of salt-tolerant, the tolerant genes reported in wild species like *H. paradoxus* and *H. mollis.*

Breeding Approaches (Methods)

Various cross-pollinated suit methods are used in sunflower improvement.

1. **Introduction:** Initially, five Russian cultivars [VNIIMK-8931 (EC-68413), Peredovik (EC-68414), Armavirskii-3497 (EC-68415), Armaverta (EC-69874) and Vashod (Sunrise)] were introduced. The genotype Armavirskii-3497 (EC-68415) was released from the University of Agricultural Sciences, Bengaluru for Karnataka state. Another popular cultivar 'Morden' was developed in 1978 through Introduction from Canada, it became popular among farmers due to high seed yield, short height, and short duration.
2. **Mass selection:** This approach is effective for the sunflower traits under the control of additive genes. The variety Surya (PKV-SUF-72-37) was developed by mass selection for Maharashtra state.

3. **Hybridization and selection methods**

   a) **Inter-varietal:** Includes pedigree, back cross and recurrent selection

      Pedigree method: It is mainly used to develop inbred and restorer lines.

      Back cross: It is used as varietal and interspecific gene transfer. Transfer of disease resistance from related species or wild species into cultivated species. To transfer male sterility trait from a donor. To develop new CMS lines and introgression of fertility restoration genes.

   b) **Interspecific:** Cross with wild species to transfer/introgression of resistant genes of various biotic and abiotic stresses, sterility and restorer genes.

4. **Population Improvement-Recurrent selection:** This approach was effectively used by Dr. Pustovoit in the USSR for oil improvement. It is also used to improve resistance against various diseases, *e.g.*, improved partial resistance to *Alternaria* leaf blight by sporophytic and gametophytic recurrent selection–development of powdery mildew-resistant restorer inbred lines.

5. **Mutation:** Mostly used in cultivated sunflower for creating genetic variability through physical mutagens like gamma rays or chemical mutagens like dimethyl sulphate.

Table 20.2. List of Mutant Cultivars Released in India

Sl.No.	*Cultivars*	*Year*	*Source of Mutagen*	*Characters Improved*
1.	Gujarat Sunflower-1 (GAUSUF-15)	1993	-	High yield, oil content 35-39 per cent
2.	CO-3 (TNAUSUF-10)	1995	Gamma rays 5KR	High yield, oil content 36-39 per cent, days to maturity 85-90 days
3.	TAS-82	2007	-	Tolerance to drought and black seed coat

6. **Heterosis breeding:** The commercial exploitation of heterosis in sunflower was possible only after the discovery of cytoplasmatic male sterility based on *pet-1* cytoplasm, followed by the identification of fertility restorer genes by Kinman in 1970 at USDA, USA. This stable system was discovered from the cross between *H. petiolaris* × *H.annus* (*var.* Armaviriskiy 9345) by Leclercq (1969) in France. This source is being used in commercial hybrid development. The first public-sector hybrid BSH-1 (CMS-234A × RHA-274) was released for commercial cultivation in 1980. Later on, several hybrids have been released from private and public sectors.

Varietal Improvement of Open-Pollinated Cultivars and Seed Production in Sunflower

Pustovit developed this method in Russia. The main aim of this method was to increase oil content and yield. This breeding method has been successfully adopted in commercially grown open-pollinated varieties in India. The steps involved in this method are:

Base population

(Consist of superior heterogeneous population, open-pollinated varieties)

↓

More than 10000-15000 individual plants are selected based on agronomic characteristics like plant height, head size, shape, uniform maturity, seed filling ability *etc.*

↓

Head-to-row progenies of selected plants

About 1000 to 1500 lines are selected based on high oil and yield

↓

Some parts of selected lines are reserved, and the rest parts are sown along with checks in sequential testing trials

↓

(Preliminary and advanced testing trials)

Based on progeny performance for oil content and yield remnant (the seeds of selected progenies (about 25 to 50 lines are bulked) are considered as nucleus seeds. These seeds are used to produce breeder and subsequent stages of seed

7. **Biotechnological approaches:** Biotechnological tools are essential to interspecific gene transfer for traits such as fertility restoration gene(s), cytoplasmic male sterility, seed quality traits, resistance to biotic and abiotic stresses, *etc.*
   a. **Genomic/Marker-assisted breeding:** Selectable gene markers and QTL has been identified and successfully used for fertility restoration, high oleic acid content, herbicide tolerance, downy mildew resistance genes (*pl* gene)
   b. **Transgenic breeding:** *Agrobacterium*-mediated gene transformation.
   c. **Genome editing:** Promising and powerful tool for accurate sunflower genetic modification and gene function identification.

Breeding Institutes

- ✰ **International Level:** The International Consortium on Sunflower Genomics.
- ✰ **National Level:**
  - i. Indian Institute of Oilseeds Research (IIOR), Hyderabad.
  - ii. Directorate of Oilseeds Development (DOD), Hyderabad.
- ✰ **State level:** Regional Research Stations-State Agricultural Universities.

Breeding Achievements

- ✰ **Variety:** Phule Raviraj, Phule Bhaskar (SS-8808), RSFV-901 (Kanthi)
- ✰ **Hybrid:** DCS-107, KBSH-53, PSFH-569, PSH-1962, PAC-334, Suryamukhi, DSFH-3, LSFH-171, COH-3, PDKVSH-952, LSFH-35 (Maruti), PSH-2080, RSFH-700, Sunflower COH 4, Tilhan Tec- SUNH-2 IIOSH-460, KBSH-88, *etc.*

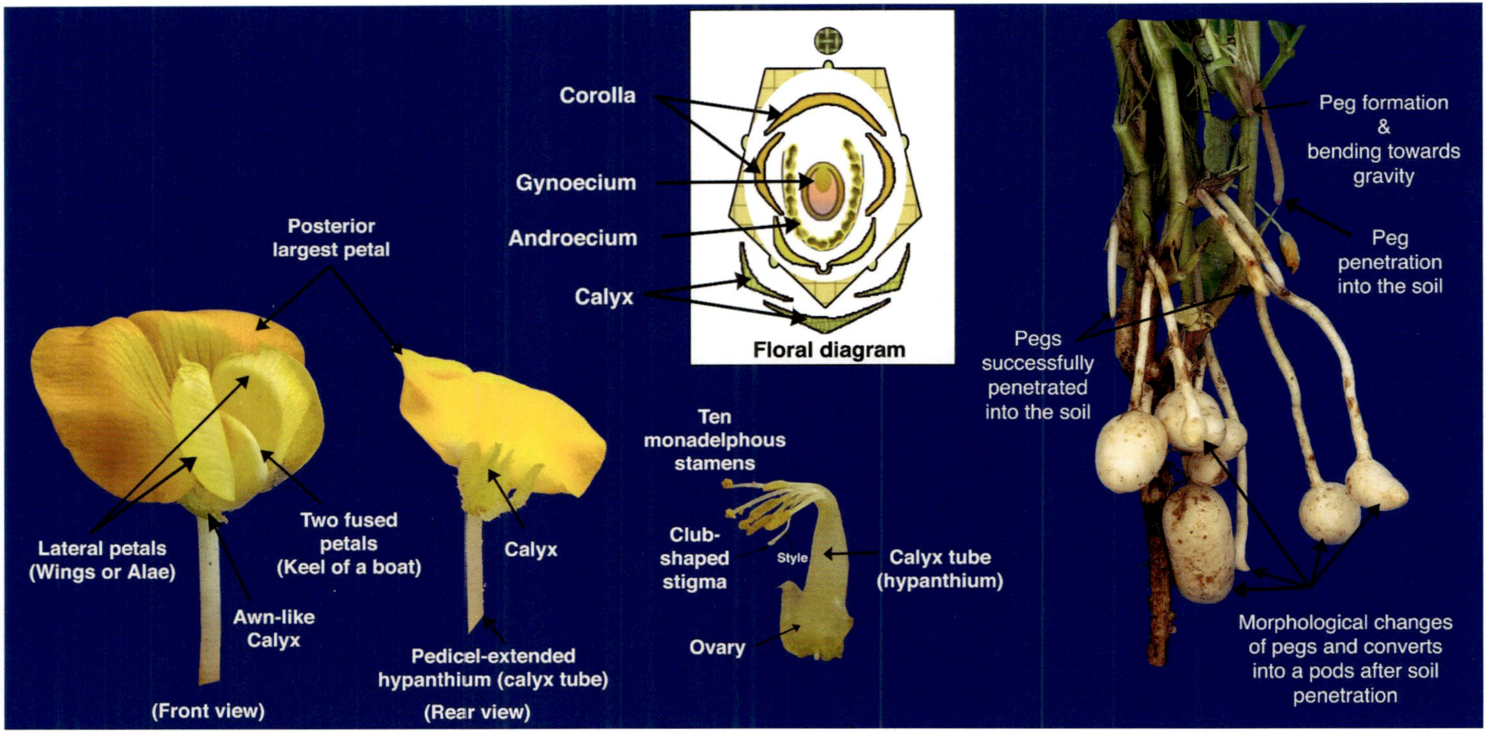

Figure 17.2. Floral Biology and Diagram of Groundnut (*Arachis hypogaea* L.). (pg. 180)

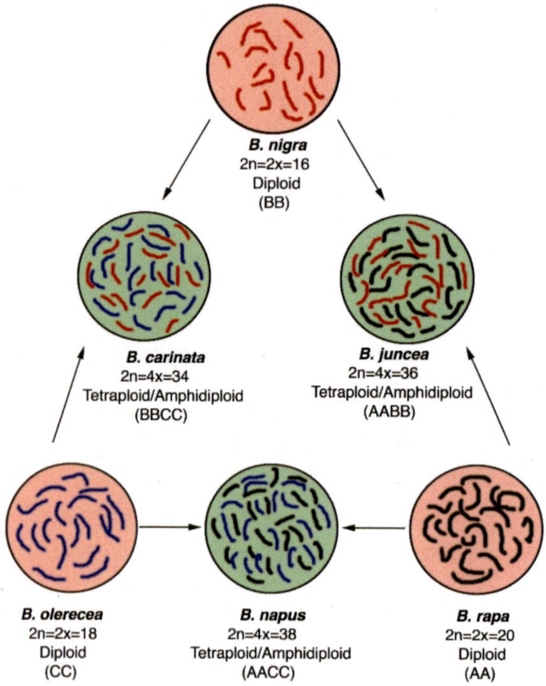

Figure 18.1. Genetic Relationship-Evolution of different *Brassica* Species. (pg. 191)

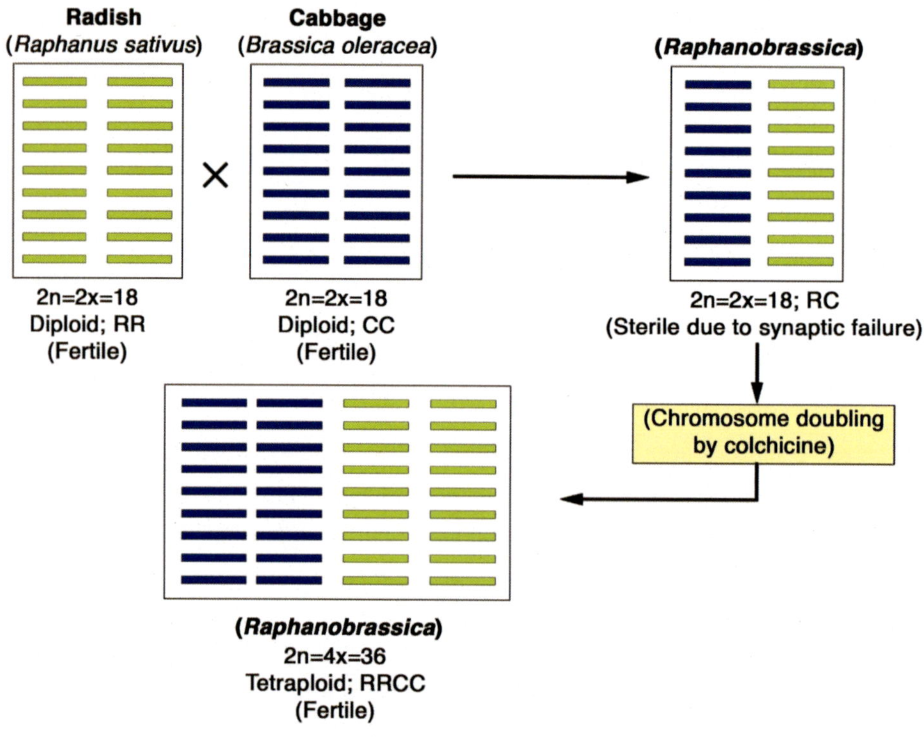

Figure 18.2. Illustration about Synthesis of *Raphanobrassica*. (pg. 192)

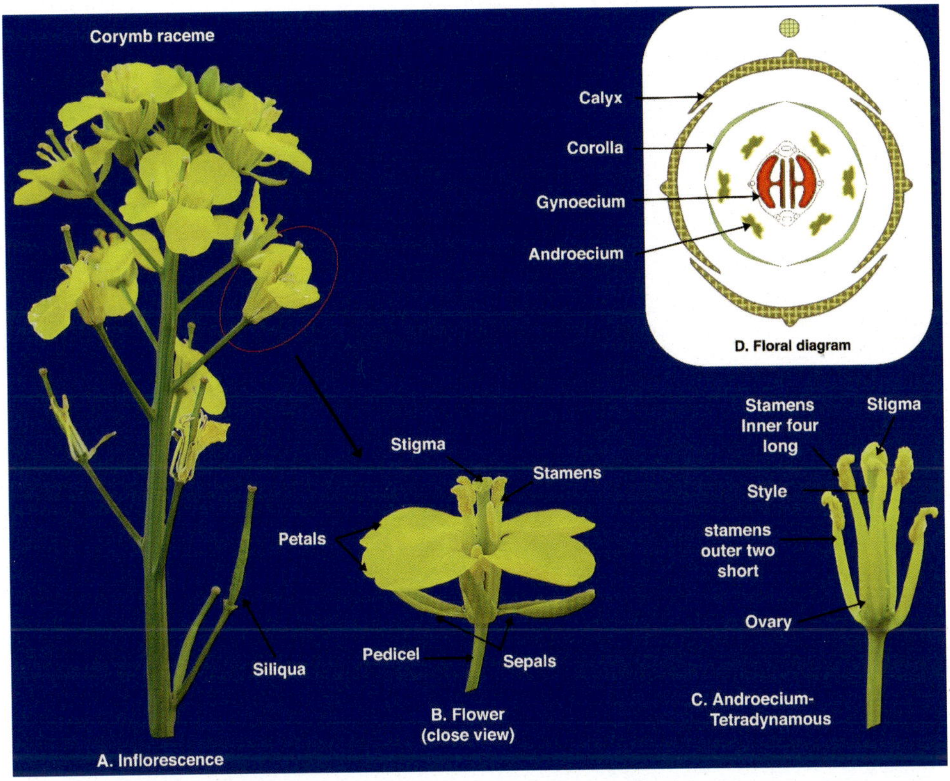

Figure 18.3. Floral Biology and Diagram of Mustard (*Brassica juncea* L.). (pg. 193)

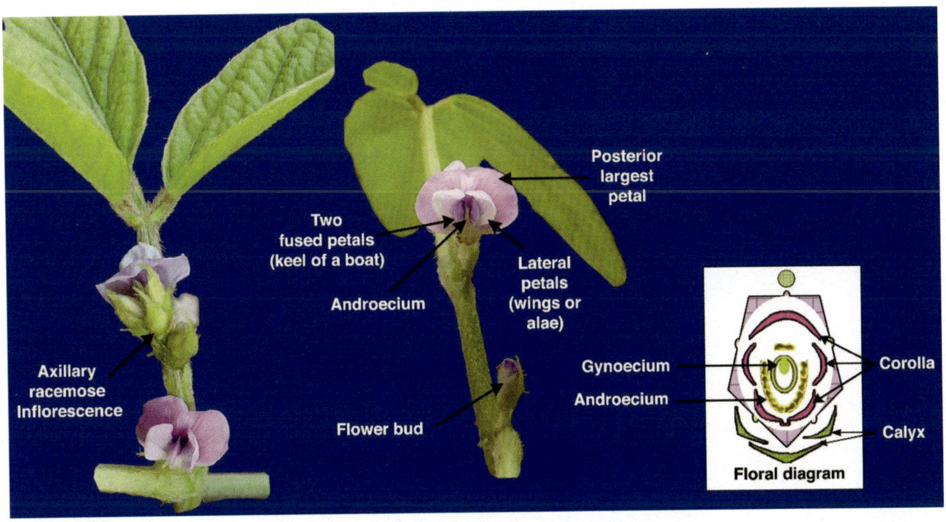

Figure 19.1. Floral Biology and Diagram of Soybean (*Glycine max* L. Merrill). (pg. 201)

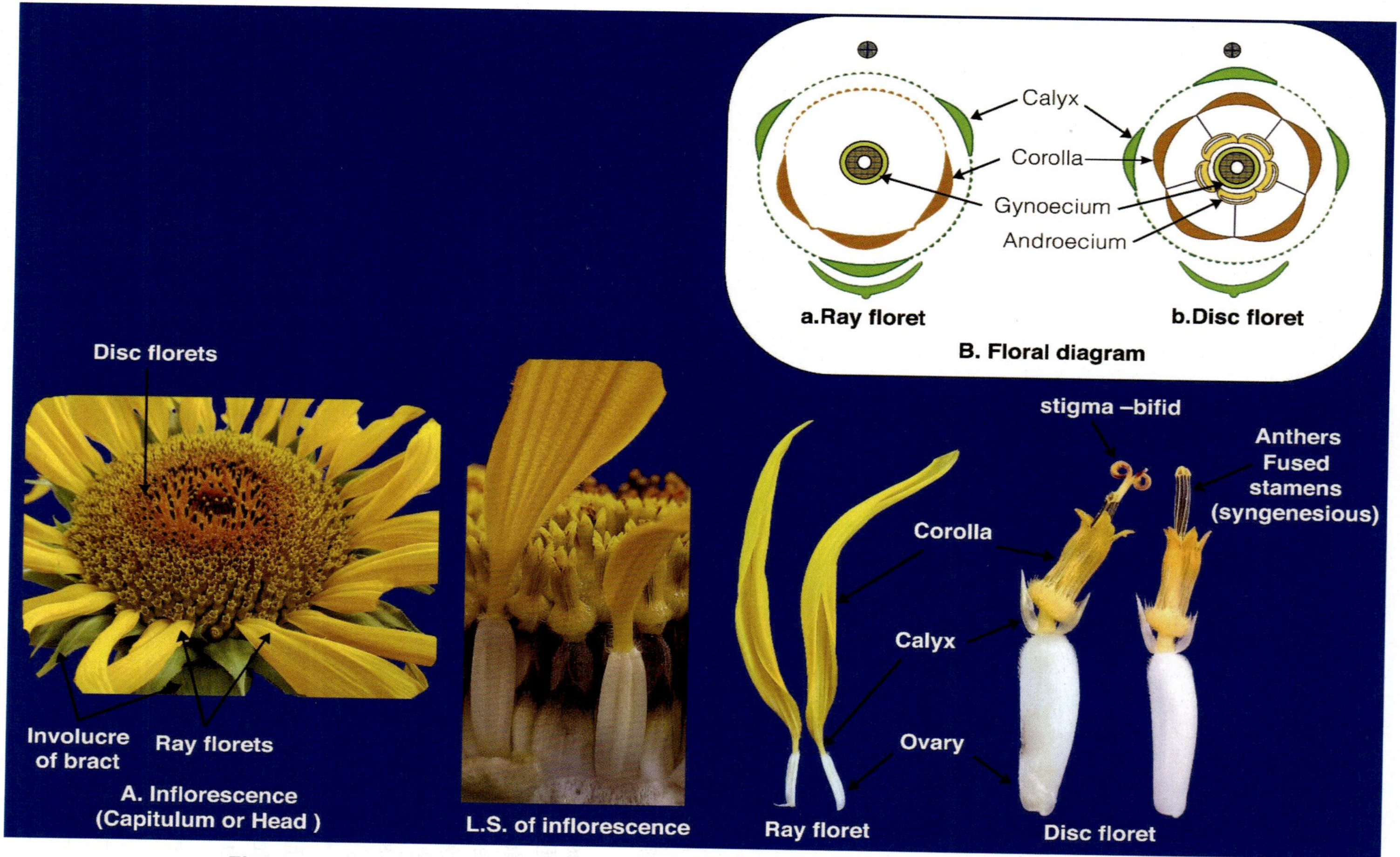

Figure 20.1. Floral Biology and Diagram of Sunflower (*Helianthus annuus* L.). (pg. 208)

Chapter 21

Breeding Sesame (*Sesamum indicum* L.)

Introduction

Sesamum is one of the most ancient and traditional oil seed crops. It is often considered the queen of oilseeds due to its high-quality edible seed oil content. Oil is a rich, high-quality protein, 85 per cent Unsaturated Fatty Acids (USFs) and endogenous antioxidants, *i.e.,* lignins (sesamin and sesamolin) and tocopherols. After roasting, sesamolin is converted into sesamol, imparting a high degree of resistance against oxidative rancidity. It is also a rich source of linoleic acid and is used as a substitute for olive oil. As per Vedas, it is excellent for human consumption. As per 'charaka', sesamum oil is the best among other oils. Worldwide it is importantly used as nutritional, medicinal, and industrial purposes. India is the first in the world in total production and area. However, productivity is one of the lowest among the sesame-producing countries in the world due to various factors, *viz.*, low yield potential of improved varieties, indeterminate growth habit, high seed shattering, low harvest index and the problem of biotic and abiotic stress.

Background Information

1.	*Common name*	Till, Tal, gingelli, benniseed, or simsim
2.	*Botanical Name*	*Sesamum indicum* L.
3.	*Family*	*Pedaliaceae*

4.	*Chromosome number*	2n=2x=26
5.	*Centre of origin*	Southern Africa to central Asia (India)
6.	*Nature of pollination*	Often cross-pollination
7.	*Pollination mechanisms*	Bisexuality, Protandry, Entomophily
8.	*List of wild relatives*	*Sesamum laciniatum, S. prostratum, S. radiatum, S. occidentale*

Cytological Group of Sesame

There are three cytogenetic groups

Cytological Group	*Ch. No*	*Species*
Group 1	2n = 26	*Sesamum indicum, S. alatum, S. africanum, S. capense, S. schenckii, S. malabaricum, S. mulayanum, S. latum*
Group 2	2n = 32	*S. prostratum, S. laciniatum, S. angolense, S. angustifolium*
Group 3	2n = 64	*S. radiatum, S. occidentale* and *S. schinzianum*

Distribution

Sesamum *spp.*, are spread throughout the tropical and subtropical areas in the Asia, Africa, and South America. The genus *Sesamum* comprised most of the wild species believed to originated in Africa, while the cultivated type *i.e., S. indicum* originated in India. Recent archaeological remnants suggest that sesame cultivation was derived from wild species at the Indian subcontinent during the Harappan civilization and spread west to Mesopotamia before 2000 B.C. Some report claims that sesame was first cultivated in Africa and later spread to India at a very early. The sesame was grown as a main oilseed crop in the Indus Valley civilization and was spread to Mesopotamia around 2500 B.C. A total of 36 species has been identified in the genus, of which 22 species are from Africa, five from Asia, seven from Africa and Asia and one species each from Crete and Brazil.

Floral Biology, Selfing and Crossing Techniques

Flower	:	Axillary, solitary, shortly pedicelled, zygomorphic, complete, perfect, hypogynous
Calyx	:	Sepals five; Gamosepalous
Corolla	:	Petals five, tubular, obliquely bell-shaped (campanulate), gamopetalous two-lipped and five-lobed and ring of hairs toward base of stamens, zygomorphic

Androecium	:	Stamens four, didynamous condition (2 short and 2 long), epipetalous, filaments are dorsifixed, and anthers bithecous, introrse, dehisce longitudinally
Gynoecium	:	Bicarpellary hypogynous, superior ovary, divided into 4 celled (tetralocular) by false septum, many ovules with axile placentation, style long filiform with forked (bifid) stigma
Fruits	:	Oblong or ovoid capsule – 2 valved with 4 chambered
Seed	:	Albuminous; obovate compressed
Floral formula	:	$\oplus$ ⚥ $2K_{(5)}\ \overset{\frown}{C_{(5)}\ A_{2+2}}\ \underline{G}_{(2)}$

Anthesis: Flowers open in the morning between 5.00 to 8.00 AM in acropetal mode; anthers dehisce very shortly around 2.00 to 4.00 AM, simultaneously stigma also becomes receptive and remains viable for about 24 h. The flower withers after 4–6 hr of anthesis.

Selfing technique: Sesame generally self-pollination; however, to achieve selfing pollination, the flower is covered under red coloured paper bags. Alternatively, selfing can be done by tying small soft threads at the tips of the flower buds about 12-16 hours before flower opening or applying semi-solid clay on the upper portion of tubular petals of unopened flowers. This method is cheap and less time-consuming but ineffective during rainy days. During rainy day fevicol is effective instead of clay.

Artificial Crossing and Pollination

Artificial crossing commences in the peak-bloom period, for emasculation select six to eight buds about to open on the next day morning. The emasculation can be done by just pulling out the corolla as such by holding it in the tip. The emasculation is carried out during evening hours between 4:00 to 7:00 PM on well-developed buds of female parents about to open the next day morning. Pollination is performed the next day morning (between 7.00 to 10.00 AM) by collecting the pollen grains from protected flowers of male parents. The pollen grains are applied on a receptive stigma by brush. The pollinated flower covered with white butter paper bags, and labelled.

Plastic straw technique: One side bent straw is used instead of a paper bag. After emasculation, the necked pistil is covered by a section of plastic straw. After pollination, straw is again inserted over the pollinated pistil.

Breeding Objectives

1. **Breeding high-yielding cultivars:** Seed yield in sesamum can be increased by breeding high adaptability genotype, early to mid-early,

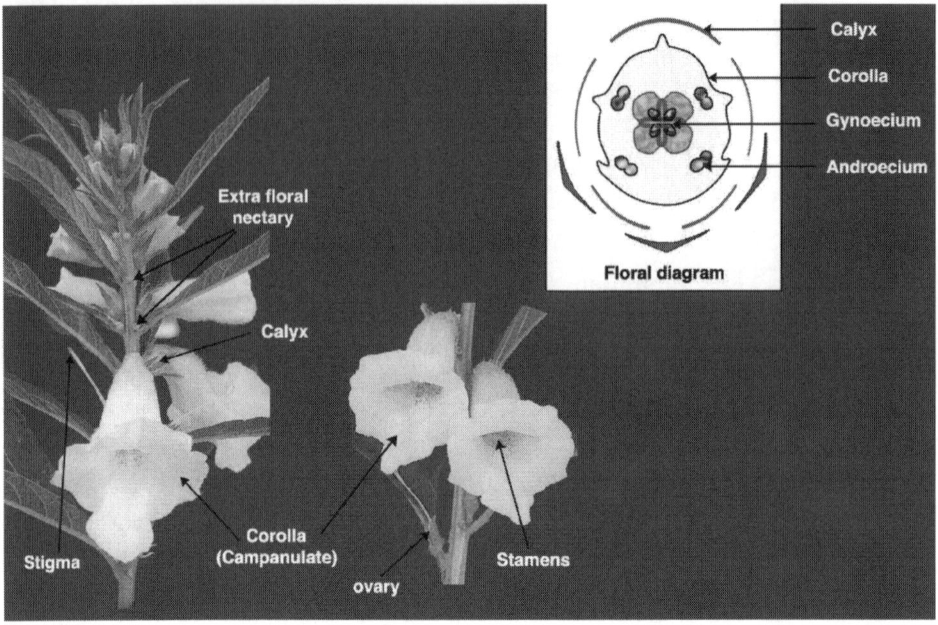

Figure 21.1. Floral Biology and Diagram of Sesame (*Sesamum indicum* L.).

earliness under low rainfall, breeding cultivars with determinate growth habit, photo insensitive, and more number of leaf axil per plant. Breeding concentrates on improving yield components like the number of capsules per plant, capsule length, Seeds per capsule and 1000-seed weight.

2. **Breeding for seed shattering and lodging resistance:** Seed shattering is a major problem in sesame during harvest; almost all cultivars are more or less susceptible–breeding of determinant growth habits and cultivars with semi-indehiscent capsules key targets. Cultivars Kalika, Amrit and Sekhar (T78) are resistant to lodging.
3. **Breeding cultivars for fit in mechanized farming:** Important plant features include medium plant height; upright stem with limited branches; determinant growth habit; strong stem/lodging resistance; uniform maturity; senescence habit, *etc.* are essential for mechanized farming.
4. **Improvement in harvest index:** The poor harvest index is one of the major issues in sesamum. It is inherited by additive mode. It can be improved by modification of plant architecture. For example, the selection of medium plant height with high-density capsules bearing starting from 15 to 20 cm above the ground level.
5. **Breeding biotic and abiotic stress resistance:** Biotic stress includes various diseases and pests:

- ☆ **Disease**: Powdery mildew, bacterial blight, phyllody, stem rot, leaf curl, capsule borer, *phytophthora* are important.
- ☆ **Insect-Pest:** Major insect pests of sesame are leaf webber/roller, capsule borer and aphids.

Abiotic stress: Sesame is mostly tolerance to drought but sensitive to salinity, waterlogging and chilling.

6. **Breeding for seed and oil quality:** Seed and oil quality includes various parameters *viz.*,
   - ☆ **Seed colour:** White, brown and black → Light colour seed content high oil than black seed cultivars. White-seeded types are most preferred for export.
   - ☆ **Seed size:** Large to medium large is preferred.
   - ☆ **High oil content:** Generally, 44-50 per cent oil content is reported in sesame. Above 50 per cent is essential. Generally, oil content is a positive correlation with seed yield but, a negatively with the protein content in sesame.
   - ☆ **Oil quality:** Oil quality can judge based on a balanced fatty acid composition with proportionately higher percentages of unsaturated fatty acid (linoleic and oleic acids) with higher amounts of antioxidants (sesamol, sesaminol) and tocopherols. Higher lignin content for long shelf life and oil colour also important. Low or zero antinutritional factors (oxalic and phytic acids) for value addition.

Breeding Approaches (Methods)

1. **Introduction:** Introduced new lines directly used as a variety or served as valuable lines for sesame breeding. For example African lines.
2. **Direct selection methods:** These include pure line selection and mass selection.
3. **Hybridization and selection**: Inter Varietal-Pedigree selection.

   e.g. Gujarat Til 3 (G. Til 3) : G. Til 1 × AHT 85
   Gujarat Til 4 (G. Til 4) : G. Til 1 × RT 125
   Junagadh Til 5 (GJT 5) : AT 90 × AT 104
   Gujarat Til 6 (G. Til 6) : AT 117 × G. Til 2
   Gujarat Til 7 (G. Til 7: Banas Gaurav) : RT 334 × G. Til 3
   Gujarat Til 11 (G. Til 11: Black seeded) : Nana Bhamo × G. Til 10

   and Inter specific: *S. indicum* × *S. malabaricum*
4. **Mutation breeding:** Mostly used in the improvement of lacking one or two traits in cultivated well-adapted cultivars and also generate variability.

Table 21.1. List of Mutant Cultivars of Sesame Released in India

Sl.No.	Cultivars	Year	Source of Mutagen	Characters Improved
1.	Kalika (BM 3-7)	1980	EMS (1 per cent for 6 hours)	Semi-dwarfness, compact growth, number of seeds per capsule and increased yield (15 per cent)
2.	Usha	1990	Arsenic-q (10 per cent)	High yield, uniform maturity and resistance to disease
3.	Uma	1990	Arsenic-q (10 per cent)	Early and uniform maturity, high oil content

5. **Heterosis breeding:** Heterosis is achieved by conventional hybridization techniques due to easy emasculation, epipetalous nature, and allogamous also favoured by insects. Heterosis can be exploited by male sterility also. The first GMS-based hybrid was developed and commercialized in China, *e.g.*, Yeti No.9 (ms 86-1 × Danbaek).
6. **Biotechnological approaches:** Biotechnological tools are used to hasten the breeding process with reliable selection of desired plant types and improvement of plant types with a combination of conventional breeding. The important tools are.
   a. **Doubled-Haploid Breeding**: It is a rapid method of genetic improvement as it considerably decreases the breeding period through early fixation of homozygosity. In sesame callus induction and androgenic plantlet regeneration from F_1s have been possible.
   b. **Genomic/Marker-assisted breeding:** Selectable gene markers and QTL has been identified for various disease and quality parameters, *viz.*, oleic, linoleic, and linolenic acids in sesame.
   c. **Genetic transformation:** *Agrobacterium*-mediated gene transformation from wild accession of sesame for biotic and abiotic stress resistance.
   d. **Tissue culture approaches:** Embryo rescue technique important for wide hybridization success.

Breeding Institutes

- ☆ **National level** Indian Institute of Oilseeds Research (IIOR), Hyderabad.
- ☆ **State level:** Regional Research Stations-State Agricultural Universities.

Breeding Achievements

- ☆ **Varieties:** Purva-1 (Brown seeded), Gujarat Til-1, Gujarat Til-2, Gujarat Til-3 (white and Bold seeded), Gujarat Til-4 (Early Maturity), Gujarat Til-7, Gujarat Til-10 (Black seeded), Gujarat Til-11 (Black seeded) Shubhra,

Smarak, Amrit, TKG-306, TKG-308, PKDS-8, PKDS-11, PKDS-12, Co-1,TSS-6, Paiyur-1, VRI-1, VRI-2, TMV-7, RT-54, RT-103, RT-125, RT-127, RT-346, RT-351, SWB–32–10–1, VRI 3, Phule Purna, Jagtiala Til-1, Sabour Til- 1, TLT- 10, Tanjila, Gujarat Til-21, *etc.*

☆ **Hybrid:** Yeti No. 9 (ms 86-1 × Danbaek) GMS-based hybrid in China

Chapter 22

Breeding Castor (*Ricinus communis* L.)

Introduction

Castor belongs to the monotypic genus *Ricinus*; the Latin meaning of *Ricinus* is 'dog tick' due to the resemblance of the seed coat colour of castor with it. It is a non-edible oilseed crop with high oil content, *i.e.*, 40-60 per cent. Oil has medicinal and industrial importance due to the only essential source of ricinoleic acid, a hydroxyl fatty acid (12-hydroxyl-cis-9-octadecenoic acid) comprising about 90 per cent of the oil. It is one of the best lubricants for high-speed engines like aeroplanes due to the viscosity of oil changing very little with temperature. The castor cake is valuable manure for field crops, but due to the presence of the poison "Ricin", it is not fed to cattles.

Background Information

1.	*Common name*	Castor bean, Divela, Erandi
2.	*Botanical Name*	*Ricinus communis* L.
3.	*Family*	*Euphorbiaceae*
4.	*Chromosome number*	2n=2x=20
5.	*Centre of origin*	Tropical Africa

6.	*Nature of pollination*	Cross-pollination
7.	*Pollination mechanisms*	Monoecious, anemophily
8.	*List of wild relatives*	*R. perciens, R. chinesis, R. maxicanus*

Distribution

Castor is cultivated under varied climatic conditions, *viz.*, tropical, sub-tropical and temperate. India, China, Brazil, USSR, and Thailand are the major countries growing castor. India ranks first in the world in respect of area and production. India, Mozambique, Brazil and China contribute about 90 per cent of world production. While India is the world's largest producer and exporter of castor oil. The major castor growing states in the country are Gujarat, Andhra Pradesh, Tamil Nadu, Orissa and Rajasthan.

Floral Biology, Selfing and Crossing Techniques

Inflorescence	:	Raceme type popularly called as spike, monoecious condition [separate male (staminate) and female (pistillate); Generally, 30-50 per cent pistillate flower born apex while 50-70 per cent staminate flowers born below]. In rare case, one or two bisexual flowers (hermaphrodite) are found in a spike.
Flower	:	Small bracteate, ebracteolate pedicellate, incomplete, regular/actinomorphic, hypogynous unisexual, either dioecious or monoecious, male flower staminate while the female flower are pistillate.
Parianth	:	Tepal Five in number in one or two whorls, aestivation-valvate, dissimilar in male and female flowers. Male flowers are apetalous with a simple perianth consisting of five sepals enclosing a stamen cluster.
(Male flower) Androecium	:	Flowers are greenish with many stamens, either tree-like or connate in one-to-many bundles, polyadelphous, and filaments branched and united to form five branches. Anthers are dithecous, globose, basifixed, introrse and dehiscing by longitudinal slits.
(Female flower) Gynoecium	:	Flowers are conical-shaped, Ovary- superior, tricarpellary, syncarpous, three-celled, three-lobed. One ovule in each locule with axile placentation. Three styles with bifid and feathery red stigmas.
Fruits	:	Regma or a schizocarpic capsule, splitting into three cocci.
Seed	:	Albuminous or dicots.
Floral formula	:	Male Flower: Br.$\oplus$ ♂ $P_{(5)}$ A_{∞} $\underline{G}_{0}$ Female Flower: Br. $\oplus$ ♀ $P_{(5)}$ A_{0} $\underline{G}_{(3)}$

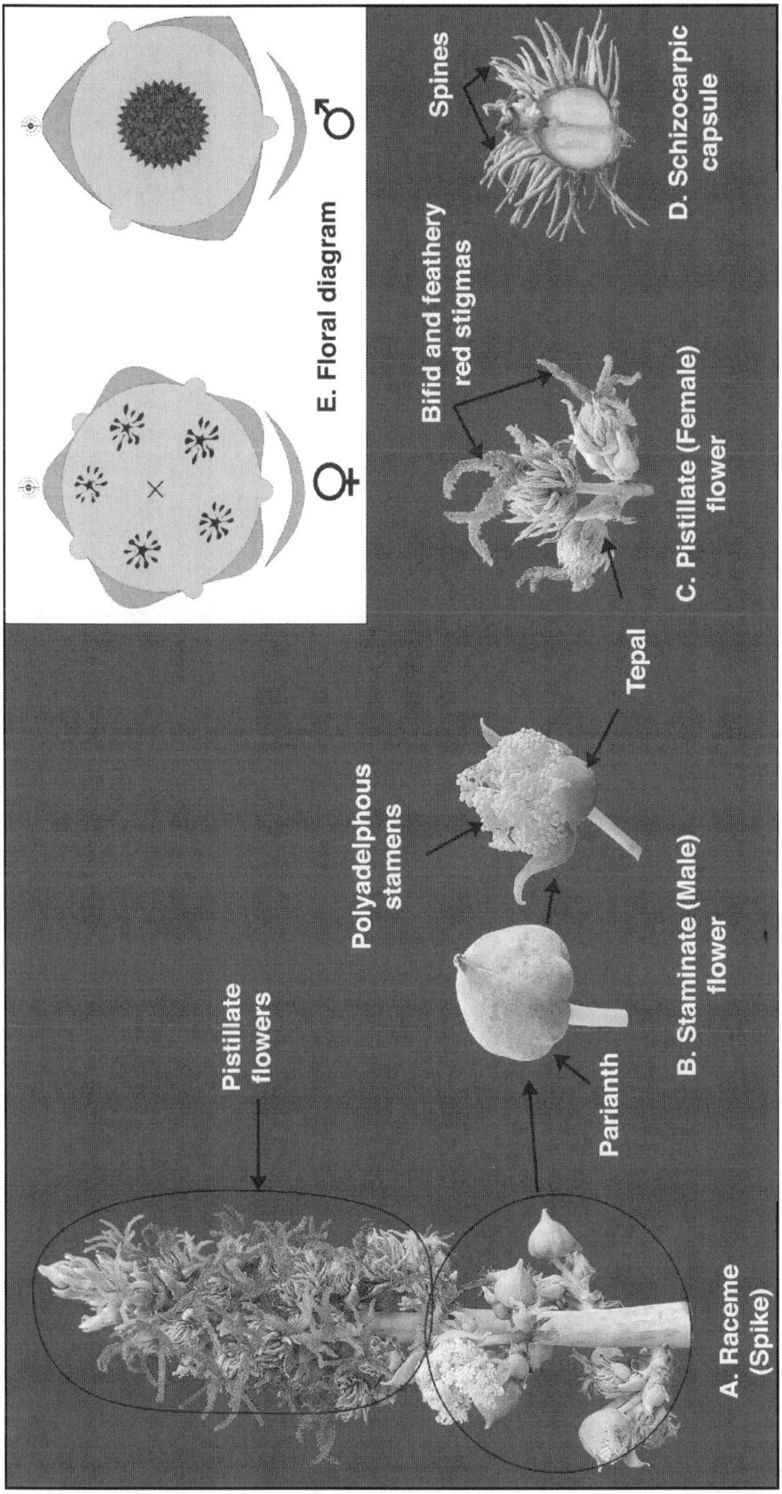

Figure 22.1. Floral Biology and Diagram of Castor (*Ricinus communis* L.).

Anthesis: The pistillate flowers start opening at 11.00 AM., but the peak opening period is from 4.30 to 5.00 PM. The pistillate flowers mature and become receptive even 2 hrs before opening the flowers and remain receptive for 2-3 days. While the male flower opens at 5.00 AM. and pollen remains viable for about 60 hours.

Selfing technique: Remove all opened female and male flowers from the raceme and cover the inflorescence before flowering under muslin cloth bag/ butter paper bags.

Crossing and Pollination Technique

The racemes of female parents are bagged before the opening of the flower. In the male parents, all opened flowers from the raceme are clipped off prior to bagging to obtain pure pollen grains for pollination. On the emergence of a stigma, pollen grains collected from bagged male racemes of desired male parent are dusted over the stigma of the flowers of the female parent. After pollination, plants are tagged and labelled. The pollen dusting is repeated three to four times to ensure sufficient seed setting on a spike of female parents. The pollen grains are viable for 3–4 days, and the stigma remains receptive for a period of 5–10 days, depending on environmental conditions.

Sex Expression (Sex forms)

There are four basic sex forms in castor

1. **Pistillate (P):** It occurs as a rare recessive mutant, with the spike having female flowers throughout the spike. It may be 100 per cent pistillate or <100 per cent (70- 90 per cent) having a few male flowers in the basal portion (Figure 22.2).
2. **Interspersed staminate flower (ISF):** A variant of pistillate form with male flowers interspersed throughout the female flowers on the spike (Figure 22.2).
3. **Monoecious (M):** The spike has basal $1/3^{rd}$ to ½ portions of male flowers, while the top portion has female flowers. In between these few whorls have both male and female flowers in an interspersed fashion (Figure 22.2).
4. **Sex revertant:** A female turns to monoecious at a later stage.

Classification of Sex Variants

According to the Shifriss (1960) the basic sex phenotypes in castor are distinguished by the kind of racemes which they differentiate during the development.

- ✰ **Type A** : Monoecious
- ✰ **Type C** : Pistillate

Figure 22.2. Basic Sex Form (Sex Expression) in Castor.

- **Type B** : Sex revertant from C to A and D to E
- **Type D** : Monoecious variant with apical interspersed raceme
- **Type E** : Uniform interspersed raceme

Among these, type A and type C are developmentally persistent forms while others are developmentally inconsistent forms.

Types of Sex Promoting Environments

The basic sex forms are classified in different ways by different authors as the expression of sex in different locations is highly influenced by environmental condition.

- **Female promoting environment:** Winter, low temperature (<30°C), young plants, early order spikes, high nutrition, less difference between maximum and minimum temperature promote female flowers and shift balance towards femaleness on a spike.
- **Male promoting environment:** Summer, rainy season, high temperature (>32°C), old plants late order spikes, low nutrition and large difference between maximum and minimum temperature promote male flowers on a spike and incline towards maleness.

Factor affecting sex expression: Sex expression is highly influenced by various environmental conditions and age of plant.

Factors	*Sex Expression*
Low temperature Young age of plant High nutrition	Promote female flower
High temperature Old age of plant	Promote male flower

Pistillate Condition and its Inheritance in Castor

There are three types of mechanisms found in castor:

1. N-Type

This type of mechanism is generally controlled by a single recessive gene. The homozygous genotype gives rise to female plants, whereas the heterozygous state gives rise to monoecious plants. It can maintain through sib matting with monoecious (heterozygous) line. This mechanism acts comparable to GMS system that exhibited a segregation ratio of 1:1 monoecious and pistillate plants in next generation. The hybrid seed production using this system is tedious since the monoecious plants are to be removed before anthesis.

- It in a backcross line that exhibited a segregation ratio of 1:1 monoecious and pistillate plants.
- The pistillate character is governed by a single sex switching gene f.
- If a plant is homozygous recessive ff for sex expression it remains as pistillate while a heterozygous recessive Ff is a monoecious plant. Thus a female plant when crossed with a monoecious, it gives 1:1 monoecious: female.
- This type can be maintained by sib mating and seed from female plants are harvested separately. The progeny from seed produced on female plants segregates in 1:1 ratio for female and monoecious plants.

  e.g. VP -1 (C) and Geeta (C)

2. S-type

The sex reversal type of pistillate developed at Israel. It is characterized by sex reverting and environmentally sensitive interspersed staminate flower expression. The plants initially 100 per cent pistillate in nature (at first order raceme). Subsequently, reversal starts from the second order that 100 per cent pistillate to monoecious condition. Such pistillate lines are developed and maintained by selecting females of late-reverted type in every generation. *e.g.*, VP-1 (Vijapur pistillate-1) [The female parent of hybrids GAUCH-1, GCH-1 and GCH-4] and Geeta (The female parent of hybrid GCH-5).

(i) Sex Reversals

Sex reversals are plant variants, which start out as female and later revert to normal monoecious at any stage after the first raceme up to 5-10 sequential order raceme. Once the change to monoecism has occurred at a given stage in a plant, it is ontogenetically irreversible. This S type pistillate line was obtained by selection within sex reversal variants at the Weigmann institute in Israel (Shifriss, 1956). This system based on sex reversal variants behave like polygenic complex with dominant and epistatic effects.

Development of stable pistillate lines from S type is based on selection from the late order revertants. On continuous inbreeding sex revertants give a spectrum of sex reversals which may be early, late and non-reverted. Early revertant on selfing gives > 50 percent pistillate plant population. Late revertants (after 4th order) on selfing give > 80 to 100 per cent pistillate populations and very less proportion of monoecious. These pistillate lines in turn are late reverted.

These sex revertant when allow to open pollinate regress to monoecism and thus have to be maintained by continuous selection for revertant pistillate plants. Among the progeny of sex revertant – female, monoecious, early revertant and late revertant, the late revertant are a potential source of pistillate line. The genetic system governing time of phenotypic reversion is unstable, but reversion as such is not associated with any serious break down in the female producing mechanism.

(ii) Non Reverted Female

Such females are developed from inbred stock of Queen 162 and Adam Mistaef and the population carry gene for interspersed staminate flower (ISF) expression at a high frequency. The penetrance and expressivity of ISF in such type of females are determined by environment. In a female promoting environment, only few ISF male flowers occur, which prematurely drop due to the competition from older female flowers that fertilized by monoecious pollen in the vicinity. This can be overcome by protecting the spikes from cross pollination by bagging. Though the initial male flowers drops, later developed male flowers to fertilize the female flowers. Shifriss (1960) developed a true breeding non reverted female line that has the gene for environmentally sensitive ISF expression. Pistillate line developed in Gujarat possesses both the systems of sex reversals and non reverted female carry the environmental sensitive gene for the expression of ISF, which offered an encourage for hybrid development programme in India.

3. NES-type

It combines both N and S types as it carries the homozygous recessive gene for pistillateness and environment-sensitive genes for ISF which are not confined to any particular raceme order and are temperature dependent. A NES type line

CNES 1 was developed by backcrossing the N-type line N-145-4 with a monoecious variety Cimarron and selecting for pistillate lines with a few interspersed staminate flowers. NES type is easier to operate and transfer a single recessive gene than both the polygenic complex, dominant and epistatic S types. The pistillate condition is temperature sensitive, the 100 per cent pistillate condition is maintained below 35°C. In contrast, temperatures above 35°C ISF (Interspersed Staminate flower) are observed at any order of raceme. In such line, seed multiplication is carried out during the summer season. This type of mechanism is similar to the EGMS type of male sterility of rice, *e.g.*, JP 65 (The female parent of hybrid GCH-6).

Further it is reported that temperature above 31°C promotes ISF while lower temperature result in fully female racemes. Thus, this type of pistillate lines can be maintained during summer, whereas hybrid seed production can be safely done during normal *kharif* planting. CENES 1 developed by Zimmerman and Smith (1996), and the female line-240 by Ankineedu and Rao (1973), JP 65 (The female parent of hybrid GCH-6) from GAU, Junagadh, and a number of pistillate lines *viz.*, VP 1, SKP 4, SKP 6, SKP 13, SKP 16, SKP 23, SKP 42, SKP 72 SKP 84, SKP 86, SKP 106, SKP 108 *etc.* has been developed from SDAU (formely GAU), Sardarkrushinagar are NES type pistillate lines.

Blooming (Waxy coat) Patterns in Castor

There are four types of blooming patterns found in castor, *viz.*,

1. **Zero bloom**: The white powder is not observed on any plant surface. *e.g.*, JP-65 (Female parent of GCH-6)
2. **Single Bloom:** The white powdery film (coat) on the stem and petiole of leaves. *e.g.*, SKI- 373
3. **Double bloom:** The white powdery film (coat) on the stem and a lower surface of leaves, *e.g.*, JI 35, VI-9, SKI 215, SH-72, DCS-89, PCS 124, GCH-5, GCH-6, *etc.*
4. **Triple bloom:** The white powdery film (coat) present on a stem as well as on the both surfaces of leaves, *e.g.*, GAUCH-1, GCH-2, GCH-4, GCH-7, GCH-8, VP-1, JI-96, SKP 84, *etc.*

Bloom trait in castor inherits as dominant or partially dominant over the zero bloom. Single bloom trait is monogenic and dominant over zero bloom, while double bloom has a dominant digenic supplementary action (*i.e.*, 9 double: 3 single: 4 zero bloom in a cross between double bloom and zero bloom). Generally, the triple bloom is always dominant over the rest bloom pattern.

Breeding Objectives

1. Breeding for high seed yield: it can improve by improving different component traits like number of racemes per plant, length of raceme, number of capsules per raceme, number of branches per plant, *etc.*

2. Breeding for high oil content
3. Breeding for short-duration cultivars with mono-branching plant types suitable for high-density planting
4. Resistance to capsule shattering and spineless
5. Dwarf to medium tall
6. Oil quality: Low ricin content
7. Breeding for disease resistance: Wilt is a major disease; GCH-7 is resistant to wilt-nematode complex
8. Breeding for insect-pest resistance: Semi-looper and sucking pest is major pest of castor, a triple booming trait of castor give resistance against it.
9. Breeding for mechanical harvesting and resistance to capsule shattering and spineless: Variety VNIIMK 165 resistances to shattering and it is suitable for mechanical harvesting due to single spike.

Breeding Approaches (Methods)

1. **Introduction:** Mauthner's Dwarf lines from Germany
2. **Mass selection:** Kavkazskaya from NVIIMK, IAC-38 from Brazil, CP-1 from interbreeding of different lines *viz.*, JI 35, HC-6, HC-8.
3. **Mutation breeding:** Mutation is used to induce variation for traits like duration, plant height and oil content. List of mutant cultivar are:

Aruna	Thermal neutrons (14 Gy)	Very early maturity (120 days instead of 270 days as the original variety) yields slightly higher
RC8	Gamma rays (400 Gy)	Late maturity, higher TGW, higher plant height, more internodes and higher yield
Sowbhagya (157-B)	Hybridization with Aruna mutant	Late maturity, dwarfness, resistance to shattering, suitable for intercropping

4. **Population improvement:** Recurrent selection method.
5. **Hybridization and selection:** Pedigree method of selection; Approach generally used to improve a polygenic trait, long duration and tall plant type variety, *e.g.*, HC 8, Junagadh-1, MC 1 developed. TMV-1 to TMV-5 series was developed at Tindivanam centre.

   Heterosis breeding: Heterosis has been significantly exploited by developing hybrid through pistillate and inbred systems. The world's first castor hybrid GCH-3 was released from GAU, Gujarat (India). In India, the first commercially successful stable S-type pistillate line VP-1(Vijapur Pistillate-1) derived from TSP 10 R (Texas Stable Pistillate 10R) was introduced in the USA. The pistillate line VP 1 was used as one of the female parents for the developed commercially successful hybrids *viz.*, GAUCH-1, GCH-2, and GCH-4. In the last four decades, India has significantly advanced hybrids development. The list of commercially successful hybrids is listed in Table 22.1. The Introduction of hybrids resulted in a remarkable increase in production and productivity. The productivity at the national level has increased from 270 kg/ha during the 1970s to 2000 kg/ha.

Steps of Hybrid Seed Production in Castor

Development of pistillate lines and maintenance of 100 per cent pistillate line:

- ☆ Diverse with 100 per cent pistillate lines with good combining ability have been developed. *e.g.*, VP-1 (Vijapur Pistillate-1), JP-65 (Junagadh Pistillate-65), Geeta, SKP 84 *etc.*, developed and maintained at SDAU, Sardarkrushinagar.

Development of inbred lines (male):

- ☆ Male lines have been developed by selfing or sibbing using generation advancement method or pedigree method.

Hybrid Seed Production

The line with 100 per cent pistillate flowers is used as female parent and inbred with ISF (Interspersed Staminate flower type) nature is used as a male parent. The hybrid seed production can be done by hand crossing in crossing block and random mating in isolation with maintaining proper isolation distance. Isolation distance for foundation stage seed is 600 m and for certified seed is 300 m.

Production of Hybrid Seed

e.g., Gujarat Castor Hybrid (GCH-7):

Parents: (SKP 84 × SKI 215)

- ☆ Two male lines should be planted as border row in seed production plots to provide sufficient pollen grains. There are one line of male interval of every three line of female.

	Female	*Male*	*Rows Ratio*
Seed rate (Kg/ha)	7.5 kg (Conventional system)	2.5 kg	3:1
	6.0 kg (Modified system)	2.5 kg	3:1

(a) Roughing

This is most crucial operation in hybrid seed production plot. It is carried out in two steps:

1. In female lines, plants showing diverse characters are uprooted before opening of the male flowers.
   - ☆ During preliminary roughing operation, look for deviations if any for various diagnostic morphological characters like stem colour, internode type, shape of leaves, blooming pattern and remove them.
   - ☆ Primary spikes are to be observed for their sex behavior. Monoecious plants beyond 3 whorls from the base are to be removed.
   - ☆ Plants with ISF, if any, should be removed.
   - ☆ The S type of female, sometimes 100 per cent pistillate line may be reverse to male flowers due to changes in temperature, as the stability of sex expression varies with environments. In such cases discard the entire plant.
2. In male lines, plants showing diverse characters are uprooted before opening of the flowers. The plant look for deviations for various diagnostic morphological characters like stem colour, internode type, shape of leaves, blooming pattern and remove them. Monoecious plants beyond 3 whorls from the base are also to be removed.

Breeding Institutes

- ☆ **National level:** Indian Institute of Oilseeds Research (IIOR), Hyderabad
- ☆ **State level:** Centre for Oilseed Research (Former Main Castor-Mustard Research Station), SDAU, Sardarkrushinagar, Gujarat

Breeding Achievements

- ☆ **Varieties:** Aruna, GAUC-1, GC-2 and GC-3, Bhagya, Jyoti, Kiran, Jwala, Pragathi, JC 24, GAC-11, *etc.*
- ☆ **Hybrids:** The important hybrids of castor are

Sl.No.	*Name of Hybrid*	*Parentage*
1.	GCH-3	TSP-10R × Junagadh-1
2.	GAUCH-1	VP-1 × VI-9
3.	GCH-2	VP-1 × JI-35
4.	GCH-4	VP-1 × 48-1
5.	GCH-5	Geeta × SH-72
6.	GCH-6	JP-65 × JI-96
7.	GCH-7	SKP-84 × SKI-215
8.	GCH-8	JP-96 × DCS-89
9.	GCH-9	SKP-84 × PCS-124
10.	GCH-10	SP-1 × SI-14

Table 22.1: Distinct morphological features of some important parental lines and hybrids of Castor

Sl. No.	*Hybrids*	*Name Parental Lines and Hybrids*	*Distinct Morphological Features*							
			Growth Habit	*Stem Color*	*Leaf Shape*	*Bloom*	*Branching*	*Capsule*	*Spike*	*Sex*
1.	**GAUCH-1**	VP-1 (♀)	Dwarf	Green	Cup	Triple	Convergent	Spiny	Semi compact	Female
		VI-9 (♂)	Medium	Green	Flat	Double	Divergent	Spiny	Compact	Monoecious
		Hybrid (×)	Medium	Green	Flat	Triple	Convergent	Spiny	Compact	Mostly male
2.	**GCH-2**	VP-1 (♀)	Dwarf	Green	Cup	Triple	Convergent	Spiny	Semi compact	Female
		JI-35 (♂)	Tall	Green	Flat	Double	Divergent	Spiny	Compact	Mostly male Interspersed
		Hybrid (×)	Medium	Green	Flat	Triple	Divergent	Spiny	Compact	Mostly
3.	**GCH-4**	VP-1 (♀)	Dwarf	Green	Cup	Triple	Convergent	Spiny	Semi-compact	Female
		48-1 (♂)	Tall	Red	Flat	Double	Divergent	Non-Spiny	Loose	Mostly male Interspersed
		Hybrid (×)	Medium	Red	Flat	Triple	Divergent	Semi-Spiny	Loose	Mostly Male
4.	**GCH-5**	Geeta (♀)	Tall	Red	Flat	Double	Divergent	Non-Spiny	Loose	Mostly female
		SH-72 (♂)	Tall	Green	Flat	Double	Divergent	Spiny	Compact	Mostly male Interspersed
		Hybrid (×)	Tall	Mahogany	Flat	Double	Divergent	Semi Spiny	Semi compact	Mostly Male
5.	**GCH-6**	JP-65 (♀)	Tall	Red	Flat	Zero/ Absent	Divergent	Spiny	Compact	Female
		JI-96 (♂)	Tall	Mahogany	Flat	Triple	Divergent	Spiny	Semi compact	Mostly Male
		Hybrid (×)	Medium	Mahogany	Flat	Double	Divergent	Spiny	Compact	Mostly Male

Sl. No.	*Hybrids*	*Name Parental Lines and Hybrids*	*Distinct Morphological Features*							
			Growth Habit	*Stem Color*	*Leaf Shape*	*Bloom*	*Branching*	*Capsule*	*Spike*	*Sex*
6.	**GCH 7**	SKP-84 (♀)	Medium	Mahogany	Cup	Triple	Convergent	Spiny	Semi compact	Female
		SKI-215 (♂)	Tall	Mahogany	Flat	Double	Divergent	Non spiny	Loose	Mostly male Interspersed
		Hybrid (×)	Medium	Mahogany	Flat	Triple	Semi Divergent	Semi Spiny	Medium loose	Mostly Male
7.	**GCH 8**	JP-96 (♀)	Tall	Mahogany	Cup	Triple	Convergent	Spiny	Semi compact	
		DCS-89 (♂)	Tall	Mahogany	Flat	Double	Divergent	Non-Spiny	Semi compact	Mostly male Interspersed
		Hybrid (×)	Medium	Mahogany	Flat	Triple	Divergent	Semi Spiny		
8.	**GCH 9**	SKP-84 (♀)	Medium	Mahogany	Cup	Triple	Convergent	Spiny	Semi compact	Female
		PCS-124 (♂)	Medium	Green	Flat	Double	Divergent	Spiny	Semi compact	Monoecious
		Hybrid (×)	Medium	Mahogany	Semi-cup	Triple	Divergent	Spiny	Loose	Mostly female/ Interspersed staminate flower
9.	**GCH 10**	SP-1 (♀)	Medium	Mahogany	Cup	Double	Convergent	Spiny	Semi compact	Female
		SI - 14 (♂)	Tall	Mahogany	Flat	Triple	Divergent	Spiny	Semi compact	Mostly male Interspersed
		Hybrid (×)	Medium	Mahogany	Semi-cup	Triple	Divergent	Spiny	Semi compact	Mostly female/ Interspersed staminate flower

Chapter 23

Breeding Cotton (*Gossypium* spp.)

Introduction

Cotton, derived from the Arabic word '*quotn*' belong to the genus *Gossypium*. The genus *Gossypium* comprises of 54 species having basic chromosome numbers (x) are 13. Among 45 are diploid (2n=2x=26) and 7 are of tetraploid (2n=4x=52). A diploid genome of genus *Gossypium* are grouped into seven types, *viz.*, A, B, C, D, E, F and G. Among genome A, B, E or F are African or Asian origin. Generally, it refers to old-world species, while C or G genomes are considered as Australian origin. D genome originated in the Western Hemisphere, generally called a New World species. Cotton is the world's most important natural spinnable fibre crop used in textile industries. Since prehistoric times, it has been cultivated in tropical and sub-tropical regions of the world. The species *G. hirsutum* is well known for its superior fibre quality, *G. barbadense* (Egyptian cotton) has fine fibre quality, whereas desi cotton has coarse and short fibre. Besides natural fibre, cotton seed is an essential source of vegetable fat and cooking oil, while seed cake is used in animal and poultry feeding.

Background Information

1.	*Common name*	Kapas
2.	*Botanical Name*	*Gossypium* spp.

3.	*Family*	*Malvaceae*
4.	*Chromosome number*	2n=2x=26 (Diploid) 2n =4x=52 (Tetraploid)
5.	*Centre of origin*	**Diploid**: India and Africa **Tetraploid**: America
6.	*Nature of pollination*	Often cross-pollinated; cross-pollination up to 5 - 25 per cent
7.	*Pollination mechanisms*	Bisexuality, Chasmogamy, Protandry, Entomophily
8.	*List of wild relatives*	**Diploid**: *G. anomalum* and *G. thurberi* (2n=26) **Tetraploid**: *G. tomentosum* and *G. mustelinum* (2n = 52)

Important Cultivated and Wild Species of Cotton

The cultivated species are divided into two major group

I. Asiatic cotton (old-world/Diploid cotton) and

II. New world cotton (Tetraploid cotton)

American (upland) cotton is the predominant species which contributes about 90 per cent to the current world production, followed by Egyptian cotton (*G. barbadense*) (8 per cent). Whereas desi cotton *G. arboreum* is mainly cultivated in India. Details of cultivated species and wild relatives, along with chromosome number, the centre of origin and common name, are depicted in Table 23.1.

Table 23.1. Cultivated and Wild Species of Cotton

Sl. No.	*Particulars*	*No. of Chromo., Genome and Ploidy*	*Chromo. Size*	*Centre of Origin*	*Common Name*
		A. Cultivated species			
1	*G. herbaceum*	2n=26A_1 (Diploid)	Large	Africa	Desi/Old World
2	*G. arboreum*	2n=26A_2 (Diploid)	Large	India	Desi/Old World
3	*G. hirsutum*	2n=52AD_1 (Tetraploid)	26-large 26-Small	North America	New World/ *American/ upland cotton*
4	*G. barbadense*	2n=52AD_2 (Tetraploid)	26-large 26-Small	South America	New World/ *Egyptian/sea island cotton*
		B. Wild species			
1	*G. thurberi*	2n=26D_1 (Diploid)	Small	America	New World
2	*G. anomalum*	2n=26B_1 (Diploid)	Medium	Africa	Old world

Sl. No.	Particulars	No. of Chromo., Genome and Ploidy	Chromo. Size	Centre of Origin	Common Name
3	*G. tomentosum*	$2n=52AD_3$ (Tetraploid)	26-Large 26-small	Hawaii	New World
4	*G. mustelinum*	$2n=52AD_4$ (Tetraploid)	26-Large 26-small	Brazil	New World

Distribution

Since prehistoric times, cotton has been cultivated in tropical and sub-tropical regions worldwide. India ranks first in terms of area among cotton-growing nations. India's major cotton-growing states are Maharashtra, Gujarat, Rajasthan, Punjab, Madhya Pradesh, Andhra Pradesh, Karnataka, and Tamil Nadu. *Gossypium hirsutum* (upland cotton), was the first cotton cultivar produced by the Mayan civilization in Mexico. In India, 90 per cent area is covered by *G. hirsutum.*

Types of Branches in Cotton

According to the development pattern, branches are classified into two types.

1. Vegetative branches (*monopodia*), and
2. Fruiting (*sympodia*)

The essential features of each branch as under

Vegetative Branches (Monopodia)	*Fruiting Branches (Sympodia)*
It, like the main stem, possesses only a single meristem	The branches from which fruiting buds arise are called fruiting branches. It produces from vegetative branches. Each branch possesses multiple meristems called axillary meristems.
Growth habit straight and erect with a single foot	Growth habit Zig-Zag with multiple foots

The Genetic Origin of Amphidiploid (Allotetraploid) New World cotton

The American cotton, an amphidiploid (allotetraploid) *Gossypium hirsutum* ($2n=4x=52, AAD_1D_1$) believed to be evolved through natural hybridization between two diploid cotton species *viz.*, African (Old-World) linted species *Gossypium herbaceaum* var. *africanum* (2n=2x=26, AA) and American (New World) lintless wild diploid species *Gossypium raimondii* ($2n=2x=26, D_1D_1$) (Figure 23.1).

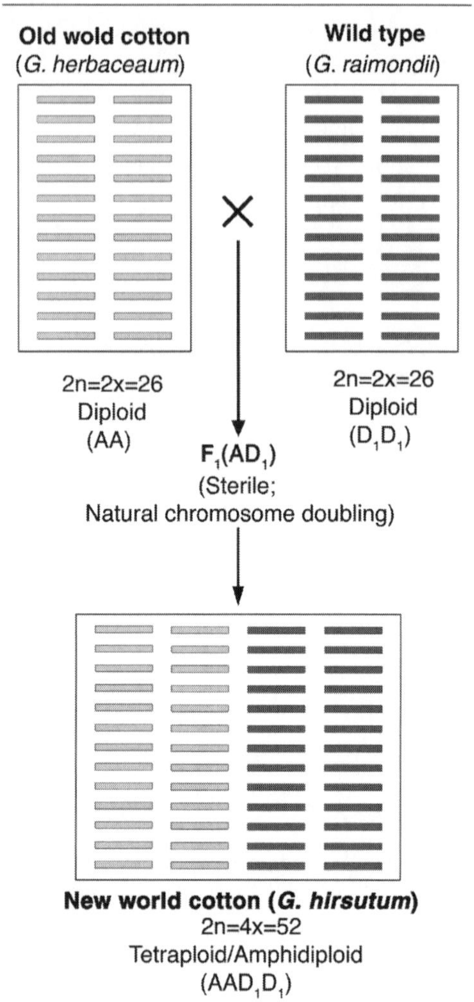

Figure 23.1. Illustration about Synthesis of Amphidiploid New World Cotton.

Floral Biology, Selfing and Crossing Techniques

Inflorescence	:	Solitary, axillary and cymose born on sympodial branches
Flower	:	Pedicellate, ebracteate, bracteolate, complete, actinomorphic (Regular), hermaphrodite (bisexual), hypogynous and pentamerous
Epicalyx	:	Three large fused
Calyx	:	Five green sepals, gamosepalous (united), persistent, cup-shaped and valvate aestivation

Corolla	:	Five yellow colour petals, polypetalous, fused at the base with staminal tube and twisted aestivation
Androecium	:	Stamen are numerous, monadelphous, epipetalous, monothecous, anther reniform and unilocular and extrose connect transversely with filaments,
Gynoecium	:	Three to five carpels, syncarpous, ovary superior and five to many locules, many ovules in each locule, axile placentation, free stigmas and simple style
Fruits	:	3 to 5 valved dry dehiscent capsules
Seed	:	Endospermic, dicot, dark coloured with epidermal hairs
Floral formula	:	$\oplus$ ⚥ Ep $K_{(5)}\ \widehat{C_{(5)}\ A_{(\infty)}}\ \underline{G}_{(5-\infty)}$

Formation of the Cotton Flower from Square to Bloom

Approximately 21 days are taken for the square (flower bud) to bloom stage. It can be recognized by different developmental stages, *viz.*,

- ☆ **Pinhead square –** The first stage at which the square (flower bud) can be identified, the flower bud visible as a small green cone
- ☆ **Match-head –** Square growth, three bracteoles (Epicalyx), joined at their bases, surround the flower bud.
- ☆ **Candle shaped –** Candle shape bud growth
- ☆ **Bloom stage –** This is said to be the actual flowering stage (white flower)

The blooming process takes several days, and the age of the bloom can be assessed by the typical bloom features, *viz.*,

- ☆ **White colour:** On the day a flower opens. Usually, pollination takes place within a few hours after the opening of a white colour flower.
- ☆ **Pinkish colour:** The day after anthesis, the flower turned pink colour
- ☆ **Reddish red colour:** Attained this colour on the third day.
- ☆ **Dried stage:** Flower is dried approximately 2 to 3 days after anthesis and falls off to reveal a developing young green boll. Sometimes a dry flower remains attached to a developing boll for extended periods, referred to as a bloom tag.

Selfing technique: To ensure the cent per cent self-pollination, The flowers, which are likely to open the following day either, should be tied with a rubber ring, mud/wire ring or covered with the proper size of a paper bag.

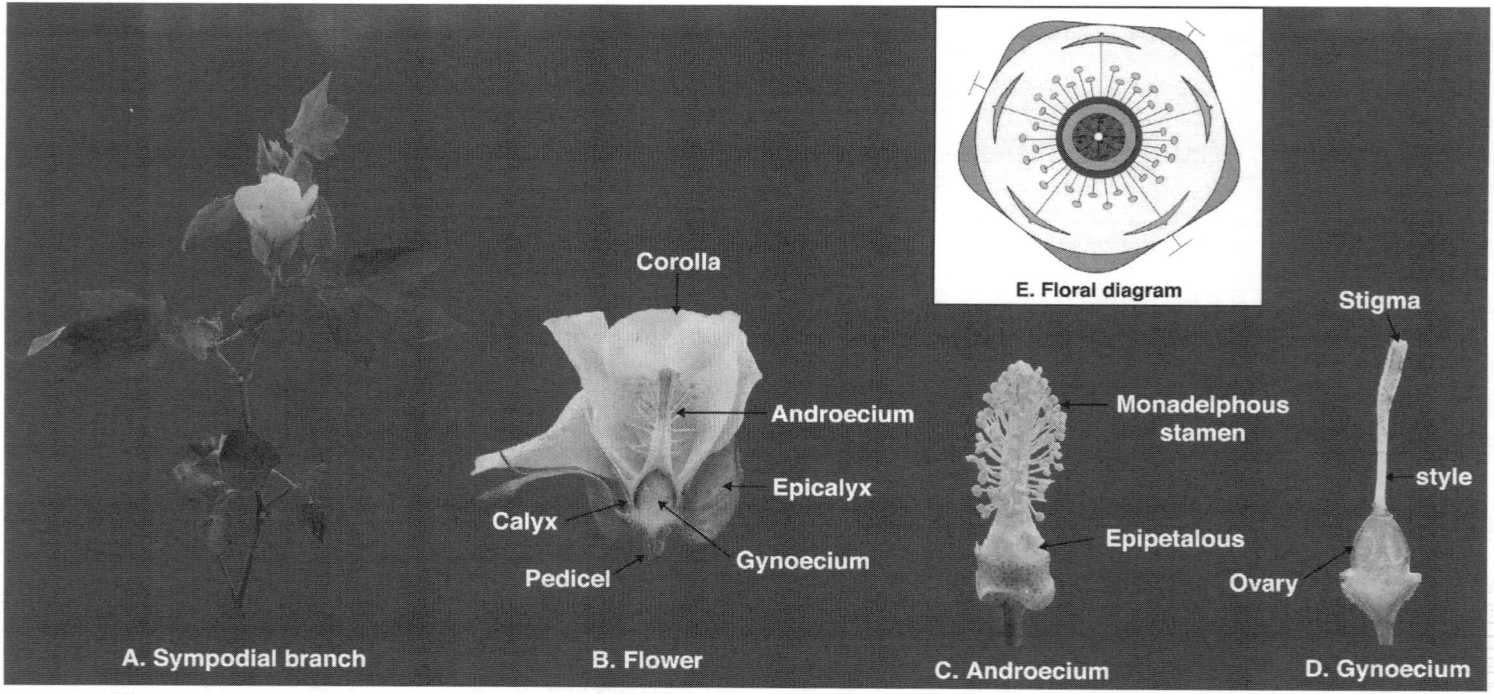

Figure 23.2. Floral Biology and Diagram of Cotton (*Gossypium* spp.).

Artificial Crossing and pollination: Hand emasculation and pollination are very easy and practically possible because of the large flower size. For the emasculation of flower buds, Doak's method is used. In this method, bracts, petals and anthers are removed by Thumbnail without damaging the ovary. Emasculation is carried out from 2.00 to 6.00 P.M. and early in the morning. After emasculation, the flower buds should be covered by a red coloured tissue paper bag or round straw tube and tied with a thread, tag, or price label. A red colour bag is helpful for identifying emasculated buds for pollination.

Pollination: Select male flower that is about to open. It is prepared for pollination by removing bracts and petals, keeping it in trays or dishes, and exposing it to sunshine to stimulate the anthers. Usually, the anthers burst at around 10.00 A.M. The red tissue bag from the emasculated female flower bud is removed, and the male flower bud is rubbed all around the surface of the stigma. One male flower is sufficient for 3 to 5 female flowers. White tissue paper bags cover the pollinated female flowers.

Breeding Objectives

1. **Fibre yield (lint fibre):** High lint fibre is the primary goal in cotton breeding. The fibre yield can be increased by a high number of bolls per meter square, boll size and medium size boll with a good opening, long reproductive growth, high lint percentage, and high lint turnout.
2. **Early maturity:** Important trait includes determinant growth habit, plant height at harvest, length of fruiting branch, effective flowering time, height to node ratio, *etc.*
3. **Breeding suitable for mechanical harvesting (picking):** The essential features that favour mechanical harvesting are early maturing, medium to compact plant type, uniform and good opening of bolls, direct fruiting branches, *etc.*
4. **Breeding for coloured cotton:** Primarily, cotton produces white lint colour. However, there is considerable lint colour variability, *viz.*, brown colour to light green found in tetraploid cotton. Coloured cotton provides great scope for the textile industry to produce cloths without synthetic dyes and chemicals and are environment friendly. However, coloured cotton fibre qualities, *viz.*, fibre length, strength, and micronaire main concerns. All colours are genetically controlled and have a monogenic inheritance.
5. **Breeding for disease resistance:** Important diseases are wilt, root rot, root-knot nematode, bacterial blight, *etc.* Breeding resistance against particular or multi-diseases resistance is essential.
   - ☆ Supriya, MCU 10→ Bacterial blight,
   - ☆ MCU 5 VT, Surbhi→ *Verticillium* wilt
   - ☆ GSHV-159 and GBHV-184→ Alternaria leaf spot

6. **Breeding for insect-pest resistance,** *i.e.*, Bollworm, pink bollworm, Minor Pest-Sucking pest, cotton mealybug, *etc.* The morphological features of plants include minimum leaf area, okra types, hairiness, glandless leaves, long frego bracts, nectariless, red canopy colour, high gossypol content *etc.*

   The different varieties of upland cotton (*G. hirsutum*) have developed for insect-pest stress resistance: *viz.*, PKV 081, B 1007, Suraj for Jassid; Kanchana for White fly; H77, F 414, Bikaneri Narma for Pink bollworm.

   Whereas *G. harbaceum* varieties: *viz.*, G Cot 11, G Cot 13, G Cot 17, G Cot-21, G Cot 23 showed resistance to Jassid and *fusarium* wilt.

7. **Breeding resistance to abiotic stresses,** *i.e.*, Drought, heat, salt. The essential plant traits favour drought tolerance, including earliness, small leaf area, high cuticular wax, longer root length, high hairiness, *etc.* Two varieties MA 9 and MCU-10 (upland cotton) were developed through mutation breeding against drought resistance. While leaf and stem angle, leaf thickness, leaf angle, ability to bear flower and retain boll at lower nodes, cell membrane stability, lower leaf senescence, plant canopy style, *etc.*, are essential for heat tolerance. Various genotypes have been identified as drought resistant/tolerant are Indore-2, SRT-1 TCH 1002, LH 900. While for salinity tolerant are MESR 17, DHY 286 and Lakshmi, *etc.*

8. **Fibre quality:** Cotton fibres are single hyper-elongated cells arising from the epidermal cells of the outer integument layer of a seed coat. The uniform growth of thousands of terminally differentiated single-celled fibres per ovule. It is considered by four overlying developmental stages: fibre commencement, primary extension, secondary cell wall (SCW) synthesis, and maturation.

   - ☆ **Lint**: The outer and longer fibre→which can easily separate from the seed in ginning and used in spinning cotton yarn and in the textile industry.
   - ☆ **Fuzz (linter)**: The inner and shorter fibre can remain attached to the seed after ginning. The fuzz proportion varies with species, *i.e.*, 1.0 per cent to 10.5 per cent. The proportion has been found to be maximum in *G. hirsutum* (up to 10.5 per cent), whereas minimum in Asiatic cotton (4.3–5.9 per cent).
   - ☆ The fuzz constitutes of pure form of cellulose. It is used to make various cellulosic products, rayon, photographic and writing papers, X-ray films, currency notes, explosives, *etc.*

     In the case of fibre cell initiation, many fibre-specific genes described, fibre elongation, or cell wall biogenesis, have been identified as candidates for genetic manipulation to improve fibre yield and/or quality. Overexpression of *GhXTH1* in cotton resulted in longer fibres than their controls without adversely affecting other fibre characters. A novel cotton sucrose synthase gene, GhSusA1, was identified from *G. hirsutum*. Silencing of GhSusA1 reduced fibre length and yield, whereas overexpression of this gene increased fibre length and strength.

Fibre is a primary product having industrial (textile) importance. Fibre quality is mainly comprised of (i) Fibre length (staple length), (ii) Fibre strength, and (iii) Fibre fineness (Fibre weight).

According to length, cotton fibre is separated into two groups:

Lint Index (Absolute weight of lint/seed)

$$= \frac{100 \text{ seed weight} \times \text{Ginning per cent}}{100 - \text{Ginning per cent}}$$

Lint per cent (Ginning per cent)

$$= \frac{\text{Weight of lint}}{\text{Weight of cotton}}$$

- ✰ High-Volume Instrument (HVI) with a specific standard is used to test quality cotton fibre samples after ginning. The HVI measures the fibre length by creating a "beard."
- ✰ **Micronaire:** It is a measure of fibre resistance to airflow, which is an indirect way to estimate fibre fineness (linear density) and maturity. High micronaire values (>5.0) indicate a coarse fibre (less fine), while low values (<3.5) indicate immature fibres, both undesirable for spinning.
    - ❖ The varieties, MCU-5, MCU-9, MCU-10, Suvin, K-9, K10, Saraswati, AKA 8401 are good identified for their quality.

9. **Seed quality:** Besides natural fibre, seed oil is an essential source of vegetable oil for human consumption. Oil quality can be deteriorated by fatty acid content, chemical composition, and undesirable pigments. The pigmented glands are present in the whole plant, *viz.*, leaves, stem, seeds, *etc.* These glands contain gossypol (a terpenoid compound), which acts as natural phytoalexins against pathogens and pests. But in the other hand, it's an important health concern which causes discolouration in cottonseed oil; Reduces the availability of lysine and other essential amino acids in cottonseed protein; Induces male sterility in humans, and poultry. Hence more consideration is required to produce oil free from gossypol content.
10. **Exploitation of wild species for various purposes:** In cotton, many wild species have been explored to widen the genetic diversity and used in pre-breeding for various stress resistance, quality, introgression

of cytoplasmic male sterility, and yield improvement. Some notable examples are given in Table 23.1

Table 23.1. Wild Species; Interspecific Cross Combination and their Utility in Cotton Breeding

Sl.No.	*Wild Species; Interspecific Crosses*	*Purposes*
1.	*G. anamolum*	Contains genes for fibre fineness and strength and resistance to insect and disease
2.	*G. barbadense*	Fibre quality traits
3.	*G. tomentosum*	Fibre quality traits-fibre elongation, length, fineness, fibre uniformity, boll weight, and boll number
4.	G. *gossypioides*	Resistant to cotton leaf curl virus disease
5.	*G. hirsutum × G. armourianum*	Profuse hairiness gives resistance to sucking pests such as Jassid, white fly
6.	Synthetic amphiploid *G. hirsutum × G. arboreum* (AADDAA)	An amphiploid resistant to *Verticillium* and drought resistance
7.	*Gossypium barbadense × G. hirsutum*	High yield, lint percentage with better lint quality traits

Breeding Approaches (Methods)

1. **Introduction and selection**: American nectariless (Parent of Hybrid-4), Russian *barbadense* SB 289 to produce Varalaxmi and SB 1085-6 (used as a parent of interspecific hybrid' CBS 156').
2. **Pure line selection:** Different varieties developed through pure line selection comprises
   - *G. hirsutum* (upland): MCU 5 VT and CO_2
   - *G. arboreum*: Cocanada 2, Gaorani 22, Gaorani 46 and Lohit
   - *G. herbaceum*: Western and Selection 69
   - *G. barbadense*: Sujata

   Mass selection- This method is now rarely used. Some varieties were developed by mass selection comprises *G. hirsutum* based are Bikaneri Narma, F 414, SRT 1, H 777, L 147, Narmada, and PRS 72 G 27, Bikaneri Narma, F 414. While *G. arboreum* based G 27, LD 133, HD 11, Cocanada white, Gaoranie, and Saraswati.
3. **Hybridization and selection:** Pedigree, bulk, back cross, *etc.*
4. **Mutation breeding:** Various mutants produced with the help of mutagens improved the agronomic features of cotton, including higher bolls yield,

earliness, biotic and abiotic stress resistance, improved quality, *etc.* The list of mutants is given in Table 23.2

Table 23.2. List of Mutant Cultivars Released in India

Sl.No.	*Cultivars*	*Source of Mutagen*	*Characters Improved*
1.	MA 9	X-Rays	Drought tolerance
2.	Indore-2	X-Rays	High yield
3.	MCU 7	X-Rays (800 Gy)	Early maturity (10-15 days), higher yield, long-staple 25 cm, tolerance to drought, resistance to black arm, root rot and leaf blight and improved spinning capacity
4.	Rasmi	Gamma rays (300 Gy)	Photoperiod insensitive, wider adaptability, superior quality and improved yield
5.	Pusa Ageti	Gamma rays (250 Gy)	Improved ginning capacity of fibre, early maturity and tolerant to jassids
6.	MCU 10	Gamma rays (300 Gy)	Drought tolerance, disease resistance and long-staple fibre
7.	DS-1	Gamma rays (200 Gy)	Semi-dwarfness and high lint

5. **Heterosis breeding:** In cotton through heterosis breeding, various types of hybrids have been developed commercially. The hybrids can be in various groups based on ploidy level or chromosome number (tetraploid and diploid) and species involved (intraspecific and interspecific).

   - **Diploid Hybrids:** Diploid hybrids are developed between *G. arboreum* and *G. herbaceum*. Diploid hybrids are of two types: intraspecific and interspecific. Intraspecific hybrids have been developed in *G. arboreum* only.
   - **Tetraploid Hybrids:** Tetraploid hybrids are developed in *G. hirsutum* and *G. barbadense*. Tetraploid hybrids are of two types: intraspecific and interspecific. Intraspecific hybrids have been developed in *G. hirsutum* only.
   - **Intraspecific Hybrids:** It is also known as inter-varietal hybrid. Intraspecific hybrids are always fertile. In cotton, intraspecific hybrids have been released for commercial cultivation in *G. hirsutum* at the tetraploid level and in *G.arboreum* at the diploid level.
   - **Interspecific Hybrids:** The development of fully fertile interspecific hybrids is possible only between species with complete chromosomal homology. In cotton, interspecific hybrids are fully fertile between *G. hirsutum* and *G. barbadense* and between *G. herbaceum* and *G. arboreum.*

Due to sound efforts of pioneer cotton breeder C. T. Patel at Main Cotton Research Station, Surat, NAU, Gujarat, India is the first country in the world to release and commercialization of hybrid with the release of Intraspecific: (G. *hirsutum* × G. *hirsutum*) hybrid H4 (Gujarat 67 × American nectariless). While Interspecific hybrids: Varalakshmi (Laxmi × SB 289E) (*hirsutum* × *barbadense*). The hybrid seed production is carried out by conventional method (hand emasculation) and male sterility system (GMS or CGMS).

Hybrids developed through conventional methods

- **Intraspecific hybrids:**
  - Intra-*hirsutum* hybrids: H 4, H 6, H 8, H 10, PKV Hy 2, JKHy 1, JKHy 2, NHH 44, Surya, Savita, Fateh, LHH 144, Maruvikas, Dhanlaxmi, Omshankar, DHH 11, CICR HH 1, *etc.*
  - Intra-*arboreum* hybrids: LDH 11, AAH1
- **Interspecific hybrids:**
  - Diploid hybrids: DH 7, DH 9, DDH 2, PHA 46, *etc.*
  - Tetraploid hybrids: Varalaxmi, NHB 12, DCH 32, HB 224, DHB 105, TCHB 213, NHB 302, Sruthi, *etc.*

Male sterility in cotton:

Male sterility is induced by transferring *G. hirsutum* chromosome into *G. harknesii* cytoplasm and restorer gene also from same species. Some available CMS sources in cotton are

CMS-D2	*Gossypium harknesii*
CMS-D8	*G. trilobum*
JA-CMS	*G. hirsutum; G. arboreum; G. thurberi*

Identifying CMS and fertility restoration sources are considered a significant development towards exploiting heterosis and commercial-scale cultivation of hybrid cotton. Based on the CMS system, various hybrids have been developed, *viz.*,

- CGMS based: PKV Hy 3, PKV Hy 4, Mech-4 (*hirsutum* × *hirsutum*)
- GMS based: Suguna (*hirsutum* × *hirsutum*)→ First GMS based hybrid released in year 1978 by Central Institute for Cotton Research and G. Cot MDH 11, AKDH-7 (*Arboreum* × *arboreum*)

Practical problems associated with the utilization of male sterility in hybrid seed production:

1. Absence of purely inherited restorer gene with stable expression over the environments
2. Absence of good combiners with CGMS and fertility restorer genes and

3. Absence of a controlled and effective pollination system for hybrid seed production

6. **Transgenic breeding:** Foreign DNA has been introduced into plants using the Ti plasmid of *Agrobacterium tumefacians.* Cotton is one of the first crops in which transgenic cultivars have been successfully commercialized for large-scale cultivation. Transgenic cotton provided resistance against the bollworms. It is transformed with resistant genes (*Cry*1Ac and *Cry*2Ab genes) encoding crystal protein δ-endotoxin (Bt toxin) obtained from *Bacillus thuringenesis* with the help of genetic engineering techniques. The Bt toxin is activated in the insect gut causing the death of the organism. Two major notable achievements have made so far, *i.e.*, 1. Hybrid for insect resistance-Bt cotton (trade name Boll Guard) and 2. Herbicide resistance (*i.e* glyphosate) (trade name Roundup ready).

   Bt cotton was the first GM technology introduced in India in 2002 as Bollgard (cry1Ac gene–Mon 531 event) and in 2006 as Bollgard II (cry1Ac + cry2Ab genes–Mon15985 event).

7. **Marker-assisted/Genomic assisted breeding:** Various QTLs and markers identified and are used in various disease resistance and oil quality improvement program. With the help of molecular markers, various cultivars, *viz.*, Abhadita, Kanchana, LK 861 and Supriya, which are tolerant to *Helicoverpa* and whiteflies, have been developed. Different lines, such as DHY 286, Mahalaxmi, MCU5, Krishna, Sujatha *etc.* have been identified for resistance against jassids. The varieties, *viz.*, Ravnaq-1 and Ravnaq-2, having higher fibre strength and improved length, have been developed through marker-assisted breeding.

8. **Genome editing tools:** Various genome editing tools such as zinc finger nucleases (ZFNs), meganucleases, transcription activator-like effector nucleases (TALENs) and the most advanced tool is clustered regularly interspaced short palindromic repeats- CRISPR-associated system (CRISPR-Cas)-mediated genome editing. Among the CRISPR-Cas9 is most promising for knockout of the genes associated with agronomical significance and improving the various stresses resistance in cotton. For example, CRISPR/Cas9 knockout of the arginase gene significantly enhanced the number of lateral roots and the root surface area in both normal and nitrogen deficiency conditions. This system successfully produced two sgRNAs in a single vector for manifold sites genome editing using Discosoma red fluorescent protein2 (*DsRed2*) and an endogenous gene *GhCLA1* in allotetraploid cotton.

Hybrid seed production is feasible in cotton because

- ☆ The large flower size, hand emasculation and pollination are very easy and become practically possible

- More number of seeds per cross
- Low seed rate for sowing

Procedure for Hybrid Seed Production in Cotton: For example, Gujarat Cotton Hybrid-4/H-4

H-4 [Gujarat-67 × American Nectariless (AN)]

- The male and female parents are grown separately but near to each other.
- Isolation distance should be maintained at 50m for foundation seed and 30m for certified seed.

Sowing of the female parent (Gujarat-67)

- Sowing is done in the last week of May to the second week of June at a distance of 1.5 m × 1.5 m or 2.0 m × 2.0 m
- A 1.5 –2.0 kg seed rate per hectare is needed to grow approximately 2500-3600 plantlets per hectare
- Extra seeds may be raised in bags for gap filling

Sowing of the male parent (American Nectariless)

- About 0.2 ha male plot is enough for one hectare of a female parent.
- The spacing required is 1.0 × 0.6 m
- The proportion of female: male is 5:1 rows

Time of sowing

- American Nectariless is early in maturity. So the male parent should be sown 20-25 days after the female parent sowing
- The male parent is usually sown on 2-3 different dates like
  - With ♀ parent sowing
  - Ten days after ♀ parent sowing; 25 days after ♀ parent sowing

Period of crossing

- The crossing begins from September onwards and continues up to 15th December. The seed setting is good during September–October
- For the emasculation of flower buds, Doak's method is used
- In Doak's method, bracts, petals and anthers are removed by Thumbnail without damaging the ovary
- Emasculation is carried out from 2.00 to 6.00 P.M. Un-emasculated open flowers should be removed

After emasculation, the flower buds should be covered by a RED coloured tissue paper bag or round straw tube and tied with a thread or tag. The red colour bag is useful for the identification of emasculated buds for pollination

Emasculation

- Emasculation is carried out from 2.00 to 6.00 PM. Un-emasculated open flowers should be removed.
- After emasculation, the flower buds should be covered by a RED coloured tissue paper bag or round straw tube and tied with a thread or tag.
- The red colour bag is helpful for the identification of emasculated buds for pollination.

Pollination

- Bracts and petals are removed from male flowers, which are about to open, and the flower is kept in sunlight in trays or dishes. The anthers burst at 10.00 AM.
- The red tissue bag from the emasculated female flower bud is removed and male flower bud is rubbed all around the surface of the stigma.
- One male flower is sufficient for 3 to 5 female flowers.
- White tissue paper bags cover the pollinated female flowers.

Picking of crossed bolls

- When the bolls open, picking should be done along with the tissue paper bags. The cotton is then taken out and filled in gunny bags after drying.
- On average, 700-1000 kg of hybrids seed can be produced per hectare.

Breeding Institutes

- **National Level:**
  i. Central Institute for Cotton Research (CICR), Nagpur, Maharashtra.
  ii. Central Institute for Research on Cotton Technology (CIRCT), Mumbai, Maharashtra.
- **State level:** Regional Research Stations-State Agricultural Universities.

Breeding Achievements

Hybrid Varieties

- **Intra *hirsutum* hybrids:** Hybrid-4, hybrid-6, hybrid-8, hybrid-10, PKVHy2, PKVHy-3, PKVHy-4, NHH-44 JKHy-1, JKHy-2, Suguna, Savita, Surya (TM 1312), DHH-11, DHB-105, Dhanalaxmi, PAU Bt 5, Nandyal Cotton 25, G.Cot-46 (Sorath Swet Hem), Phule Satpuda, GN. Cot. 27, G. Cot. 31, G.Cot.10 Bt, H Bt Cotton 65, H Bt Cotton 66, *etc.*

- ✰ ***Hirsutum* × barbadense hybrids:** Varalaxmi, JKHy-11, NHB-12, MHB-302, CICR HH-1, Sruthi, HB 224, CBS 156, TCHB-213, DCH-32, *etc.*
- ✰ **Bt hybrids:** Boll Guard-I, Boll Guard-II and Fusion Bt, RCH-2, RCH-138, RCH-118, RCH-144 (Rasi Seeds, Attur, TN), MECH 12, MECH 162, MECH 184, MRC 6301 (MAHYCO, Jalana, MH), Ankur-651, Ankur 09 (Ankur Seeds, Nagpur, MH), Bunny and Mallika (Nuziveedu Seeds, Hyderabad, AP) VSCH-369 BG II, ARCH 3106 BGII, Mahabeej-124 BGII, CICR- H Bt Cotton 62, CICR- H Bt Cotton 63, CCH-03 BG II, ARCH 3106 BG II, Ajeet 9-2 BG II, GTHH-49 BG II, Hy. Daftari 1579 BG II, DC 5100 BG-II, US 711 BG II, *etc.*

Chapter 24

Breeding Sugarcane (*Saccharum officinarum* L.)

Introduction

The word sugarcane is derived from the Sanskrit word 'sharkara', meaning sugar. It is a tropical plant belonging to *Poaceae* family. It has very efficient in converting solar energy into chemical energy due to C_4 photosynthetic physiological apparatus. It is an important sugar crop in all tropical Asian countries. In India, it is one of the essential sugars-yielding crops. India is the largest sugar producer in the world. It comprises three cultivated species *viz., S. officinarum* (2n = 80), *S. barberi* (2n = 111 – 120), and *S. sinense* (2n = 81 – 124) among cultivated species *S. officinarum* Tall, handsome, large-barrelled and colourful canes known as noble cane. Dutch scientists in Java gave the term noble due to its attractive appearance and quality. Sugarcane is primarily used for producing sugar, jaggery and ethanol. Secondary products are molasses, bagasse and press mud. Its green leaves and tops are used as cattle fodder.

Background Information

1.	*Common name*	Shirdi, Ganna, Eekh
2.	*Botanical Name*	*Saccharum officinarum*
3.	*Family*	*Poaceae*
4.	*Chromosome number*	2n=80

5.	*Centre of origin*	New Guinea and North India
6.	*Nature of pollination*	Cross-pollination
7.	*Pollination mechanisms*	Protogyny, self-male and female sterility, anemophily
8.	*List of wild relatives*	*S. spontaneum, S. robustum* and *S. edule* Hassk

Species of Sugarcane

The *Saccharum* species are highly heterozygous due to high chromosome numbers and complex ploidy levels (due to the hybridization of several genomes). Six sugarcane species (*Saccharum* spp.) are recognized among three cultivated and three wilds. A detail description is given in Table 24.1.

Table 24.1. Detail of Sugarcane Species

Sl. No.	*Name of Species*	*Chromosome No. (2n)*	*Descriptions*
		Cultivated species	
1.	*Saccharum officinarum* (Noble canes)	80	Popularly known as Noble canes. It is supposed to have resulted from complex introgression between *S. spontaneum, Erianthus arundinaceus* and *Miscanthus sinensis*. However, some reports support it originating from *S. robustum* through normal and human selection in New Guinea. It migrated from its centre of origin to other places and was modified by natural hybridization. It is an essential source of genes for sucrose accumulation, all milling qualities such as high cane yield, juiciness, thick, erect stem and low fibre, easy detrashing due to soft rind, resistance to smut, *etc.* Sucrose content: High (18-25 per cent) and Fibre content: Low (5-15 per cent).
2.	*S. barberi* (North Indian canes)	111–120	This species is also known as North Indian canes with dark green leaves, short, cylindrical internodes, hard rind, low-to-medium juice and sucrose content suitable for jaggery. Sucrose content: Medium (13-17 per cent) and Fibre content: High (10-15 per cent). It is mainly cultivated in subtropical India; well adapted to adverse growing conditions such as drought, waterlogging, and cold weather (winter). This might have originated by natural hybridization between *S. officinarum* and *S. spontaneum* .

Sl. No.	*Name of Species*	*Chromosome No. (2n)*	*Descriptions*
3.	*S. sinense* (Chinese canes)	81–124	This species is also known as Chinese cane. Morphological characteristics are comparable of S. *barberi*. However, it has a spindle internode shape, light green or yellowish leaves and is taller than *S. barberi.* Sucrose content: Medium (12-15 per cent) and Fibre content: High (10-15 per cent)
			Wild species
4.	*S. spontaneum*	40–128	It is a most primitive species, originated and widely distributed in India, adapted to drought and water-logged conditions. It lacks of sucrose (sweetness) but produces several thin shoots. Sucrose content: Very low (1-4 per cent) and Fibre content: Very high (25-40 per cent)
5.	*S. robustum*	60, 80	Possibly evolved in New Guinea by natural hybridization between *S. spontaneum* × *Erianthus*, *Sclerostachya* and *Miscanthus* species. The closest wild relatives of *S. officinarum*. It has thin-to-thick bare hardy stalk without rhizome. Sucrose content: Low (3-7 per cent) and Fibre content: Very high (25-35 per cent)
6.	*S. edule* Hassk	60–80	Closely resembles to *S. robustum* except inflorescence features. The inflorescence is compact, remains unopened and enclosed inside the leaf sheaths. The aborted inflorescence is used as edible and so it is known as edule. Sucrose content is low

Gene Pool of Sugarcane

A gene pool of sugarcane is grouped into four categories

Sl.No.	*Gene Pool*	*Species*
1.	Gene pool 1 (GP-1)	Commercial cultivars and near-commercial clones/genetic stocks of crosses *S. officinarum* × *S. spontaneum*; *S. barberi* × *S. spontaneum*; *S. sinense* × *S. spontaneum*
2.	Gene pool 2 (GP-2)	Clones of *S. officinarum*, *S. barberi* and S. *sinense*.
3.	Gene pool 3 (GP-3)	Wild species *viz.*, *S. spontaneum* and *S. robustum* and *S. edule*
4.	Gene pool 4 (GP-4)	Different allied genera *viz.*, *Erianthus*, *Miscanthus*, *Narenga* and *Sclerostachya* (Species Sorghum, *Bambusa* and Zea)

Nobilization

In India, Indian noble cane (*S. officinarum*) did not survive well because of red rot disease. However, it established well in New Guinea, where the disease was not a problem. In sugarcane breeding, a general practice has been carried out to cross between a noble cane (*S. officinarum*) with other desirable species. The aim of these crossings to combine the high sugar content of the noble cane's clone with the hardness and disease resistance of the other desirable species through a backcross breeding programme. This technique is commonly known as nobilization.

The breeding success in yield, quality and resistance improved by the significant efforts of eminent sugarcane breeders, namely C. A. Barber, T.S. Venkataraman and others at CBI, Coimbatore. Co 205 was the first inter-specific sugarcane variety released commercially in 1918. The tropical cane (*Saccharum officinarum*), commonly known as noble cane, has essential attributes like high sugar content, thicker stem, *etc.* These components were transferred to sub-tropical cane (Indian cane; *S. spontaneum*) through a breeding process popularly known as Nobilization. With the efforts of Nobilization, the variety Co 312 was released in 1928. It remained dominant for many years in sub-tropical farmers' fields. In addition, to yield and sugar advantage, the hybridization is made between normal cane and wild species (*S. spontaneum* var. Kans; Kans grass) to transfer disease-resistant attributes.

The variety of high cane yielding with high sugar recovery is popularly known as wonder cane. Recently a wonder variety Co 238 has released at Karnal centre in 2009 for the sub-tropical region. However, it has gained popularity in both sub-tropical and tropical regions of India.

Floral Biology, Selfing and Crossing Techniques

Inflorescence	:	The racemose type, it is an open-branched panicle or tassel. Based on appearance, it is known as 'Arrow'. It is long, featherlike and tapering–central stem and lateral (side) branch (Figure 24.1). The oldest flowers are at the bottom, and the youngest are at the top. The lateral branches bear rames (delicate, spikelet-bearing branches) and each rame has many pairs of spikelets.
Flower	:	The spikelets in sugarcane are primarily in pairs. However, the spikelets at the ends of the branches are in groups of three. Each spikelet pair includes one pedicellate (stalked) spikelet and one sessile (stalkless) spikelet; groups of three spikelets have one sessile and two pedicellate spikelets. Each spikelet has one fertile floret. Fertile lemma present or absent, Palea absent or minute, the fertile floret is **bisexual**, means it has both stamens (male or pollen-producing structures) and a pistil (female or ovule-producing structure).

		Flower Lodicules usually absence of hairy structure, rarely with a few sticky hairs 2–3 times as long as the spikelet, variable in shape, ranging from wedge shaped-bilobed to wedge-shaped-short edge.
Androecium	:	The fertile floret has three stamens
Gynoecium	:	Single superior ovary with bifid style and feathery stigma
Fruits	:	Small Caryopsis (about 1mm long)
Seed	:	Monocots
Floral formula	:	Br. ⊕ ⚥ $P_{2(Lodicules)} A_3 \underline{G}_{(1)}$

Figure 24.1: Sugarcane Inflorescences: An Arrow.

Anthesis: Anthesis commences early in the morning, between 5:00 AM and 6:00 AM . Maximum anthesis occurs between 6:00 AM and 8:00 AM. Protogyny condition - stigma protrudes out first, and anthers dehisce afterwards. The opening of flowers occurs from top to downwards. 7–10 days are required to complete the anthesis of an Arrow.

Selfing technique: In sugarcane, selfing is archived by covering an arrow with a lantern. Lantern is a farmwork or cage made from bamboo, covered with a muslin cloth or polythene paper. It prevents accidental cross-pollination. The lantern has to be supported by bamboo poles. The lantern has to be opened once a day, mostly during the afternoon hours, to reduce the temperature that my build up inside during the day. Lantern has to be retained in position till the seeds are set (Figure 24.2).

Figure 24.2. 'Selfing Technique' in Sugarcane.

Hybridization Technique

The crossing is challenging in sugarcane:

1. Spikelets are minutes. So, hand emasculation is not possible. Alternative methods of emasculation like hot water treatment can not possible
2. Self-sterility of both male and female organs is predominating in most cultivars.
3. Vegetative propagation is common practice. Some varieties seldom flower outside the tropics. Some varieties flowers once in 6 to 8 months and is highly controlled by photoperiods.

Emasculation is not practicable in sugar cane

Various morphological environmental factors include

- ✰ The small size of the flower
- ✰ A large number of flowers present in arrows
- ✰ Tall plant height
- ✰ A particular range of photoperiods is essential for flowering. It flowers round the year at equator, where the day length is constant for 12 hours.

Crossing Techniques

The crossing can be successful due to protogyny conditions. The crossing performed in sugarcane by various techniques, each described below:

1. Lantern Technique

Lantern covers the female plant before the initiation of anthesis. After initiating the anthesis, the arrow is cut from the desired male parent. This arrow is used to hybridize by shaking up with a female arrow inside Lantern. This may repeat for 2-3 days to achive more seed sets.

The difficulties that occur in field crossing in Lantern methods can be overcome by Scaffolding [a robust structure called 'Elevated Hybridization Runways (EHR)] (Figure 24.3).

Figure 24.3: 'Lantern Technique' a Robust Structure called 'Elevated Hybridization Runways (HER)'.

2. Rooted Stalk (Marcotting) Technique

At the time of flowering, cut around the stem and tie a polythene bag with a growth medium (nutrients). The roots arise near the cut area of the bud. This can be cut and used for hybridization purposes. This technique is called marcotting. It is used in TNAU, Coimbatore and Sugarcane Breeding Institute, Coimbatore.

3. Sulfurous Acid Technique (Hawaiian solution method)

A sulfurous acid solution keeps the inflorescence alive for several weeks. Cut both male and female flowering arrows along with a small portion of the stem. These cut ends are immersed in a bowl containing 0.01 per cent sulphuric acid and 0.01 per cent phosphoric acid. The cut end in the lab is brought nearer and performs cross-pollination. They absorb weak acids. Add weak solution daily to change the acid taken by the stem. Once a week, entirely change the solution. This is done for 25-30 days until the seed becomes mature.

The female plant is cut and kept in weak acid in the improved technique. At the time of flowering, the male parent can be brought nearer, and the pollen can be shed by shaking, as done in the Lantern method.

Sugarcane Breeding Work is Carried Out at Coimbatore Instead of Lucknow. Why?

Day length is nearly constant for 12 hours; warm nights, humid conditions and high rainfall favour flowering in sugarcane. Such conditions are found in Coimbatore. In contrast, cool weather and high altitude where the day length varies is a common situation of Lucknow that affects the flowering. So, all the breeding activities have been carried out at Coimbatore instead of Lucknow.

Flowering is Undesirable in Commercial Canes

Sugarcane stalk becomes hollow due to the uses of energy and sucrose; plant growth ceases and matures rapidly after the appearance of the flowering stalk. Cane and sugar yield reduction as the food is diverted towards growing inflorescence. The loss of apical dominance and subsequent development of side shoots leads to a reduction in the sucrose content in the stalk.

Breeding Objectives

1. **High cane yield and sugar content:** High cane yield and sugar content are the primary objectives of sugarcane. The yield attributes include stem height, thickness, tillering capacity, number of viable canes per plant, weight of individual cane, *etc.*
2. **Variety suitable for early to full season maturity:** When varieties attained juice quality, *i.e.*, juice brix (per cent) concentration >18 per cent, Polarization per cent (Pol per cent) in juice is >16 per cent and above and purity of juice >85 per cent levels attained ten months after planting is called early varieties, and those attaining sucrose maturity at

11–12 months are called mid-late varieties. Early maturing high sugar content varieties are ideal for crushing in the early phase of the crushing season. Some known examples of early varieties are CoC 671, Co 6806 and Co 997 for tropical regions, while for sub-tropical regions, are CoJ 64, Co 89003 and CoS 8436. To decrease the crop duration of sugarcane from 12 to 8 months, the sugarcane breeder attempted crosses between sugarcane and sorghum, and hybrids were also produced.

3. **Breeding for diseases resistance:** In India, major prevailing diseases are Red rot, Smut, Wilt, Mosaic, Ratoon: stunting disease and Grassy – shoot.
   - **Red rot:** It is a major problem in sub-tropical countries. The Resistance sources have been found in genotypes of *Saccharum officinarum*, *S. sinense*, *S. barberi*, *S. robustum* and S. *spontaneum* and related genera, *viz.*, *Erianthus*, *Sclerostachya*, *etc.* The numbers of resistance verities have been released, *viz.*, Co 98014, Co 1148, Co 1336, Co 6304, Co 0118, Co 5659, CoS 698, Co 0237, Co 0238, Co 0239 and Co 05009.
   - **Smut:** Serious disease in many sugarcane-growing countries. The resistance is governed by two dominant genes (*S1* and *S2*) also influenced by inhibitor and anti-inhibitor genes. The resistant sources have been identified from *S. spontaneum* and *S. officinarum* clones. The resistant commercial varieties in India are Co 449, Co 527, Co 853, Co 1148, and Co 1336.
4. **Breeding resistance/tolerance to insect pests:** The major prevailing pests in sugarcane are Shoot borer, Cane borer, Pyrilla, Mealybugs, White flies, Termites and White grub. Breeding multi-pest resistance is essential. *S. barberi* and *S. sinense* are key sources of pest resistance.
5. **Breeding tolerance to abiotic stresses:** Since the long growth cycle, *i.e.*, 10–12 months and prolonged and increased under varying climatic conditions, sugarcane crop gets more chances to be exposed to various abiotic stresses, such as drought, salinity, flooding, cold/frost, and high temperature.
6. **Breeding for quality improvement:** Factors in sugarcane quality are juice yield, millability, sugar content in juice, quality of juice *etc.* Sugar yield can be determined through juice yield and sugar content in juice, Total solids, a non-sugar fraction of the juice, and solids per cent in the juice (brix).
   - **Millability:** It refers to the characteristics of the canes that make it possible to recover sucrose from the stalk by normal extraction methods.

     Desirable characteristics associated with good milliability:

     1. Moderate hardness of rind (bark)
     2. Low-fibre: Sucrose ratio

3. Good fibres length
4. Long internodes
5. Good fibre quality for paper making

- **Bagasse:** The fibrous portion of sugarcane remains after the juice extraction. It comprises of two types of fibre: the long fibres in the rind and the shorter, softer fibres in the pith of the cane stem.

7. **Breeding variety for specific purposes:** Besides sugar production, sugarcane also breeds for the production of bioenergy, high biomass, and bioethanol. Sugarcane is an important bioenergy crop and source of affordable raw material for producing bioethanol and bagasse-based electricity. Stalk yield, raw sugar juice (molasses) and juice sugar content in sugarcane varieties are key components for ethanol production.

   Molasses: The thick syrupy residue left after the sucrose is removed from the clarified sugar juice (syrup). Molasses is the key source for ethanol production. According to quality variation for a specific purpose, sugarcane (energy cane) of two types

   - **Type I energy cane:** Moderate-to-high fibre content (<25 per cent) and low sucrose (>15 per cent brix). Used for dual purposes canes.
   - **Type II energy cane:** High biomass yield with high fibre (>25 per cent) and low brix (<15 per cent).

   Recent attempts have been made in India to produce Type I and II energy cane through a cross between *S. officinarum* × *Erianthus arundinaceus* and *S. officinarum* × *S. spontaneum*.

Breed Cultivars as per Needs

- ***Saccharum officinarum*:** For chewing due to the soft nature of the fibre
- **White sugar:** Hybrid having higher sucrose and moderate fibre content is preferred
- **Jaggery purpose:** Variety having high sucrose content and better cleanability are suitable
- **Ethanol production:** Variety having total sugar content are more suitable

Breeding Approaches (Methods)

1. Germplasm Collection, Introduction and Clonal Selection

Saccharum barberi is the native species of sugarcane in India. Several plants including various kinds of sugarcane were introduced from China. An introduced a new species of sugarcane called *S. sinense*. The earliest efforts to utilize the wild germplasm of sugarcane were undertaken in Java and India.

2. Hybridization

In sugarcane selection performed (*i.e.*, F_1 population) after hybridization, since the sugarcane plant is heterozygous, segregation occurs in F_1 generation.

Different types of crosses are used in sugarcane breeding programmes *viz.*,

i. **Biparental crosses:** Crosses between two known desired parental clones.

ii. **Area crosses:** In this programme, cross between male sterile female clones and one male parent in an isolated area.

iii. **Poly crosses:** Also known as Melting pot crosses, crosses several superior/potential parental cultivars in an isolated area through natural cross-pollination.

iv. **Recurrent selection:** Used for improving cane yield and quality separately or simultaneously

The interspecific hybridization in the *Saccharum* genus was initiated by Dutch plant breeders in the Java Island, around 1885. Initially, interspecific hybridization between *Saccharum officinarum, S. spontaneum.* Released of Co 205, the First Interspecific Sugarcane Hybrid. *S. barberi* proved successful during the early years of sugarcane breeding in India, leading to the development of several early varieties like Co 213, Co 244, Co 312 and Co 313.

3. Mutation Breeding

According to Heinz, X-ray-Irradiation to induce mutations in sugarcane were carried out in 1927. Several mutation breeding programmed with X-ray and gamma rays were initiated in India's early sixties. Mutation breeding in sugarcane aims to create economic mutants for higher cane yield, non-flowering and resistance to various diseases such as red-rot, smut, downy mildew and various insect borers. Gamma rays and chemical mutagens such as EMS are applied mostly on buds. The list of mutant varieties released in India are depicted in Table 24.2.

Table 24.2. List of Mutant Varieties in Sugarcane

Sl.No.	*Cultivars*	*Source of Mutagen*	*Characters Improved*
1.	Co 6608 mutant	Gamma rays (30-50 Gy)	Resistance to red rot
2.	Co 997 mutant	Gamma rays (30-50 Gy)	Resistance to red rot
3.	Co 8153	Gamma rays (150 Gy)	Improved juice quality and high yield
4.	Co 85035	Gamma rays (150 Gy)	Resistance to *Ustilago scitaminea*, high cane yield and high sucrose percentage

4. Biotechnology and Genetic Engineering Approaches

☆ **Tissue culture techniques:** Micropropagation: To *invitro* regeneration of plantlets well established and standardized in sugarcane. The callus induction from leaf whorl independent of genotype is well established in sugarcane. The in vitro culture system is mainly established for isolating somaclones, producing disease-free plants, and genetic transformation. Spontaneous occurrence of somaclonal variation was reported for the first time in sugarcane. A successful example is **'Ono'** variety developed from **Pindal** variety using Somaclonal selection. It is resistant to Fiji disease.

- **Genetic transformation:** Development of transgenic: Their success rate has increased with the advancement of genetic transformation. For example, insect resistance *via* the δ-endotoxin gene; disease resistance *via* genes encoding antimicrobial; peptides resistance to sugarcane mosaic virus (SCMV); Modified sucrose content *via* down-regulation of pyrophosphate-dependent phosphofructokinase, *etc.*
- **Molecular marker technology:** The molecular approaches larges used in sugarcane for the evolution of better varieties through selection of parents for crossing, use as an effective selection tool in the identification progenies and for characterization of germplasm and introgression of wild germplasm, DNA fingerprinting, varietal documentation and depicting genes of economic importance and their selective transfer to the desired genetic background.
- **Genomics and genome editing tools:** Due to polyploidy (euploid and aneuploid chromosome sets) and heterozygous constitution of genome complexity, the huge genome size (~10 Gb), make it hard to study at a genetic level. However, recent advancement of genome sequencing tools makes it possible to study sugarcane genomic level and comparative analysis. The accessibility of a whole genome sequencing (WGS) platform, genome sequencing and advanced genetic analysis models, such as genome-wide association studies (GWAS) and genomic estimate, presents prospects to develop complete datasets in sugarcane to hasten genetic gains in the sugarcane breeding programmes. In the case of genome editing tools, the utility of TALEN has been demonstrated in sugarcane to reduce lignin content for enhanced biofuel production.

Breeding Institutes

- **National level:**
    i. Sugarcane Breeding Institute (SBI), Coimbatore, Tamil Nadu
    ii. Indian Institute of Sugarcane Research (IISR), Lucknow, Uttar Pradesh
- **State level:** Regional Research Stations-State Agricultural Universities.

Breeding Achievements

Co-205, Co-791, Co-775, Co-6806, Co-6304, Co-7602, CoC-671, Co-7527, Co-62175, Co-8338, Co-87263, Co 86249 (Bhavani), Co 2001-13 (Sulabh), Co 238

(Karan 4), Co 98014 (Karan-1), Co 0118 (Karan 112), Co 99004 (Damodar), CoM 265 (Phule 265), CoC 671 (Vasant-1), Co 09004 (Amritha), Co 86032 (Nayana), Co 0238, Co 15023, Gujarat Sugarcane-1, Gujarat Sugarcane-2, Gujarat Sugarcane-3, Gujarat Sugarcane-4, CoLk 15201, CoLk 15207, Co 11015, Divyashi (GNS-12), Genda (CoSe 11453), Karan 17 (Co-17018), Roshan (CoS-16233), IKHSU-16, *etc.*

Section-V

Breeding Forages (Lucerne, Berseem and Oat)

Chapter 25

Breeding Forage Crops

Introduction

Forages, also termed fodder crops, are mostly cultivated for livestock feed. These crops are an essential component of agriculture, providing a reliable and cost-effective source of animal feed. Indian sub-continent is one of the world's mega centres of crop origin and crop plant diversity. They are grown for their high protein content, digestibility and palatability, which are essential for the health and productivity of livestock. Depending on the anticipated use, they can be grown as a mono-crop or as a mixed crop, along with cereals or pulses or as a cover crop. Many diverse crops are considered forage crops, and the primary group fall under the *Poaceae* and *Leguminosae* families. The Indian gene centre has rich genetic diversity in native grasses and legumes. The list of essential forage crops is depicted in Table 25.1.

Table 25.1. List of Essential Forage Crops

Sl.No.	*Crops–Common Name*	*Scientific Name*
1.	Lucerne/Alfalfa	*Medicago sativa* L
2.	Oat	*Avena sativa* L
3.	Berseem/Egyptian clover	*Trifolium alexandrinum* L
4.	Forage Sorghum and Sorghum × Sudan hybrids	*Sorghum bicolor* L. Moench
5.	Fodder Bajra/Pearl millet	*Pennisetum glaucum* L

Sl.No.	Crops–Common Name	Scientific Name
6.	Bajra Napier Grass-Hybrid	*(Pennisetum americanum* L. Leek) × (*P. purpureum* Schum.)
7.	Fodder Maize	*Zea mays* L
8.	Cowpea	*Vigna unguiculata* L. Walp
9.	Cluster bean/Guar	*Cyamopsis tetragonoloba* L. Taub
10.	Rice bean	*Vigna umbellate* Thumb Ohwi and Ohashi
11.	Indian bean/Field bean/Lablab bean	*Lablab purpureus* L. Sweet
12.	Guinea Grass	*Panicum maximum* Jacq
13.	Dharaf grass/Dhawalu grass	*Chrysopogon fulvus* Spreng. Chiov
14.	Anjan grass/Buffel grass	*Cenchrus ciliaris* L.
15.	Dhaman grass/Birdwood grass	*Cenchrus setigerus* Vahl
16.	Marvel grass	*Dichanthium annulatum* Forsk Stapf
17.	Barley/Jau	*Hordeum vulgare* L

Breeding Forage Crops and Challenges

Considering the prominence of forage crops in the agriculture and livestock industry, it becoming appropriate to pay more consideration to forage breeding and its improvement. In India, more than 50 plant species are being used as forage, of which nearly 30 species only pay attention to forage breeders for developing new potential cultivars. However, in the last few decades, a multi-disciplinary approach has led to the development of numbers of varieties in forages. The breeding procedure of forage crops are common like other field crops; however, some breeding constraints, *viz.*, diverse pollination behaviours and mechanisms, small flower size, irregular fertilization, and poor seed setting, and the perennial nature of crops.

The major breeding challenges are

- Poor seed setting and low seed viability.
- Most forage species are cross-pollinated in nature.
- Non-synchronous flowering and spikelet maturity, abscission of spikelets after maturity in grasses and the occurrence of a large number of sterile glumes in many grasses.
- It is more challenging to maintain true breeds, and there are problems associated with propagation due to heterozygosity.
- Vegetative and reproductive growth phases lapping.
- In many forage legumes, irregular pod setting, non-synchronous maturity and seed shattering is major problems.

- In some cases, self-incompatibility encounters a problem with producing homozygous lines.
- Artificial hybridization is tedious due to small flowers.
- Problems encountered in crossing and obtaining gene recombination are due to reproducing essentially through apomictic means.
- The forage yield performance may differ in the initial evaluation of selected plants or lines in the breeding trials and farmer fields.
- In some forages, plant stands are not quickly established due to weak seedlings.
- Some forage species may perform differently with different grazing management systems.
- Many forage species are polyploids, which increases their genetic complexity.
- Many years are required to evaluate the persistence and productiveness of new varieties/genotypes as the most forage crops are perennials in nature.

Essential Characteristics of Forage Crops Considered while Breeding

- **Better persistence:** The persistence of the herbage is markedly affected by biotic and abiotic stresses and excessive grazing. It is also associated with the vigour and growth habit of crops. However, in many crop species, it can be increased by agronomic management.
- **Better quality forage yield:** Besides biomass yield, quality is essential for forage crops. It comprises crude protein, *in-vitro* dry-matter digestibility, carbohydrate, fat, fibre, minerals and vitamin content. The low percentage of neutral detergent fibre and toxicity depends on the species' nature, agronomic practices, growth and harvesting stage.
- **Reproduction Conveniency:** The negative correlation between seed setting and high forage yield. Suitable breeding strategies need to be taken up for strategic forage improvement.
- **Management easiness:** Easy to establish in the field due to high seedling vigour. The ease of mix cultivation means there is no need to separate management practices.
- **Palatability:** Though it is not associated with nutritive value, it is one of the most essential features of forage crops. It decides the digestion of forage/fodder. Plant leafiness and succulents' traits are more important for better palatability.

Common Breeding Objectives of Forage Crops

1. Suitable growth habit – For grazing grasses dwarf types, while for hay, making tall growth habit more suitable.
2. Greater aggressiveness/fast growth and competing ability as solo cropping and in association with other crops, *i.e.*, legumes.
3. Resistance to biotic and abiotic stresses.
4. Ease of reproduction, *i.e.*, both seeds and vegetative type.
5. Elimination of undesirable features: Zero or minimum content of undesirable chemical constitutes or undesirable features/toxicity. For example HCN content in sorghum and Sudan grass, coumarin content in sweet clover, presence of awns and leaf shedding, dry and pithy culms, steaminess in grass, *etc.*

A. Breeding Lucerne (*Medicago sativa* L.)

Introduction

Lucerne, or alfalfa (Rajka), is an ancient crop and one of the oldest cultivated perennial forage legumes. It is often termed the 'queen of forages' as it covers major areas of cultivated grassland of the world compared to the other grasses and legumes due to its nutritional superiority. The cultivated species of lucerne is an autotetraploid (2n=4x=32) crop. It grows mainly in the temperate agro-climatic regions of the world. Due to its high protein content, it is mainly preferred for grazing, hay and silage production, and quality pasture for livestock. It has the highest productivity of feed protein per unit area. Also, it contains a good amount of all the macro and trace minerals and all the fat and water soluble vitamins, making it ideal for feeding to animals.

Background Information

1.	*Common name*	Lucerne, Alfalfa
2.	*Botanical Name*	*Medicago sativa* L.
3.	*Family*	*Fabaceae* (*Leguminosae*)
4.	*Chromosome number*	2n=4x=32
5.	*Centre of origin*	North-western Iran and north-eastern Turkey
6.	*Nature of pollination*	Cross-pollination (> 80 per cent)
7.	*Pollination mechanisms*	Herkogamy, Entomophily- Melittophily
8.	*List of wild relatives*	As per the list (Table 25.2)

Table 25.2. List of Wild Relatives and their Chromosome Number

Sl.No.	*List of Wild Relatives*	*Chromosome Number*
1.	*Medicago caerulea* (Progenitor of *Medicago sativa* L.)	2n=16
2.	*Medicago disciformis*	2n=16
3.	*Medicago laciniata*	2n=16
4.	*Medicago falcate*	2n=16,32
5.	*Medicago rugosa*	2n=30
6.	*Medicago arborea*	2n=32
7.	*Medicago prostrata*	2n=32
8.	*Medicago glomerata*	2n=32
9.	*Medicago glutinosa*	2n=32
10.	*Medicago hemicycle*	2n=32
11.	*Medicago polychroa*	2n=32
12.	*Medicago strasseri*	2n=32
13.	*Medicago scutellata*	2n=32
14.	*Medicago citrina*	2n=48

Different Varietal Types of Lucerne/Alfalfas

Based on morphological features and location/distribution, it is classified into different groups

Sl.No.	*Types*	*Features*
1.	Common alfalfas (*Medicago sativa* ssp. *sativa*)	✰ Flower Purple colour ✰ Mostly upright types ✰ Differ in winter hardiness
2.	Chilean alfalfas (Spanish)	✰ Cold tolerance and in fall and spring growth habits
3.	Turkistan alfalfas	✰ Distributed in southern Russia, Iran, Afghanistan, and Turkey ✰ Susceptible to leaf and stem diseases ✰ Resistant to some insects and crown and root diseases. ✰ Winter hardiness: Moderate to high.
4.	Flemish alfalfas	✰ Developed in northern France ✰ Vigorous and steamy, recover quickly after cutting and early type ✰ Resistant to certain foliar diseases, susceptible to root and crown diseases

Sl.No.	Types	Features
5.	Yellow-flowered alfalfas (*Medicago sativa* ssp. *falcata*)	☆ High resistant to cold ☆ Growth habits prostrate upright ☆ Broad crowns, creeping root habit ☆ Resistance to foliar diseases
6.	Variegated alfalfas (*Medicago sativa* ssp. × varia)	☆ Resulted due to natural hybridization between the common and yellow-flowered types ☆ The flower colour is variegated (includes brown, purple, green, greenish-yellow, *etc.*). ☆ High resistant to cold

Floral Biology, Selfing and Crossing Techniques

The flowers are typical legume types: Calyx 5 united; Corolla 5 – Standard 1, Wing 2, keel 2 free, stamens 10 (9+1 or 8+2, Staminodes in few), Gynoecium superior, monocarpellary. The petals partially join and enclose the stamens and pistil. Most legumes are self-pollinated and nectar is secreted at the base of the corolla tube. But in lucerne (*Medicago sativa*), pollen is dispersed by an short-tempered action known as ***tripping***.

Inflorescence	:	Raceme-Racemose type: axillary oblong, dense, short, pyramid-shaped clusters of 8–30 flowers.
Flower	:	Pedicellate, zygomorphic, irregular, complete, perfect, and pentamerous except for pistil, slightly perigynous, and papilionaceous.
Calyx	:	Tubular with five hairy lobes (sepals), persistent, gamopetalous, calyx tube oblique and sepaloid.
Corolla	:	Five petals [1+2+(2)], polypetalous, unequal size, vexillary aestivation, Papilionaceous
Androecium	:	Diadelphous stamens – (9) +1; one posterior stamen is free, but the remaining nine are fused to form an incomplete stamina tube that encloses ovary, anther-dithecous, basifixed and dorsifixed–the bursting of anther longitudinally by a short-tempered action commonly known as tripping.
Gynoecium	:	Monocarpellary, ovary- superior, unilocular, single ovule, Placentation- marginal, style – short and curved, feathery stigma covered with thin waxy membrane.

Fruits	:	Spirally coiled legume or pod (dehiscence formed by marginal placentation).
Seed	:	Ex-albuminous with two cotyledons.
Floral formula	:	⊕ ⚥ $K_{(5)}\, C_{1+2+(2)}\, A_{(9)+1}\, \underline{G}_{1}$

Selfing techniques: Generally, pollen is dispersed by a short-tempered action known as ***tripping***. Tripping action is essential for the seed set, it is accomplished by physical disturbance of flower by insects mostly honey bee and fast wind. Therefore, self-pollination is accomplished by covering the whole plant under a wire cage with honey bees.

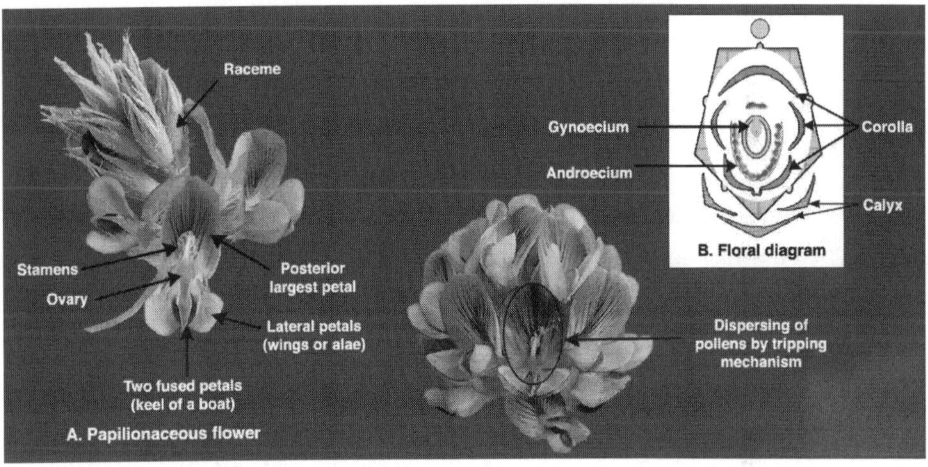

Figure 25.1. Floral Biology and Diagram of Lucerne (*Medicago sativa* L.).

Artificial Crossing and Pollination

- **Hand Emasculation:** The emasculation and hybridization are tedious due to the small flower structure. In the case of artificial hybridization, Select the healthy raceme and discard open flowers. Keep the flower bud (hooded stage), owing to open the next day. The bud is held at the base between the thumb and forefinger. The flower opened with the help of a pointer or forceps. The anterior sepal is exposed with the help of forceps, and the keel is pushed down or a small cut at the base of the keel petal and makes a cut along the central axis down to the distal end. The exposed anthers are then softly removed from the filaments. Tagging of emasculated bud by tying around the internodes below the emasculated bud with coloured cotton thread or a tag.
- **Alcohol treatment:** Select the healthy raceme and discard open flowers from it. Dip it in 57 per cent alcohol for 10 seconds. Then, wash it with tap water and rub on pollen grains from the desired plants.

☆ **Cage method:** Grow the male and female parents close or in different pots and tie both plants in one wire cage. Keep bees inside the cage to facilitate the outcrossing. Harvest seeds from both the parents.

The seed setting is higher in cross-pollination than self-pollination due to

1. Self-pollen is partly incompatible
2. Pollen tube development is faster with foreign pollen
3. In self-pollination, embryos abort and fail to grow customarily compared to cross-pollination.

Breeding Objectives

1. **Breeding for better forage yield**: Vigorous tall growth habit, high tillering and fast regeneration abilities after cutting with good persistence over the years are selected.
2. **Breeding for earliness:** The early flowering is essential for per-day fodder productivity.
3. **Breeding for high seed yield**: Besides better forage yield, cultivars with good seed yield are also essential for seed production.
4. **Breeding improves persistence:** It is markedly affected by biotic and abiotic stresses. The cultivar with disease and insect resistance, cold tolerance, drought tolerance, and salinity tolerance enhanced persistence ability. For example, perennial cultivar AL-3 (Anand Lucerne 3) gives palatable fodder continuously for 2-3 years.
5. **Breeding variety suitable for multicut:** Produce forage round the year, give the number of cuts, and give the juicy plant stature–for example, perennial cultivars AL-4 (Anand Lucerne 4) and TNLC-14 as CO-3.
6. **Breeding improves quality parameters**: It comprises high nutritive values, *viz.*, a good amount of crude protein content (18-22 per cent) and crude fibre (25-35 per cent) with high palatability. Better digestibility and lower saponins content (anti-nutritional factor, in which they cause a reduction of nutrient utilization and feed conversion efficiency).

Breeding Approaches (Methods)

1. **Plant Introduction:** The introduced genotype is directly or indirectly used in lucerne improvement. *e.g.*, 'Ladakh', a cold and bacterial wilt-resistant genotype, was introduced from USA to North India. It has the capacity for renewing growth after prolonged drought.
2. **Selection:** Natural and various artificial selection approaches, *viz.*, Mass selection and Recurrent selection, top cross (half-sib) selection, full-sib selection, and progeny selection (open pollination), have been used in lucerne improvement. For example, Cultivars CO-1, 'Grimm

451' developed from the mass selection. LL Composite-5 (LLC-5) was synthesized from 125 downy mildew resistance clones selected from Kutch areas of Gujarat state by recurrent selections followed for three cycles of mating. The cultivars Anand Lucene-2, Anand Lucern (AL-3), T-9, *etc.*, developed through open pollination mass progeny selection.

3. **Multiline varieties:** Multiline is a line developed by composite (mixing) seeds of isogenic lines or near-isogenic lines (NILs) in desired proportion. Original lines are maintained separately.
4. **Synthetic and composite varieties:** An open-pollinated population derived from an intercrossing of selfed inbred, maintained through routine mass selection procedures. Due to lack of pollination control the synthetic varieties are more important than hybrid. The poly cross approach is adopted to establish the desired synthetic. The cultivar' LL Composite-3 (LLC 3)' is synthesized from 20 local genotypes collected from Gujarat, and CO-2 is a poly cross derivative of CO-1. The 'Krishna (RRB-07-1)' variety was developed by selecting superior plants from diallel crosses made among Anand – 2, AL-3, RL-97-1, RL-88, Chetak, Anand-21 and Annad-22.
5. **Hybridization- Hybrid cultivar:** Intra-varietal and Interspecific hybridization (*e.g.*, a cross between *Medicago sativa* × *M. Falcata* to exploit winter hardness and vigour); introgression of *M. arborea* genome into *Medicago sativa* for disease resistance, *e.g.*, Anthracnose.
6. **Mutation breeding:** Mutagenesis has improved specific traits and adaptation. The mutant lines' Xinmu 1' and 'Nongjing 1' are identified in China.
7. **Biotechnological approach:**
   - ☆ **Transgenic approach:** Genetically modified cultivar-Roundup Ready alfalfa has been developed in California for resistance against broad-range herbicide glyphosate. Another example is regulating Pro-Anthocyanidins (TaMYB14) in leaves, which enhances dry matter yield.
   - ☆ **Marker-assisted selection and Genomic selection:** Offer a more rapid breeding cycle and precise selection over conventional selection.
   - ☆ **Somatic hybridization:** This approach is used to introduce traits from wild relatives, for instance, synthesis of somatic hybrids of *M. sativa* with the annual wild species *M. rugosa* and *M. scutellate* through the protoplast electrofusion technique.

Breeding Achievements

AL-3 (Anand Lucerne-3), Anand Lucerne-4 (AL-4), Anand-2 (GAU-L-1), Anand-3, Chetak (S-244), CO-1, CO 2, GAU-L-2 (SS-627), LL Composite-3 (LLC-3), LL Composite-5 (LLC-5), RRB-07-1, RL-88 (RLS-88), Type-9 (T-9) and TNLC-14 as CO-3, BAIF Lucerne-5, IGFRI-DL-5, *etc.*

B. Breeding Berseem (*Trifolium alexandrinum* L.)

Introduction

Berseem is an important long-duration *rabi* legume fodder in India. It is also considered the 'King of fodder crops.' It is well adapted to semi-arid conditions. It is the principal green forage for animals. Due to its high regeneration ability, it yields four to eight cuts, enabling a constant feed supply to animals. Its fodder is succulent and has high palatability with rich nutritional values. The fodder quality is better as it contains 20 per cent protein and has good digestibility (>65 per cent).

Background Information

1.	*Common name*	Berseem/Egyptian clover
2.	*Botanical Name*	*Trifolium alexandrinum L.*
3.	*Family*	*Fabaceae* (*Leguminosae*)
4.	*Chromosome number*	2n=2x=16
5.	*Centre of origin*	Mediterranean region
6.	*Nature of pollination*	Predominantly Self- pollination, but cross-pollination also occurs (<10 per cent)
7.	*Pollination mechanisms*	No self-sterility, but tripping is essential for a good seed set
8.	*List of wild relatives*	As per the list (Table 25.3)

Table 25.3. List of Wild Relatives and their Chromosome Number

Sl.No.	*List of Wild Relatives*	*Chromosome Number (2n)*
1.	*Trifolium dubium*	14
2.	*Trifolium incarnatum*	14
3.	*Trifolium pratense*	14
4.	*Trifolium fragiferum*	16
5.	*Trifolium hybridum*	16
6.	*Trifolium resupinatum*	16
7.	*Trifolium subterranum*	16
8.	*Trifolium repens*	32
9.	*Trifolium ambiguum*	16,32,48

Breeding Objectives

1. Breeding for high adaptability.
2. High persistence in annual multicut forages, *e.g.* Jemini variety of berseem.

3. Variety with high dry matter yield: The variety gives a good yield when cut in the early stage of development, *e.g.* the genotype of Italian origin.
4. Breeding cultivars with better quality parameters include crude protein content, in-vitro dry-matter digestibility (IVDMD), low percentage of neutral detergent fibre and toxicity. However, the nutritive value is influenced by the nutrient concentration of morphological fractions, which is an important cutting treatment.
5. Breeding cultivar tolerance to abiotic stresses *viz.*, adverse soils – acidic/ saline soils, high rainfall and low moisture system.
6. Breeding variety resistance to various biotic stresses, *viz,* disease and insect pests, for example, genotypes B1, B2 and B3 are resistant to bean yellow mosaic potyvirus).

   Berseem cultivars are susceptible to the various diseases, *viz.,*

   - ✰ Root rot caused by *Rhizoctonia solani* and *Fusarium semitactum*.
   - ✰ Stem rot is caused by *Sclerotinia trifoliorum*; leaf blight is caused by *Epicocum* sp.
   - ✰ Powdery mildew is caused by *Oidium* spp.; Downy mildew is caused by *Perenospora trifolif.*

   These diseases can affect the health and productivity of berseem cultivars.
7. Breeding cultivar for superior summer persistence.
8. Breeding variety that has a high response to input.
9. Breeding good quality of fodder having physical, chemical, high nutritional and marketing value.

Breeding Approaches (Methods)

1. **Introduction** – The crop was introduced in India. Initially, two stable genotypes *viz.* Mescavi and Fahli were introduced in 1904. JBSC-1 Selection from EC 318954 introduced from Germany.

   The main weak point in berseem genetic improvement is the lack of available genetic variability. A narrow genetic base and although various breeding approaches like selection, polyploidization, mutation, wide hybridization, *etc.* have been used.
2. **Selection methods:** Initially, many cultivars were developed through selection methods. Examples: BL1, BL 2, Jawahar Berseem 1 (JB 1), JBSC-1, Hisar Berseem 1 (HFB 600), *etc.*

   - ✰ Mass selection: Superior single plants are selected based on the first cut, *e.g.* Bundel Berseem 2, Bundel Berseem 2.
   - ✰ Recurrent selection: It is a cyclic selection approach. The variety "JB 5" was developed.

☆ Composite selection: The variety developed by composite different lines, *viz.*, UPB-101,102,103,105,106,107 and 109.

3. **Polyploidy Breeding** - An essential revolution in berseem breeding in India was achieved by the induction of polyploidy. "The varieties developed through polyploidy breeding are "Pusa Giant" and "Bundel Berseem 3 (autotetraploid variety).
4. **Wide Hybridization:** *T. alexandrinum* (2n = 16) was successfully hybridized with *T. berytheum* (2n = 16) and *T. salmoneum* (2n = 16) and found the utmost feasible parent. Interspecific hybrids progenies exhibited introgression of various desirable characteristics, *viz.*, late flowering and resistance to diseases like root rot and stem rot.
5. **Mutation Breeding:** The genetic variation in berseem also genereated through various chemical and physical mutagens. Cultivars, *viz.*, BL 22, BL 42, BL 180, HB 2 were developed through mutation breeding.

Breeding Achievements

BL-1, Mescavi, Jawahar Berseem-1, Pusa Giant, BL-10, BL-180, Bundel Berseem – 2, BL-2, BL-22, BL-42, (JHB-146), Wardan (S - 99 - 1), Bundel Berseem – 3 (JHTB-96-4), Hisar Berseem – 1 (HFB – 600), HB 2, JB-5, UPB – 110, JSBC-1, BL 45, Bundel Berseem-7, BL 46, Jawahar Berseem 07-15

C. Breeding Oat (*Avena sativa* L.)

Introduction

Oat (*Avena sativa* L.) is an important winter cereal worldwide. The crop is widely cultivated for food, feed, and fodder. The bulk of the oats produced worldwide are used extensively as feed for livestock (cattle, sheep, poultry and horses), and only 17 per cent of the world's production (grain) is used as human food. The fibre of oats grain has unique physiological properties. Fibre content in oats reduces serum cholesterol and helps to promote a more normal blood glucose level in people with diabetes. Oat straw has a similar proximate analysis to other cereal straws. However, it has a higher metabolizable energy (ME), which confirms the traditional view of its superior feeding value. It is a quick-growing, palatable, succulent and nutritious crop. Due to the congenial climate, it is an important rabi cereal fodder crop mainly cultivated in North-western, Central and Hilly parts of India. Other forages like berseem clover and alfalfa remain dormant in hilly areas during winter, whereas oat can be grown as a main alternative forage crop because it has drought and cold tolerance ability. Its popularity as a fodder crop in the region is increasing because of outstanding growth habits, fast re-growth and high nutritive value for animals.

Background Information

1.	*Common name*	Oat
2.	*Botanical Name*	*Avena sativa* L
3.	*Family*	*Poaceae (Gramineae)*
4.	*Chromosome number*	2n=6x=42
5.	*Centre of origin*	Asia and Europe
6.	*Nature of pollination*	Predominantly Self- pollination
7.	*Pollination mechanisms*	Bisexuality and cleistogamy
8.	*List of wild relatives*	As per the list (Table 25.4)

Table 25.4. List of Wild Relatives and their Chromosome Number

Sl.No.	*List of Wild Relatives*	*Chromosome Number (2n)*
1.	*Avena atlantica*	14 (2x)
2.	*Avena canariensis*	14 (2x)
3.	*Avena agadiriana*	28 (4x)
4.	*Avena insularis*	28 (4x)
5.	*Avena fatuva*	42 (6x)
6.	*Avena sterilis*	42 (6x)

Breeding Objectives

1. Breeding for high yield.
2. Breeding for winter hardiness.
3. Breeding for wide adaptability under varying soil and abiotic stresses.
4. Breeding for lodging and shattering resistance.
5. Breeding for disease resistance.
6. Breeding for insect resistance.
7. Breeding for better fodder quality.

Breeding Approaches Methods

1. **Introduction** – The most basic method in oat breeding. Initially, many lines were introduced from different countries. For example, the Algerian genotype is from Algeria. The famous cultivar 'Kent' was selected from the USA-introduced material.
2. **Selection** – Pure line and mass selection performed. The cultivar Palampur-1 pure line selection is from Algerian materials, and the Phule Haritha (RO-19) was selected from the Kent variety. Bundel Jai -

851 (JHO-851) is a single plant selection from the material introduced from Japan. The variety UPO – 94 was a single plant selection from the American material (OGP73 – M 94). A mass selection approach was used for developed the cultivar Narendra Jayee-1 (NDO-1).

3. **Hybridization:** Pedigree and interspecific hybridization have been followed. Many cultivars have been developed by hybridization followed by pedigree selection. For example: Haryana Javi-8 (HJ-8), Jawahar Oat 03-93, JO 2, NDO 1101, Pant Forage Oat-4 (UPO-06-2), *etc.* Bundel Jai-2004 (JHO-2000-4) was developed from a derivative of the cross JHO 851 (*A. sativa*) × *A. maroccana*-16/30B [Interspecific hybridization + polyploidy]
4. **Multiline**: Multiline breeding approach used for developing disease resistance lines. *e.g.* crown rust resistance.

Breeding Achievements

Algerian, Kent, Bundel Jai -2001-3 (JHO-2001-3), Haryana Javi-8 (HJ-8), Bundel Jai-2004 (JHO-2000-4), JO-2, NDO -1101, Bundel Jai-991 (JHO 99-1), Jawahar Oat 03-93, Narendra Jayee-2 (NDO-2), Palampur-1, Sabazar (SK0-7), UPO – 212, Bundel Jai 2010-1 (JHO-2010-1), JHO-2009-1, OS-346, OS-377, RO-11-1, OL-1769-1, SKO-225, Pant Forage Oat-3 (UPO-06-1), SKO-96, OL-1802, Phule Haritha (RO-19), Pant Forage Oat-4 (UPO-06-2), HFO 806, OL 1874-1, OL 16, Narendra Jaypee- 1101, JWGO-01, HFO 906, Jawahar Oat- JO 07-310, *etc.*

Forages Breeding Institutes

- **International level:** The Alliance of Bioversity International and the International Center for Tropical Agriculture (CIAT), Rome, Italy.
- **National level:** Indian Grassland and Fodder Research Institute (IGFRI), Jhansi, Uttar Pradesh.
- **State level:** Regional Research Stations-State Agricultural Universities.

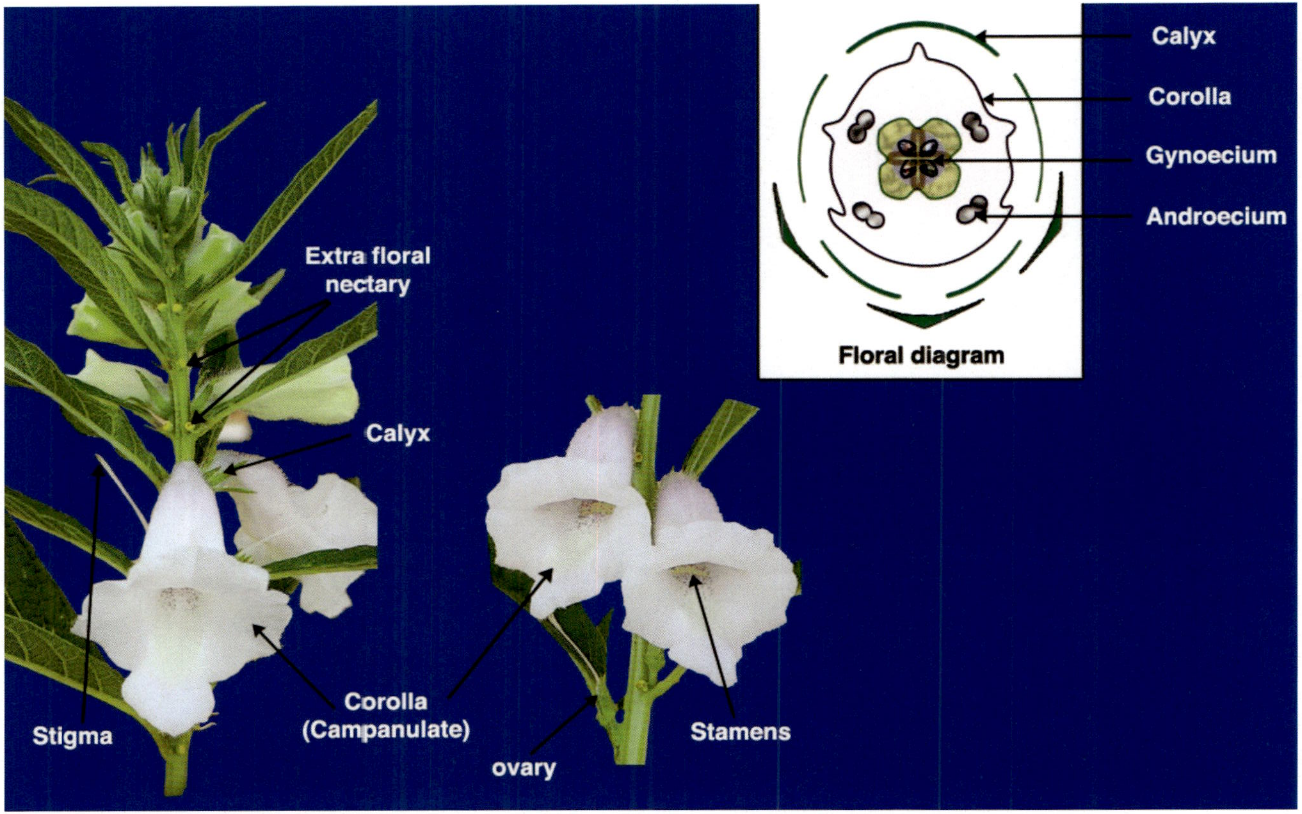

Figure 21.1. Floral Biology and Diagram of Sesame (*Sesamum indicum* L.). (pg. 218)

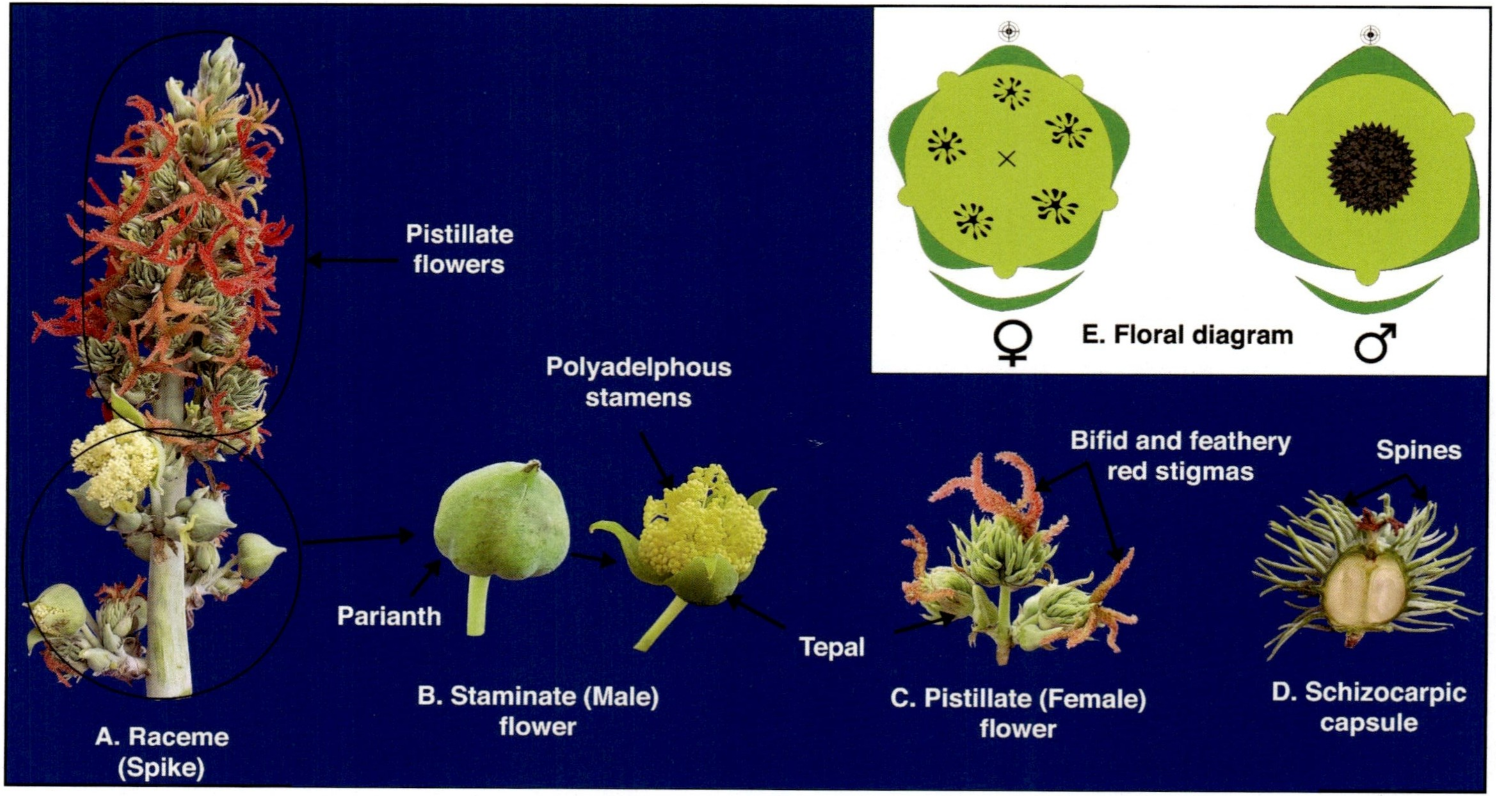

Figure 22.1. Floral Biology and Diagram of Castor (*Ricinus communis* L.). (pg. 225)

Figure 22.2. Basic Sex Form (Sex Expression) in Castor. (pg. 227)

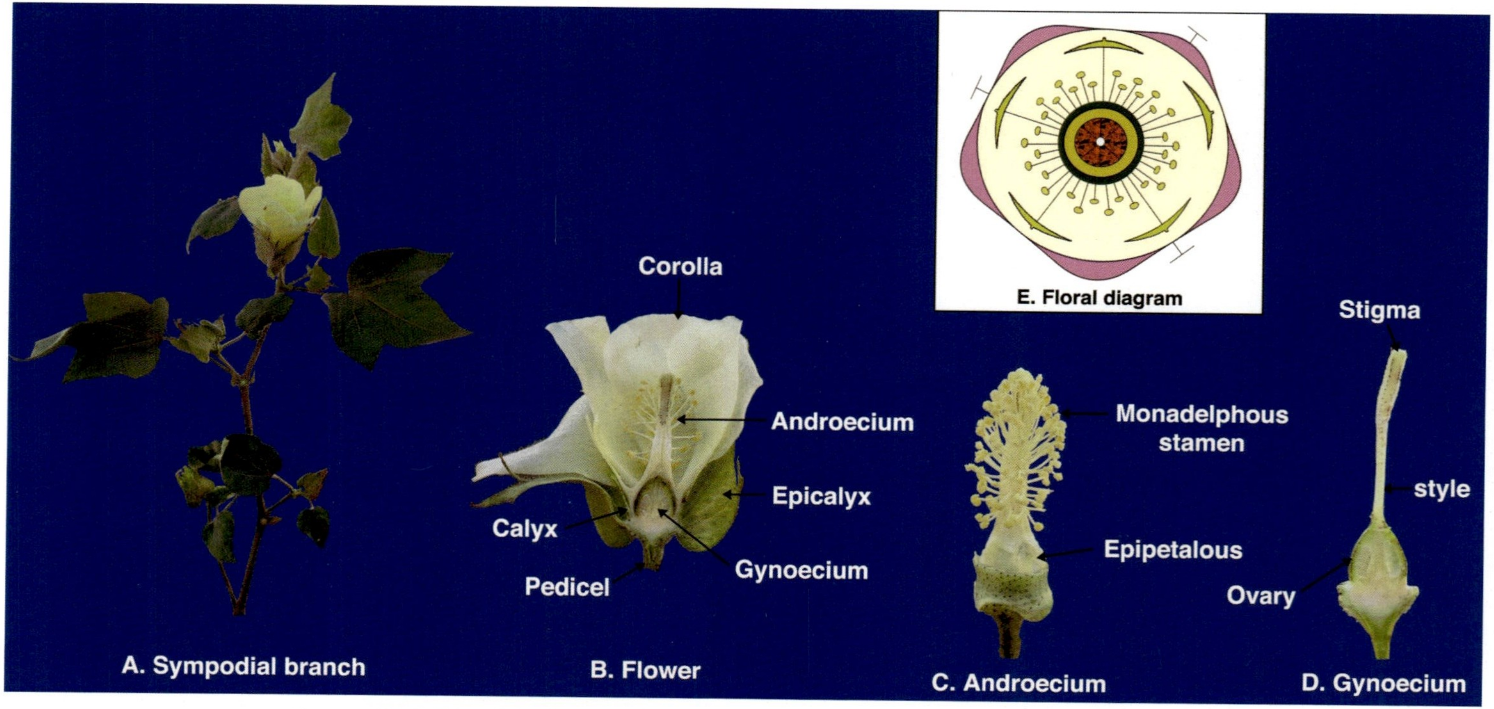

Figure 23.2. Floral Biology and Diagram of Cotton (*Gossypium* spp.). (pg. 242)

Figure 24.1: Sugarcane Inflorescences: An Arrow. (pg. 257)

Figure 24.2. 'Selfing Technique' in Sugarcane. (pg. 258)

Figure 24.3: 'Lantern Technique' a Robust Structure called 'Elevated Hybridization Runways (HER)'. (pg. 259)

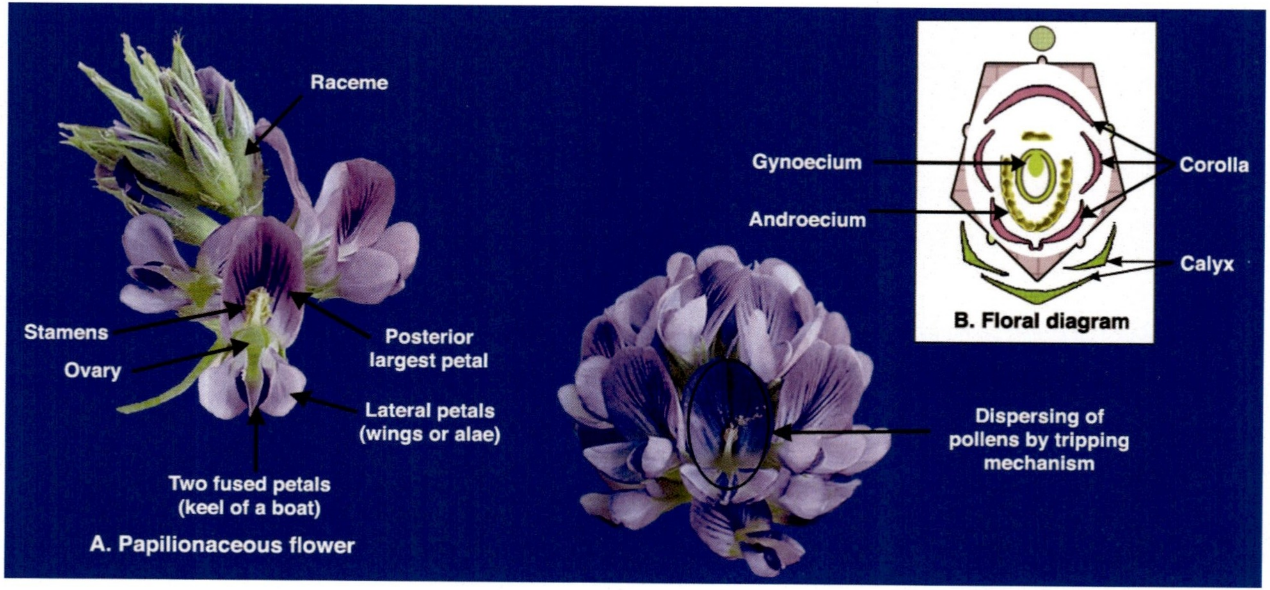

Figure 25.1. Floral Biology and Diagram of Lucerne (*Medicago sativa* L.). (pg. 275)

Appendices

Appendix-I

Important Terminologies of Crop Breeding

A

Abiotic stress: Any unfavourable condition for a plant's normal growth and development caused by non-biological agents such as high/low temperature, water deficiency, excessive moisture, toxicity or deficiency of element, salinity, alkalinity, acidity, and various pollutants.

Accession: A discrete germplasm sample (*i.e.,* cultivar, population, mutant line, breeding lines, *etc.*) which represents the total genetic variation of plant species maintained for conservation and evaluation.

Acclimation or Acclimatization: A process by which a plant species adapts, adjusts or modifies itself to survive and reproduce in a variable environments or stressful conditions over generation or locations. In other words, The ability of new plant material to adjust itself and naturally perform in the new environment.

Active collection: In germplasm conservation, seed storage for medium-term (usually 10 to 15 years) through 5-8 per cent moisture. It is used for regeneration, multiplication, evaluation, documentation, and distribution. Further, it can be maintained in sufficient quantity to be accessible on request.

Adaptability: The ability for adaptation; capacity of individual plant or populations to adjust to varying environmental conditions and perform typically like in local (normal) environmental conditions.

Adaptation: The adjustment of an organism or plant species to different environment or stressful condition.

Advance breeding line: The parental line developed from germplasm stock by various breeding approaches. It is further used to develop new varieties or hybrids.

Advance trait: A trait that differentiates a plant from an ancestral trait or kind, *i.e.,* brittle rachis of wheat.

Adventitious embryony or adventive embryony: A form of apomixis in which an embryo develops from the somatic tissue of the ovule, *e.g.,* as nucellar, integumentary or chalaza tissue.

Agamic complex: A form of asexual reproduction where seed formation occurs partially or entirely by asexual means.

Agamogenesis: Lack of participation of male gametes in the reproduction process. So, reproduction takes place without male gametes.

Agamogony: It is one type of apomixis where abnormal meiosis occurs during spore formation in the ovule, resulting in a diploid embryo-sac instead of a normal haploid.

Agamospermy: It is a form of apomixis where seed development occurs through sporophytic tissue, *i.e.*, adventitious embryony.

Albino: Plant lacking pigmentation, *i.e.*, chloroplasts showing white or light-colour.

Alien germplasm: The sum of genes introduced from a wild relative or distinct or non-adapted species.

A-line: In a male sterility system, the male sterile line which is used as a female line (seed-bearing parent) in hybrid seed production.

Allogamous: Cross-pollinating or fertilizing plant species

Allogamy: Pollen grain transfers from one flower to another flower of different plants. It is also referred to as cross-pollination/cross-fertilization. In other words, Sexual fertilization followed by seed development through pollen belonging to the different flowers borne on different plants.

Alloplasmic CMS: Cytoplasmic male sterility which arises due to the fusion of diverse cytoplasm through intraspecific, interspecific, or intergeneric hybridization.

Allopolyploidy: Combination of chromosome sets (genome) from two or more different but compatible species—generally produced by chromosome doubling in F_1.

Allotetraploidy (4x): A particular species having four replicas (copies) of two different but compatible chromosome sets (genome) derived from two different parental species.

Ambisexual or Bisexual or Ambosexual: The plant possessing reproductive organs of both sexes (male and female).

Amphidiploid: The interspecific hybrid which received a diploid set of a chromosome from each of two diploid species, and such species behave like normal diploids. *e.g.*, *Brassica* species.

Amphimixis: Normal fusion of male and female gametes in sexual reproduction. It is the opposite of apomixis.

Ancestor species: The original species of plant from which progenies have descended or evolved.

Androdioecious: A plant species in which hermaphrodite and male flowers are borne on different plant.

Androecious: Plant bearing only male flowers.

Androgenesis: Male nucleus (sperm nucleus) is responsible for forming a haploid embryo, while the female nucleus (egg nucleus) is inactive or degenerated. In other words, the haploid embryo arises directly from male gametes shortly after piercing into the female gametophytes.

Andromonoecious: A plant species in which hermaphrodite and male flowers are borne on the same plant.

Angiosperm: It covers all flowering plants in which seeds develop inside ovary.

Anther culture: It is one of the tissue culture techniques where haploid plant or callus formation occurs through anther or pollen grain *via* tissue culture techniques.

Anther: In the flowering plant, the terminal portion of a stamen bearing pollen.

Anthesis: One of the events in a flowering plant where dehiscence of a pollen grain occurs from anther.

Antixenosis: The host is unsuitable or unattractive for insect pests for the purpose of shelter, colonization, reproduction, *etc* and thereby exhibit resistance against such insect pests.

Apomeiosis: The formation of the embryo sac without the process of meiosis.

Apomictic: A plant that forms asexual progenies.

Apomixis: A modified form of asexual reproduction where no mixing (union) of gametes occurs, and the seed develops from unfertilized embryos.

Apospory: A non-reduced embryo sac originating from a somatic cell, *i.e.*, the integument or nucellus instead of the embryo sac mother cell.

Artificial (induced) mutation: Mutation generated by treating with particular mutagenic agents.

Artificial selection: Selection, which favours those traits which are valuable and desirable to humankind.

Asexual reproduction: A form of reproduction where male and female gametes do not unite to form a zygote, *i.e.*, vegetative propagation and apomixis.

Asexual: Male and female gametes do not unite in the reproductive process.

Autogamy: Transfer of pollen grains from anther to a stigma of the same flower or different flower of the same plant. In other words, Sexual fertilization followed by seed development through pollen belonging to the same flower or of different flowers borne on the same plant. It is also referred to as self-fertilization or self-pollination

Automixis: After accomplishing megasporogenesis, the embryo developing due to the union of two haploid nuclei.

Autonomous apomixis: Seed formation by asexual means without pollination. It is also referred to as parthenogenesis.

Autoplasmic CMS: Cytoplasmic male sterility which arises due to spontaneous mutation in the mitochondrial genome of a species.

Auto-polyploidy: Polyploidy originating by multiplying identical sets of chromosomes (genome) of the same species.

Autotetraploidy (4x): four replicas (copies) of the same sets of chromosomes (genome).

Autotriploidy (3x): Three replicas (copies) of the same sets of chromosomes (genome).

Average heterosis: Superiority of F_1 (hybrid) over mid-parent (mean value of both parents involved in hybridization) value. It is also termed mid-parent heterosis or relative heterosis.

Avirulent: A pathogen strain incapable of causing disease in a host plant due to specific resistance.

Avoidance: It is one of the mechanisms of stress resistance. An escape from various stresses can be accomplished either due to earliness or changing the planting date.

B

Backcross breeding: A breeding approach in which repeated back crosses are made in F_1 and subsequent generations to improve specific character in agronomically well-adapted variety.

Backcross: In plant breeding, the cross of F_1 generation or segregating generation with either of parents (dominant or recessive). *e.g.*, Rr × RR or Rr × rr. The back cross may or may not be considered as a test cross.

***Barnase*:** A gene derived from *Bacillus amloliquefaciens* encoding RNase to induce male sterility in the plant.

***Barstar*:** A gene derived from *bacillus amloliquefaciens* encoding a specific inhibitor of barnase gene and ultimately produces fertility restoration in barnase-inducing male-sterile plants.

Better parent heterosis: The hybrid is superior to the better parent. It also refers to heterobeltiosis.

Binucleate: Cells having two nuclei.

Biofortification: Plant breeding plays an essential role in improving nutritional status or modifying quality traits by enriching essential micronutrients, vitamins, antioxidants, reducing anti-nutritional factors, bitterness, or improving the flavour in several foods.

Biotechnology: The application of scientific and engineering principles in manipulating biological agents at the cellular-molecular level for altering its trait for advantageous use of humankind.

Biotic stress: Any unfavourable condition for a plant's normal growth and development caused by biological agents such as fungi, bacteria, viruses, phytoplasma, insect pests, and parasitic weeds.

Biparental mating design: One of the population improvement approaches in which plants are randomly selected from segregating generation (F_2 or onward) of a cross between two lines possessing contrasting traits.

Bisexuality: The flower is said to be perfect as it has both sex organs, *i.e.*, male and female organs. It is also referred to as hermaphrodite.

B-line: In the male sterility system, a fertile male line is used as a pollen parent to maintain the male sterility in A-line (male sterile). It is also known as the counterpart of the A-line. It lacks a fertility restoring gene.

Bloom: The white-waxy powdery substances present on the surface of leaves, stems, or ears, *e.g.*, castor, cereal crop. It is often used as a marker trait.

Blooming: The periods during florets are open, and anther comes out from it, *i.e.*, grasses.

Breeder seed: Seed or vegetative propagating material produced by a concerned breeder or the institute. It is a progeny of the nucleus or breeder seed stage-I, and it is used for source of foundation seed.

Breeder: The person who engages in research associated with breeding activity.

Breeding line: Group of plants with similar traits or combination of traits selected for breeding a new variety or hybrid or directly released as a new variety and may be further used in the breeding program.

Breeding method: Various breeding approaches which are used to improve or breed new plant varieties or hybrids.

Breeding: It is a combination of art with science and technology that deals with modifying and improving hereditary traits of an organism.

***Bt* gene**: A gene isolated from bacteria *Bacillus thuringiensis* used to transfer in various crops, *i.e.*, cotton, tobacco, brinjal, maize, *etc.*, thereby making it resistant against various insect pests.

Bud pollination: Selfed-seeds can be produced in self-incompatible species by temporarily suppressing incompatible effects by pollinating at the bud stage.

Bud selection: A form of clonal selection where mutant bud is selected for further propagation.

Bulbils: Reduced, modified flower (axillary fleshy bud). It is a vegetative body that skips fertilization, followed by seed formation.

***Bulbosum* technique:** The haploid production technique used in barley through an interspecific cross between *Hordeum bulbosum* × *Hordeum vulgare,* commonly known as *bulbosum* techniques developed by Kasha and Kao in 1970.

Bulk breeding or bulk method: The method used to handle segregating generation in the self-pollinated crop where plants of F_2 and subsequent generations are bulked with natural selection. This is done until the desired level of homozygosity is achieved. At the end of bulking period after individual plant selection is carried out to produce pureline variety.

Bulk segregants analysis (BSA): A molecular breeding approach used to reveal specific markers for particular traits, *e.g.*, a gene for disease resistance.

Bulk selection: A selection procedure of self-pollinated crops where the segregating population continues bulk with or without selection up to F_6 or F_7 generation.

C

Canola: A rapeseed cultivar having less than 2% erucic acid and less than 30 m mol/g glucosinolates content in oil. It is also referred as double zero rapeseed to denote both low erucic acid and low glucosinolates.

Centre of diversity: The geographical area or local region where cultivated plant species and their wild relatives exhibit more extraordinary genetic diversity than anywhere in rest of the world.

Centre of domestication: The geographical area where it is believed that a particular plant species was cultivated for the first time.

Centre of origin: The geographical areas where crop plants have originated.

CGIAR: The Consultative Group on International Agricultural Research is the world's largest global agricultural innovation network.

Chalaza: The opposite part of the micropylar end and acts as the basal part of the ovule.

Character/Trait: Any morphological, biochemical, behavioural or anatomical attributes (features) of an organism resulting from G × E interaction and may display heritable variation.

Chasmogamy: A situation where bisexual flowers open only after pollination is over.

Check (control): Row or plot of standard variety included in particular crop trials to evaluate genotypes/lines under testing.

Chemical hybridizing agent: Also referred to gametocides or pollen sterilant. Chemicala which induces male sterility by killing or inactivating male gametes without disturbing or adversely affecting the female reproductive organ.

Chemically induced male sterility: Male sterility induced by chemical hybridizing agents (CHA), various chemicals that kills pollen or male gametes.

Cleistogamy: It is one of the mechanisms of autogamy in which pollination and fertilization occur in an unopened flower, which means the flower never opens at all, *i.e.*, no anthesis occurs.

Clonal degeneration: Due to mutations or some infection, the vigour and performance of a clone declines with time.

Clonal selection: The method of selecting clones from mixed vegetative propagules population. It skips sexual reproduction, *e.g.*, Sugarcane, potato *etc.*

Clone: A group of plant-produced through asexual means from a single plant, so it is an exact copy of an identical genome from where it is derived.

Combination breeding: The breeding method used to transfer one or more traits into a single variety/genotype from another variety/genotype, *i.e.,* backcross.

Commercial hybrid: First generation of hybrid that is commercially cultivated on farmer's field.

Composite variety: Variety constituted by a seed mixture of several phenotypically promising open-pollinated superior lines without testing combining ability.

Conventional hybrids: The hybrid developed from breeding two or more homozygous lines (inbred/pure lines) through conventional means.

Crop improvement or Plant breeding: Different techniques of plant breeding comprising scientific principles, art and technology, used to improve or modify the heredity of crop plants as per human needs.

Cross incompatibility: A functional pollen of one species cannot pollinate or fertilize the female of another species.

Cross pollinated crop: Crop plants representing up to 90 % or more cross-pollination; it also represents the allogamous crop.

Cross: Mating or hybridization between two or more individual lines with different genotypes to achieve genetic recombination, *i.e.,* F_1 or segregating generation.

Cross-pollination: Transfers of pollen grain from one flower to another in different plants. It is also referred to as allogamy.

Cultivar: A population of cultivated crop plants related by ancestry is produced by scientific breeding or standard selection method and has a common genetic trait for distinctiveness and reproducibility.

Cytoplasmic Genic/Genetic Male Sterility (CGMS): Male sterility controlled by cytoplasmic genes, while the nuclear gene, often termed as the restorer gene, is available to restore fertility.

Cytoplasmic inheritance: Inheritance governed by extra-chromosomal gene carried by cytoplasmic organelles, *i.e.,* mitochondria, plastid. Also known as non-Mendelian inheritance.

Cytoplasmic male sterility (CMS): Male sterility controlled by cytoplasmic genes, *i.e.,* mitochondrial genome. It is maternally inherited.

D

Dee-geo-woo-gen: Semi-dwarf rice varieties containing a recessive gene '*sd1*' responsible for dwarfness in rice

Derivative hybrid: Hybrid developed by a particular cross between two hybrids.

Detasseling: During hybrid seed production of maize, mechanically removing/ detaching tassel from the female parent.

Determinate growth: An inflorescence, the terminal flower becomes open first, which ends the further growth of the floral axis.

Dichogamy: A situation in plants where both male and female organs mature at different times.

Dioecism (Dioecious): A condition in which the Staminate and pistillate flowers are borne on a different plant.

Dioecy: A form of the unisexual condition that represents the dioecious state of the plant.

Diploid (2n): Somatic chromosome number of any individual species. It is represented by '2n'.

Diploid cell (2n): Cell containing a pair of chromosomes, each received from male and female parents.

Diploid parthenogenesis: Embryo develops from a diploid egg cell without fertilization.

Diploidization: It is a natural or induced mechanism in which polyploid species act as diploid ones.

Diplospory: The non-reduced embryo sac develops from a megaspore mother cell (an archespore cell), bypassing the process of meiosis.

Direct embryogenesis: A tissue culture technique where an embryo develops directly from somatic embryos excluding the callus phase.

Direct introduction: A kind of plant introduction where an introduced plant species is directly adopted in the new environment without any changes.

Directional selection: Natural selection favouring only an extreme phenotype over other types, *i.e.,* tall and dwarfness in the plant.

Disassortative mating: Mating between different plants.

Distant hybridization (Distant/wide cross): A hybridization (cross) performed among different species of the same genus or between two different genera of the same family.

Domestication: Wild species brought under human supervision and converted to the usual cultivated crop via a selective breeding approach for many generations according to human needs.

Donor parent: It also refers to non-recurrent parent. The parent which is used only once in the backcross method for transferring desirable genes into the recipient parent.

Double cross: Cross made between two F_1 hybrids or two single crosses; four inbred lines are used.

Double fertilization: The synchronized union of one sperm with the nucleus and a second with two polar nuclei to form a diploid zygote and triploid endosperm, respectively.

Double haploid lines: A homozygous lines developed by doubling chromosome numbers of haploids through colchicine treatment.

Double haploid method: Breeding approach which is used to produce a homozygous line by shortcut way to generate a haploid plant through parthenogenesis or by anther culture, followed by chromosome doubling by colchicine treatment.

Double haploid: Diploid homozygous plant produced due to spontaneous or induced chromosome doubling.

Double top cross hybrid: A hybrid generated by a cross between F_1 hybrid (single cross) with an open-pollinated genotype/variety.

Drought avoidance: Under moisture stress, plants maintain a favorable internal water balance.

Drought escape: Plant completes its life cycle before water-deficiency commences.

Drought resistance: Plants can show minimum yield loss under water-deficient conditions.

Drought tolerance: Ability of a plant to withstand low tissue water content.

Drought: Stress which occurs due to water deficit.

Dwarfing gene: One class of genes that control plant height, *i.e., Norin-10* in wheat, *Dee-Geo-Woo-Gen* in rice.

E

Ear to row selection: It is the simplest form of progeny selection. Selection is practiced from progenies/lines that rise by a single plant/ear.

Early generation testing: Evaluation cum selection scheme where poor lines/ recombinants are discarded in the early generation, *i.e.,* F_2 and F_3

Economic heterosis: Superiority of the hybrid (F_1) over the standard, commercial, or check variety. It is also known as standard or useful heterosis.

Economic yield: The dry weight of a plant from total biological yield, which has economic value.

Egg cell: Female gamete found between two synergid cells at the micropylar end in the embryo sac, giving rise to embryo after fusion with the male gamete.

Egg: Female gamete or germ cell.

Elite genotype: A genotype or line which is truly superior over other materials of the same categories.

Emasculate or Emasculation: A process to remove or kill pollen/anther from bud or flower before anthesis without damaging female reproductive parts.

Embryo culture: A tissue culture technique in which artificial culturing of the young embryo excised from seed under aseptic condition and growing in an artificial nutrient medium under the *invitro* state.

Embryo rescue: A technique used to save embryos from aborting during wide hybridization through embryo culture.

Embryo sac: Female gametophyte formed by megagametogenesis in plants generally containing eight-nucleate.

Embryogenesis: The formation of an embryo; in tissue culture, forming a somatic embryo from the callus.

Embryoid: An embryo-like structure found in cell and tissue culture.

Endopolyploidy: Cells of diploid organisms having multiples of the 2n gametes (*i.e.,* 4n, 8n, *etc.*) result from endo-mitosis.

Endosperm: Triploid tissue, which arises from the triple fusion (one sperm nucleus + two polar nuclei of embryo-sac), act as reserved food material to nourish the developing embryo.

Euploidy: The increase of complete chromosome set (genome) of a particular species. The chromosome sets are an exact multiple of original chromosome sets (genome).

Evolutionary breeding: A breeding procedure in which the variety is developed from the naturally selected progeny of cross or multiple crosses that have undergone evolutionary changes.

Exotic collection: The germplasm is considered exotic when it collected from outside the country.

Exotic plant: Plant introduced from outside of the country.

Explant: A detached portion of a tissue or small piece of the plant organ used to initiate an *invitro* culture.

Exploration: A expedition set to collect germplasm of cultivated and wild relative species from various sources.

Ex-situ conservation: The conservation of plant genetic resources carried out outside of (away from) their natural habitat, in other words, outside the area of their origin. It also refers to the out-site preservation of germplasm, *i.e.*, Seed gene bank.

Extinction: The permanent loss of a species, maybe due to overexploitation.

F

F_1: Symbol used to designate first filial generation.

F_2: Symbol used to designate second filial generation or self-fertilized progeny of F_1 generation.

F_3: Symbol used to designate third filial generation or self-fertilized progeny of F_2 generation.

Facultative apomixis: Seeds are produced through apomixis and sexual reproduction, *e.g.*, *Sorghum bicolor*.

Fertilization: During sexual reproduction, union of two gametes (sperm + egg cell) to produce a zygote (2n).

Field gene bank: It is *in-situ* conservation and mostly suitable for recalcitrant and vegetatively propagated species—such conservation concentrates on preserving cultivated plant species on farmers' fields.

Filial generation: The progeny resultant from the first cross between two parents in the first filial generation later can advance as self-fertilization of respective F_1 (hybrids) is also considered filial generation F_2, F_3, *etc.*

Flower: Reproductive part of the angiosperm (flowering plants) which facilitates reproduction and is a rich source of nectar.

G

Gametocide: It also refers to pollen sterilant, a chemical hybridizing agent. A chemical agent used to induce male sterility by killing or functionally inactivating male organs/ gametes.

Gametogenesis: Forming male and female gametophytes from the microspores and megaspores, respectively, through mitotic division.

Gametophyte: A multicellular structure enclosing male and female gametes produced through mitosis, *i.e.*, male gametophyte (pollen grain) from microspore, whereas female gametophyte (embryo sac) from megaspore.

Gametophytic apomixis: Asexual seed formation where in an embryo sac develops from an unreduced cell, *i.e.*, diplospory or apospory.

Gametophytic self-incompatibility (GSI): A kind of homomorphic self-incompatibility where haploid genetic constituent determines (genotype of pollen itself) pollen incompatibility.

Gamma garden: A specific area where gamma rays emitted from a particular source are subjected to induce mutation in a particular plant grown in this area.

Geitnogamy: Fertilization accomplished by pollen belonging to the different flowers borne on the same plant.

Gene bank: It also refers to germplasm bank. A suitable storage condition where different viable forms of germplasm collection, *i.e.,* pollen, seed, vegetative propagules, DNA, *etc.*, are maintained and conserved.

Gene pool: Sum total of genes and their alleles (combined genetic information) present in a population. In other words, the Interbreeding group in a random mating population sharing a common set of genes.

Gene pyramiding: Incorporating two or more genes in a particular variety to achieve multi disease-pest resistance.

Gene redundancy: The presence of multiple copies of a gene due to polyploidy or any other chromosomal modifications.

Gene sanctuary: A particular area protected from human interferences to save or maintain available plant diversity in this area.

Gene: According to the classical concept, the gene is a hereditary factor generally liable for governing traits. As per the molecular concept, a gene is a nucleotides sequence of DNA that codes for a specific peptide chain. W. Johannsen first coined the term gene (Gen) in 1909.

Generative nucleus: During microgametogenesis, haploid nucleus of microspore (pollen grain) which divides mitotically to produce two sperm nuclei.

Genetic disharmony: In wide hybridization, the embryo's demise occurs due to genetic disparity.

Genetic emasculation: Prevention of self-pollination through genetic factors either due to male sterility or self-incompatibility.

Genetic erosion: The loss of genetic variability from cultivated species and their wild relatives or wild forms.

Genetic gain: The genetic improvement through the selection of variation with the heritability of a particular trait from generation to generation.

Genetic homoeostasis: The buffering capacity of a genotype/line against environmental fluctuation.

Genetic male sterility: A kind of male sterility governed by a nuclear gene.

Genetic polymorphism: An occurrence of two or more genotypes in a genetic population.

Genetic resistance: Resistance against pests or pathogens due to general or specific gene action.

Genetic resources: The sum total of all the genes and their allele present in a particular species or wild relatives/wild form.

Genetic slippage: The genotype fails to express its actual potential to reach an optimal phenotypic level due to G × E interaction.

Genetic vulnerability: The susceptibility of a genetically uniform crop variety to different biotic or abiotic factors due to narrow genetic base.

Genetically modified organisms (GMO): Plant which carries new genes that have been inserted through genetic engineering techniques, *e.g.*, *Bt* cotton, *Bt* maize, flavour-saver ("Flavr-Savr") tomato, *etc*. It also refers to the transgenic organism.

Genic (Nuclear) Male Sterility (GMS): Male sterility governed by nuclear genes.

Genome: The total amount of DNA or RNA (genetic material) present in a particular organism.

Genotype: It represents the genetic constitution of an organism. It remains stable generation after generation.

Genotype-environment interaction: Differential performance of a genotype in different environments.

Genotypic frequency: The frequency of individuals with a specific genotype in a population. It ranges from 0 to1.

Genotypic variation: The genetic variation which remains unaltered by environmental factors. Such variation in the population is essential for effective selection—It is measured in terms of genotypic variance.

Germ pore: It is a minute pore on pollen grain from which a pollen tube emerges, exine wall is absent in this region.

Germplasm conservation: See conservation.

Germplasm: See genetic resources.

Gossypol: A natural polyphenolic aldehyde-dark pigment, $C_{30}H_{30}O_8$, derived from unrefined cottonseed oil.

Gynodioecious: Plant species in which hermaphrodite and female flowers are borne on different plants.

Gynoecious: Plant bearing only female flowers.

Gynoecium: Female reproductive organ of a flower, comprising one or more carpels.

Gynomonoecious: Hermaphrodite and female flowers are borne on the same plant.

Gynophores: The stalk that pushes groundnut flowers into the soil after pollination.

H

Half sibs: Progeny with one parent common.

Half-diallel cross: Crossing made between series of selected genotypes/lines in all possible combinations exclusive of reciprocal.

Haplodiploidy: It is a condition of sex differentiation where the male is haploid, and the female is diploid.

Haploid (n): Gametic chromosome number of any individual species. It is denoted by 'n'.

Haploid breeding: Various breeding approaches/techniques are being used to produce haploid, *i.e.,* androgenesis or gynogenesis, ultimately producing a homozygous line (true-breeding line).

Haploid cell (n): A cell containing a single chromosome set. Also called gametic chromosome number.

Haploid parthenogenesis: The condition where the unfertilized egg produces a viable male whose cells are haploid.

Hardy-Weinberg law: Gene and genotype frequencies (Genetic structure) in a random mating population (Mendelian population) persist constant generation after generation in the absence of evolutionary forces, *viz.*, mutation, natural selection, migration, or random genetic drift.

Harvest index: The proportion of economic yield to the biological yield.

Harvest-Plus program: It is a global institution & scientists association mainly looking to boost human nutrition by breeding varieties of new staple food crops with higher micronutrient level.

Heat tolerance: It is an ability of plant or genotype to perform normal and yield normal against the high-temperature condition.

Heredity: The inheritance of the characters from parents to their offspring, in other words, the transmission of characters from generation to generation.

Herkogamy: One of the mechanisms that control self-pollination and promote cross-pollination where stigma remains covered by some physical barrier (transparent membrane) which remains unbreakable until and unless any external force applied, *i.e.,* wind, visit of honey bee *etc.*

Hermaphrodite: Both male and female reproductive organs are present on the same flower of the same plant, *e.g.*, mustard, cotton.

Heterobeltiosis: See better parent heterosis.

Heterogametic sex: The male or female sex with one X and one Y chromosome and gives rise to two gametes during meiosis.

Heterogamy: (*Hetero* = Dissimilar and *Gamy* = Fusion or Union). Fusion of morphologically dissimilar male and female gametes.

Heterogeneous: A population possess a mixture of different types (different genotypes).

Heteromorphic: A plant having different floral morphology within the same species.

Heterosis breeding: Various breeding approaches are used to produce hybrids to exploit heterosis for yield, quality, fertility, stability, *etc.*

Heterosis: The superiority of hybrid in terms of overall growth-vigour, yield potential and fertility compared with its parents. It is associated with increasing heterozygosity.

Heterostyly: A genetic condition found in some species where hermaphrodite flowers have variations in style and length of filaments, *e.g.*, Linseed.

Heterozygote: An individual comprising unidentical alleles for a specified trait at the same locus of a homologous chromosome. It produces two different types of gametes, *e.g.*, Rr or Pp.

Hexaploidy (6x): A particular species having six replicas (copies) of two or more different but compatible chromosome sets (genome) derived from two or more different parental species.

Homoeostasis: Tendency of a genotype to remain biologically stable to environmental fluctuation.

Homogamy: A situation in plants where both male and female organs of a bisexual flower mature simultaneously. In other words, anthesis and stigma-receptivity occur at the same time.

Homogeneous: A population possessing genetically like individuals (e.g., identical genotypes).

Homomorphic system: A kind of self-incompatibility system caused due to internal factors, *i.e.*, genetic or physiological factors of a flower. In other words, a plant has the same floral morphology within the same species.

Homozygosity: The presence of identical alleles at one or more loci in homologous chromosomes.

Homozygote: An individual comprises identical alleles for a specified trait at the same locus of the homologous chromosome. It produced only one type of gamete, *e.g.*, PP or RR and pp and rr.

Horizontal Resistance (HR): It governed by polygene or quantitative gene to impart resistance against different races of pathogens.

Hybrid breakdown: It is a kind of reproductive failure where the fertile hybrid becomes weak, sterile, or inviable in further generations (*i.e.*, F_2).

Hybrid inviability: Inability of hybrid zygote to grow into a normal embryo under the normal conditions of development.

Hybrid: Progeny of cross between two or more genetically unlike strains, inbreds, purelines, *etc*.

Hybridization: Mating or crossing between two diverse-homozygous parents.

I

Ideotype: It is a biological model of a plant with specific traits expected to perform and provide maximum economic yield over the defined environmental condition.

Imperfect flowers: Unisexual flowers that contains only one sex, either male or female.

***In situ*:** In the case of a plant, the particular area where it is naturally occurring.

***In situ*-conservation:** The conservation of plant or plant material/germplasm under original habitat or natural condition where species originally evolved in the natural environment and are protected from human interference. It also refers to on-site conservation or preservation of germplasm.

Inbred: A homozygous line developed by continuous selfing (inbreeding) in a cross-pollinated crop. It is a true-breeding homozygous line of a cross-pollinated crop.

Inbreeding depression: Reduction in vigour, overall growth, survival, and fertility of individuals due to inbreeding.

Inbreeding: The mating of a closely related individual resulting a decrease in vigour and fertility.

Incompatibility: Inability to fertilize and produce normal seed in self-pollination or cross-pollination.

Indeterminate: An inflorescence where the terminal flower becomes open at last, resulting in the continuous growth of the floral axis.

Indigenous collection: The germplasm is considered indigenous when collected within the country's area. Which is genetically correlated with that trait.

Indirect selection: Improvement of a particular trait (yield) by selecting another related or component trait.

Intergeneric cross/hybrid: A cross/hybrid between two plants belonging to different genera of the same family.

Interspecific cross/hybrid: A cross/hybrid between two plants belonging to different species of the same genera; it is also referred to as the intrageneric cross.

Inter-varietal cross/hybrid: A cross/hybrid between two plants belonging to a different variety of the same species; it also refers to the intraspecific cross or hybridization.

Intra-varietal hybridization: Crossing performed between two plants belonging to the same variety.

Introgression: Transfer of genes of one species into the gene pool through backcross breeding or somatic hybridization.

Introgressive hybridization: Introgression of a novel gene through repeated backcrossing or somatic hybridization.

Isogamy: (*Iso* = Similar and *gamy* = Fusion or Union). Fusion of morphologically similar male and female gametes.

Isogenic lines: Lines identical to each other except for a single gene.

L

Landrace: A traditional population of crop cultivars or clone produced, selected and maintained by farmers for many generations with great diversity and wider adaptability due to more resistance to stresses.

Landraces: Locally adapted varieties also known as traditional local varieties.

Latent trait: In wild species, a trait which expresses only in limited hybrid combination when hybridizing with some crop species.

LD_{50}: A specific concentration level of chemical (mutagen) at which 50 per cent (50%) of cells or tissue gets killed.

Lethal doses (LD): A specific concentration level of chemical (mutagen) at which certain cells or tissue gets killed.

Line: A group of individuals sharing common parents.

Linebreeding: A variety form by combining lines or genotypes based on test performance.

Luxuriance: In some cases of wide hybridization, the hybrid shows superiority in terms of vegetative growth and vigour but declines reproductive ability (fertility).

M

M_1 seeds: Treated seed with the specific mutagen.

M_1, M_2, M_3: Symbolic representation of a first, second and third generation in mutation (generation after mutagenic treatment), respectively.

M_2 population: The self-progeny of M_1 plants, plant population expressed mutation in a homozygous state.

Macro-mutations: Mutations which cause enormous morphological modifications on phenotype, usually observed in qualitative characters.

Male fertile: Plant producing normal functional pollen.

Male gametocides: Any chemical substance that kills or inactivates the male reproductive organ without any adverse effects on female reproductive organs. Also known as pollen sterilant, CHA; It is mainly used to induce non-genetic male sterility or chemical induce male sterility.

Male sterile: A plant producing non-functional or even absent or lacking male organ/pollen.

Male sterility: Characteristic in flowering plants where pollen grain remains non-functional or unable to produce functional pollen grains. May occur due to cytoplasmic, genetic or interaction of both.

Marker gene: A gene whose function is known and its location can be traced quickly. It may be due to morphological, biological or DNA or molecular base.

Marker: Any properties (traits) of an individual, *i.e.,* morphological, biological, physiological, DNA based (molecular) that can be easily recognized and selected.

Marker-aided selection or marker-assisted selection: Selection strategies based on information obtained at the molecular level using various molecular markers.

Mass emasculation: Emasculation carried out mechanically, *e.g.,* maize.

Mass pedigree method: A modified pedigree method in which the selection is not performed when the environmental condition is unsuitable for selection at that time plant population is maintained as bulk until the condition becomes favourable for pedigree selection.

Mass selection: Forms of selection method of breeding in which individual plant selected on phenotype basis and seeds of selected plants are maintained in bulk for next generation.

Mating: Also termed as crossing or hybridization.

Megasporogenesis: Production of functional megaspore from archesporial cell.

Metaxenia: The influence of pollen grain on maternal tissue of fruit (endosperm), *e.g.,* fruit quality of date palm, also termed Zenia effects.

Micro-centres: A small area within the main centre containing dense crop plant diversity. The rate of natural evolution is more.

Mid–parent heterosis: See average heterosis.

Molecular marker: Unique sequence or segment of DNA closely related to the gene of interest.

Molecular plant breeding: An application of various molecular techniques which indirectly help in breeding techniques for selecting suitable line/genotype at the genetic level.

Monoecism (Monoecious): A condition in which the Staminate and pistillate flowers are borne on the same plant.

Multiline variety: A variety or cultivar that comprises many isogenic lines of the self-pollinated crop (resistance to a different pathogen race).

Multiline: A line developed by composite (mixing) of isogenic lines or near-isogenic lines (NILs).

Multilocation testing: To identify stable newly developed strain/variety by breeder over different geographical areas/locations by testing its performance over several selected locations.

Mutation breeding: A special breeding approach to utilize variation created by mutagenesis for crop improvement.

Mutation: Unexpected genetic alterations in the DNA sequences of an organism giving rise to variable phenotypes.

N

Natural selection: Selection of an individual by natural means without the interference of humans.

Near-Isogenic lines (NIL): The line produced by repeated backcross breeding and selection. It is a genetically similar except for a specific region of the genome or gene introduced from donor parent.

Negative mass selection: A kind of mass selection where undesirable off-type (disease-pest infested) plants are removed, and the rest of the plant grows. It is used for varietal purification.

NES pistillate: In castor femaleness in the monoecious plant due to homozygous recessive gene.

Nobilization: The tropical cane (*Saccharum officinarum*), commonly known as noble cane, has essential attributes like high sugar content, thicker stem, *etc.* These components were transferred to sub-tropical cane (Indian cane; *S. spontaneum*), through a breeding process.

Non-recurrent apomixis: Apomixis occurs occasionally; the embryo-sac contains haploid cells, which is usually sterile. It is developed due to haploid parthenogenesis.

Non-restorer line: The male fertile line cannot restore the male fertility in hybrid, *i.e.,* B-line.

Norin 10: A wheat variety containing semi-dwarfing gene.

N-pistillate: In castor femaleness in the monoecious plant due to heterozygous gene for pistillate nature.

Nucellus: It is a mass of diploid cells enclosed by integuments. Its cells contain abundant reserve food material. An embryo sac is found inside the nucellus.

Nuclear (Genic) male sterility (GMS): Male sterility governed by a single recessive nuclear gene.

Nuclear centre: It refers to the area where agricultural practices began or is considered to have done so.

Nucleus seed: Source of breeder seed; usually developed by concerned plant breeder to maintain cent per cent genetic purity.

Nutri-Cereals: Various millets and other crops have the potential in delivering food and nutritional security as they are the powerhouse of nutrients and health benefit properties.

O

Obligate apomixis: A kind of apomixis where the seed is produced fully by asexual means.

Obsolete (outdated) varieties: Old cultivars which were popular earlier. Later it became replaced by other improved and high yielding cultivars.

Off type plant: A plant in seed production plots, even though it belongs to the same cultivars or species but deviates from normal morphological or other related traits.

Opaque-2: A mutant line of maize rich in lysine content.

Open pollination: Crossing occurs naturally or randomly within cultivars.

Organ culture: Regeneration of plantlets from a different organ, *i.e.,* shoot, root, meristem, bud, *etc.*, under *invitro* condition.

Orthodox type seed: The seed is considered as a dry category having longer viability; their viability generally does not get affected even up to below 5 per cent moisture content.

Outbreeding: see allogamy.

Outcross: Progeny results due to natural means of cross-pollination between two different lines/genotypes. It differs concerning the genetic constitution of both parents.

P

Parthenocarpy: Seedless fruit or fruit production without seed, *e.g.*, banana.

Parthenogenesis: Embryo develops autonomously without fertilization. In other words, Embryo developed from an embryo sac (megagametophyte) without pollination or any role of a male gametophyte.

Participatory plant breeding: Genetic achievement can archive by involving or close collaboration between breeders, researchers, farmers, *etc.*

Passport data: It refers to basic information *viz.*, an accession-Latin name, local name, name of the genus, family, name area of origin, procurement/collection date, unique identity number if available, and institute code for maintaining it, *etc.* of germplasm in question.

Pedigree method/ breeding: A method of breeding used to handle segregating generation where individual plants are selected from segregating generation concurrently pedigree record is maintained for all selected individual plants.

Pedigree records: An ancestry record of individuals, families, genetic line, or variety.

Phenotype: It represents an external feature or outer appearance of an organism. It remains unstable from one generation to the next due to interaction of a genotype and environment. Examples are flower colours, seed colour, seed shape, plant height, eyes colour, *etc.*

Photoperiod-sensitive genetic male sterility (PSGMS): Male sterility regulated by the range of photoperiod (day length).

Pin flower: Flowers having short stamen and long style.

Plant breeding: It is an art and science of improving the heredity of plants for the economic benefit of humankind.

Plant exploration: Tours of plant collection and acquaint with available flora and plant species at an area of concern.

Plant genetic resources (PGR): Any living plant material, either reproductive or vegetative propagules of modern or old cultivars, wild-weedy species or any other plant genetic stock that has potential value for plant breeding and agriculture technological advancement.

Plant introduction: Acquisition of superior plant varieties by importing them from a cultivation area to a newly grown area.

Plant quarantine: A lawful restriction on movement or keeping plant materials in isolation to eliminate or inhibit the spread of harmful exotic insects-pests and diseases in the new area.

Pollen grain: The male gametophyte originating from microspore via microsporogenesis.

Pollen mother cell (PMC): A male sporophyte gives rise to haploid microspore by meiotic division.

Pollination agents: The pollen transferred from anther to stigma is accomplished through various external physical agencies, called pollination vectors.

Pollination: In a flowering plant, transfer of pollen grain from anther to the stigma, either on the same plant or a different plant.

Polygenes: Gene has small and cumulative effects on the expression of a particular trait. Their expression is greatly affected by the environment.

Polymorphism: The occurrence of two or more diverse phenotypes in a genetic population.

Polyploidization: The spontaneous or induced multiplication of the haploid or diploid genome of a cell.

Polyploidy: Multiplication or increase of chromosome sets (genome) in a particular species.

Population genetics: A subdiscipline of genetics merely deals with gene and genotype frequencies considering one or few gene loci in a large group of individuals (Mendelian population).

Population improvement: Improvement of Mendelian population (random mating population) through various selection approaches with or without progeny test.

Pre-breeding: All the activities belonging to research and screening of plant material for biotic-abiotic resistance traits or nutritional aspect before it is used in further breeding process, *i.e.*, evaluating exotic material before initiating breeding programmes.

Primary Centre of diversity: Uninterrupted areas like forests, mountains, hills, valleys, rivers, *etc.*, where plants evolved from wild species—having vast genetic diversity of crop plants.

Primary centre of origin: The geographical region where particular crop species presume to be originated.

Primary gene pool (GP1): The gene pool comprising all the cultivated and spontaneous (*i.e.*, wild or weedy species) races of particular crop species.

Primary plant introduction: The exotic plant material (*i.e.*, cultivated or wild species) is directly introduced into a new environment for commercial cultivation without altering its original genotype.

Progeny selection: Selection of the superior plant is made from a heterogeneous population based on the progeny test.

Progeny test: Determine the genotypic value of an individual parent based on their offspring performance.

Progeny: It also refers to offspring; a plant is grown from the cross seeds produced by parental plants.

Protandry: A dichogamous condition where the male organ (*i.e.*, stamens) releases the pollen grain before the female organ (*i.e.*, stigma) comes in a receptive state. It is one kind of mechanism which promotes cross-pollination.

Protogyny: A dichogamous condition where female organs (*i.e.*, stigma) become receptive earlier, compared to a male organ (*i.e.*, stamens) dehiscence of pollen grains. It is one kind of mechanism which promotes cross-pollination.

Protoplast culture: Regeneration of whole plantlets by isolation and culture of plant protoplast under *invitro* condition.

Protoplast fusion: A somatic hybrid obtained by fusing two protoplasts under *invitro* culture.

Pure line: The progeny of a single, self-pollinated, homozygous (true-breeding) plant.

Pureline selection: Develop a new plant variety by selecting and isolating the single best pure line progeny from a mixture of pure lines.

Q

QPM: Quality Protein Maize (QPM) biofortified maize with improved protein quality by increasing tryptophan and lysine content.

Qualitative inheritance: Inheritance of trait that is governed by oligogene.

Qualitative traits: Traits governed by one or a few genes. They are expressed in few but distinct phenotypic classes and shows discontinuous variation.

Quantitative genetics or Biometrical genetics: A sub-discipline of genetics which deals with quantitative inheritance.

Quantitative inheritance: Inheritance of trait that is governed by polygenes.

Quantitative traits: Traits governed by polygenes. It does not inherit in a simple Mendelian fashion. Instead, each gene has small and cumulative effects that depict continuous variation and are markedly affected by environmental factors.

Quasi-quantitative trait: Polygenic trait showing discontinuous variation, *e.g.*, tillers/ plant.

R

Random mating (panmixia): Mating in which each female gamete is equally likely to unite with any male gamete. Also referred to as panmictic.

Raphanobrassica: In 1928, Karpenchenko produced an amphidiploid from a cross between a radish and cabbage having combo traits from both, *i.e.,* leaves of radish and root of cabbage.

Raphe: It is a ridge-like structure formed by the union of funicles to the ovule wall.

Recalcitrant type seed: The seed is considered a moist category having short viability and loses its viability rapidly if moisture content declines below 12-13 per cent.

Recipient parent: In the backcrossing programme, a parent is repeatedly used. It is also termed the recurrent parent. It received the desired gene from a non-recurrent parent.

Reciprocal cross: Cross in which line is used as female, formerly used as male.

Reciprocal recurrent selection: A form of recurrent selection used to improve both GCA and SCA of the population for a trait simultaneously using a heterozygous tester.

Recurrent selection for GCA: A form of recurrent selection that improves the GCA of a population using a heterozygous tester (open-pollinated variety).

Recurrent selection for SCA: A form of recurrent selection that improves the SCA of a population using a homozygous tester (inbred line).

Recurrent selection: A population improvement approach in which the genetic recombination generates through repeated cyclic selection over generations with the interbreeding of individuals.

Regeneration: Development of plant either through vegetative mean or *in-vitro* condition.

Registered seed: Progeny of foundation seed having good genetic purity and identity.

Relative heterosis: See average heterosis.

Release of variety: After a successful breeding trial, the best strain/line/genotype is permitted to reproduce and for commercial use.

Reproduction: An organism gives rise to similar kinds of progeny by the process of either sexual or asexual means.

Residual heterobeltiosis: The extent of heterosis exhibited by F_2 and succeeding segregating generations over the better parent.

Residual heterosis: The extent of heterosis exhibited by F_2 and succeeding segregating generations.

Residual standard (economic) heterosis: The extent of heterosis exhibited by F_2 and succeeding segregating generations over the standard parent.

Resistance: A host plant response to avoid the development or growth of biological agents (*i.e.*, pathogen, insect, and pest).

Resistant: Trait of a host plant capable of suppressing or retarding the development of insect-pest, pathogen or other factors.

Restorer gene: Usually, the dominant gene can restore fertility by suppressing effects of cytoplasmic male sterility and produce functional gametes.

Restorer line (R-line): The line contains the restorer gene, restoring fertility in the cytoplasmic male-sterile plant.

Roguing: During the hybrid seed production programme, removal of undesirable off-type plants for maintaining genetic purity.

S

Secondary centre of diversity: Cultivated or interrupted areas where certain crop species or valuable crop plants show considerable diversity although they did not originate there.

Secondary gene pool (GP2): The gene pool comprising all those related species in which crossing with species included in GP1 is tricky and leads to partially fertile and less vigorous F_1 hybrids.

Secondary plant introduction: The exotic plant material (*i.e.*, cultivated or wild species) is used as cultivars only after selection or altering its genotype by using it into a hybridization program for desirable gene sources.

Seed bank: The plant seeds or obsolete varieties are stored, usually under vacuum-packed and cold conditions to maintain viability.

Seed technology: The science that deals with the production, harvesting, processing, testing, packaging, storing and marketing of seeds.

Seed vigour: Property of seed which determines its potential for quick and uniform emergence and development of normal seedlings under various field conditions.

Segmental allopolyploids: The combination of two or more partial homologous genomes.

Segregating generations: F_2 and onward generation obtained through selfing.

Segregating population: F_2 and onward generation in which genetic differences are detectable.

Segregation: Separation of alleles during meiosis, resulting in each gamete containing one allele from a pair.

Selection: Differential proportion of reproduction of genotype.

Self-incompatibility: Failure of functional self-pollen to fertilize the ovule of the same flower (hermaphrodite) or another flower of the same plant. It may be due to morphological or physiological reasons.

Selfing: Achieving self-pollination through protecting the inflorescence from receiving foreign pollen.

Self-pollination: Transfer of pollen to the stigmas of the same plant.

Self-sterility: Failure to attain fertilization and seed set after self-pollination.

Semi-dwarf: A common term used in wheat breeding, it designates individuals or a variety showing medium short stature.

Semigamy: The embryo develops without fusion of male and female gametes. The nuclei of the female and male gametes divide and develop independently in the embryo sac.

Shuttle breeding: In segregating generation, the segregating materials are grown in alternative generations at two different locations to select appropriate stress tolerance for yield and its component.

Single or simple cross (hybrid): Hybrid generated by a cross between two parental lines (inbred lines).

Single-seed descent (SSD) method: A modified version of the bulk selection method where only a single seed is selected randomly from each plant to advance each generation.

Somaclonal variation: Heritable or non-heritable variation occurs in the cell when plant culture is under *invitro*.

Somatic hybrid: A Hybrid (cybrid) derived from the fusion of genetical different somatic cells by protoplast fusion.

Specific combining ability: Deviation in performance of a cross from that predicted based on general combining ability.

Sporogenesis: A form of haploid microspores (male spore) and haploid megaspores (female spore) by meiotic division.

Sporophytic apomixis: The embryo developed directly from diploid somatic (sporophytic) ovule cells.

Sporophytic self-incompatibility (SSI): A kind of self-incompatibility where the pollen incompatibility is determined by the diploid genome of the pollen-producing (parent) plant (sporophytic generation).

Standard heterosis: The kind of heterosis estimated over standard/ commercial variety.

Sterility: Incapable to fertilize the stigma either by self or foreign pollen, resulting in no seed set.

Syn_0: The parental line of the synthetic variety.

Syn_1: The first generation of the synthetic variety.

Synergid cells: Two cooperative or accessory cells situated on either side of the egg cell inside the embryo sac play a vital role in pollen tube attraction, guidance, and reception.

Syngamy: See sexual reproduction.

Synthetic variety: In cross-pollinated crops, variety developed by crossing in all possible combinations among the number of lines having good general combining ability.

T

Tassel: The inflorescence of the male flower present at the top of a maize stalk.

Tasseling: The phenomena of emerging out of tassels from maize stalk.

Tertiary gene pool (GP3): The gene pool comprises all those species in which crossing with cultivated plant species (GP-1) is challenging and leads to sterile F_1 hybrids.

Testcross: The cross of an unknown heterozygous (F_1) with its homozygous recessive parent. *e.g.*, Rr × rr. Test cross may be considered as a back cross. It is used to test the linkage and genotype of an unknown parent.

Thermo-sensitive genetic male sterility (TSGMS): Male sterility markedly influenced by temperature range.

Three-way cross (hybrid): A hybrid generated by a cross between a single cross and an inbred line. In other words, this cross/hybrid involves three parental genotypes (inbred lines).

Tissue culture: A biotechnological approach to grow a plant or cell or tissue in a suitable artificial medium under *in vitro* conditions.

Tolerance: A form of resistance in which the ability of crop plants to give optimal yield compared to other varieties in the same stress level.

Top cross hybrid: Hybrid resulting from a cross between an inbred with a common pollen parent (generally an open-pollinated variety).

Transgene: A foreign gene isolated from taxonomically distant organisms and inserted into desired organisms with the help of biotechnological approaches.

Transgenic Male Sterility: Male sterility is imparted by inserting a foreign gene (transgene) with the help of genetic engineering tools.

Transgenic plant: Plants that carry transgene. Also termed as genetically modified organisms, *i.e.*, Bt cotton.

Transgressive breeding: Various breeding methods are used to handle transgressive segregants produced during transgressive generation.

Transgressive segregation: Segregants generated in F_2 or onward generation of a cross between two diverse parents produce a more extreme phenotype outside of both parents.

Trihybrid: A hybrid heterozygous for three pairs of alleles.

Trimonoecious: Male, female and hermaphrodite flowers borne on the same plant.

Trioecious: Plant with male, female, and hermaphrodite flowers on three different plants.

Triploid: Individual having three sets of chromosomes (3n), *e.g.*, Banana.

Trisomics (2n+1): Represents the presence of one extra chromosome.

Triticale: First human-made artificial allopolyploid produced from a cross between wheat (*T. aestivum*) and rye (*Secale cereal*).

True breeding: A set of similar individuals producing progeny with identical phenotypes by intercrossing.

U

Unisexual flower: A flower bears either stamen (male organ) or carpels (female organ). In other words, only one sex is present.

Useful heterosis: See Economic heterosis.

V

Variability: The sum of variation (genetic or phenotypic level) present among a group of a species.

Variation: The differences among a group of an individual for a trait.

Varietal deterioration: Decline of genetic or agronomic traits of a well-developed variety.

Variety: In plant breeding, strain or genotype distinguishes at least one trait (retain such trait in the next generation when reproduced) from other such groups. It is released for commercial cultivation by the variety release committee.

Vegetative reproduction: A form of a new individual from the vegetative part without fertilization.

Vertical Resistance (VR): The host display resistance against a specific race—the resistance is governed by oligogene.

Vertifolia effects: The resistance cultivar losing resistance against a particular pathogen race due to the evolution of such virulence race.

W

Wide hybridization: See distant hybridization.

Wild type: Also termed as a normal, the most often observed phenotype in the population.

X

Xenia: Effects of pollen on seed and fruits of the fertilized plant. This phenomenon is commonly observed in maize.

Z

Zygote: A diploid (2n) cell produced by fusion of male and female gametes (fertilized egg).

APPENDIX-II

Recipient of World Peace Prize and World Food Prize and their important contributions in crop breeding

A. Recipient of World Peace Prize

Year	*Name of Recipient and Country*	Contribution
1970	**Dr. Norman Ernest Borlaug; United States**	✰ He is known as the Father of the green revolution in the World. ✰ Developed dwarf varieties of wheat, resulting in a green revolution in the World. ✰ To feed a hungry world.

B. Recipient of World Food Prize

Year	*Laureate(s) and Country*	*Contribution*
1987	**Dr. M. S. Swaminathan; India**	✰ He is known as the Father of the green revolution in India. ✰ Introduced, developing, and distributing high-yielding wheat and rice varieties in India during the 1960s initiated the green revolution in India.
1988	**Dr. Robert Chandler, Jr.; United States**	✰ The founding director of the International Rice Research Institute (IRRI), Philippines and the World Vegetable Centre, Taiwan. ✰ A fundamental role in forming a CGIAR system in the world enhances world food production.
1991	**Dr. Nevin S. Scrimshaw; United States**	✰ The reduce malnutrition in developing nations. Studies combat protein, iodine, and iron deficiencies and tutor the people and experts. ✰ Developing nutrition-rich supplements (foods).
1992	**Dr. Edward F. Knipling, and Dr. Raymond Bushland; United States**	✰ To develop an environmentally approachable effective way to (the sterile insect system) to control pests that impend crops and livestock production.

Year	*Laureate(s) and Country*	*Contribution*
1994	**Dr. Muhammad Yunus; Bangladesh**	✰ Encouraging the economic and social empowerment of the poorest peoples of Bangladesh, specifically through an innovative no-collateral microloan approach, so the poor people are enable to access sufficient food and nutrition.
1996	**Dr. Henry M. Beachell; United States**	✰ The advances in rice breeding would help maintain sufficient food supplies against the rising populations in Asia and around the World.
	Dr. Gurdev Khush India	
1998	**B.R. Barwale; India**	✰ Transformed the face of Indian agriculture through privatizing the development and distribution of high-yielding varieties and hybrid seeds throughout India.
2000	**Dr. Surinder K. Vasal; India**	✰ For the research and guidance in refining maize's productivity and nutritional content by developing Quality Protein Maize (QPM).
	Dr. Evangelina Villegas; Mexico	
2004	**Dr.Monty Jones; United Kingdom**	✰ He developed NERICA – "New Rice for Africa"–by crossbreeding high-producing Asian genotypes with African genotypes, which thrived in poor soils and drought conditions.
	Prof. Yuan Longping; China	✰ For the development of the World's first hybrid rice varieties and associated technologies known by the father of Hybrid Rice.
2009	**Dr. Gebisa Ejeta; Ethiopia**	✰ Improved food security for hundreds of millions of people in sub-Saharan Africa by developing Africa's first high yielding sorghum hybrids resistant to drought and the parasitic weed-Striga.
2010	**David Beckmann andJo Luck; United States**	✰ For their leadership in grassroots efforts to improve the conditions of hunger and nutrition.

Year	*Laureate(s) and Country*	*Contribution*
2013	**Dr. Marc Van Montagu; Belgium**	✰ Their modernization accomplishments in founding, developing and applying modern agricultural biotechnology supported sustainability and global food security through improved yields, insect-disease resistance, and tolerance to climate resilience.
	Dr. Mary-Dell Chilton and Dr. Robert Fraley; United States	
2014	**Dr. Sanjaya Rajaram; India and Mexico**	✰ To improve the lives of hundreds of millions of people by developing 480 high-yielding and disease-resistant wheat varieties that increase the world production by 200 million tons.
2016	**Dr. Maria Andrade; Cape Verde**	✰ The development and application of biofortification, breeding critical vitamins and nutrients into staple crops, thus intensely reducing "hidden hunger" for millions. ✰ Develop disease-resistant, drought-tolerant, high yielding varieties of orange-fleshed sweet potato (OFSP) rich in Vitamin A.
	Dr. Howarth Bouis and Dr. Jan Low; United States	
	Dr. Robert Mwanga; Uganda	
2020	**Dr. Rattan Lal; India and United States**	✰ Developing and mainstreaming a soil-centric approach that increases food production, preserves natural resources, and alleviates climate change.
2024	**Geoffrey Hawtin and Cary Fowler; United Kingdom/ Canada and United States**	✰ For preserving and protecting the world's heritage of crop biodiversity (conservation of over 6,000 species of crops and culturally important plants) and mobilizing this critical resource to protect against threats to global food security.
2025	**Mariangela Hungria Brazil**	✰ For innovative work in biological nitrogen fixation (BNF) and development of microbial technologies which reduce the requirement for chemical fertilizers.

APPENDIX-III

List of Genome Sequenced in Crop Plants

Sl.No.	Common Name	Scientific Name	Haploid Genome Size (Mbp)	Year
1.	Model plant	*Arabidopsis thaliana*	125	2000
2.	Rice	*Oryza sativa*	430	2002
3.	Grapes	*Vitis vinifera*	505	2007
4.	Papaya	*Carica papaya*	372	2008
5.	Cucumber	*Cucumis sativus*	367	2009
6.	Sorghum	*Sorghum bicolor*	818	2009
7.	Maize	*Zea mays*	2300	2009
8.	Castor	*Ricinus communis*	320	2010
9.	Soybean	*Glycine max*	1115	2010
10.	Apple	*Malus domestica*	742	2010/17
11.	Date palm	*Phoenix dactylifera*	658	2011
12.	Potato	*Solanum tuberosum*	844	2011
13.	Pigeonpea	*Cajanus cajan*	833	2011
14.	Rapeseed	*Brassica rapa*	485	2011
15.	Banana	*Musa acuminate*	523	2012
16.	Barley	*Hordeum vulgare*	5100	2012
17.	Cassava	*Manihot esculenta*	760	2012
18.	Sweet orange	*Citrus sinensis*	380	2012
19.	Tomato	*Solanum lycopersicum*	900	2012
20.	Foxtail millet	*Setaria italica*	490	2012
21.	Melon	*Cucumis melo*	450	2012
22.	Chick pea	*Cicer arientium*	738	2013
23.	Sugar beet	*Beta vulgaris*	714-758	2013
24.	Tobacco	*Nicotiana sylvestris*	2636	2013
25.	Capsella	*Capsella rubella*	130	2013
26.	Common bean	*Phaseolus vulgaris*	520	2013
27.	Kiwifruit	*Actinidia chinensis*	616	2013
28.	Peach	*Prunus persica*	265	2013
29.	Pepper	*Capsicum annum*	3500	2014

Sl.No.	*Common Name*	*Scientific Name*	*Haploid Genome Size (Mbp)*	*Year*
30.	Mango	*Mangifera indica*	450	2014
31.	Pine apple	*Ananas comosus*	526	2015
32.	Sunflower	*Helianthus annus*	3600	2017
33.	Lettuce	*Latuca sativa*	2500	2017
34.	Quinoa	*Chenopodium quinoa*	1400	2017
35.	Bitter gourd	*Momordica charantia*	339	2017
36.	Coconut	*Cocos nucifera*	~24200	2017
37.	Pearl millet	*Pennisetum glaucum*	~1790	2017
38.	Bread wheat	*Triticum aestivum*	14500	2018
39.	Peanut	*Arachis hypogaea*	~2700	2019
40.	Cauliflower	*Brassica oleracea* L. var. *botrytis*	584.6	2019
41.	Upland wild rice species	*Oryza granulata*	736.6	2020
42.	Annona	*Annona muricata*	799.1	2021
43.	Blackgram	*Vigna mungo*	475	2021
44.	Medicinal herb	*Picrorhiza kurrooa*	~ 1700	2021
45.	Faba bean	*Vicia faba L.*	13000	2023

Source: ncbi.nlm.nih.gov/and http://www.nature.com, Michael and Jackson, 2013.

References

A. K. Roy., R. K. Agrawal, Shahid Ahmad, R. V. Kumar, A. K. Mall, Nitish R. Bhardwaj, Ritu Mawar, D. N. Singh, S. R. Kantwa and S. A. Faruqui (2018). Database of Forage Crop Varieties: 2018. AICRP on forage crops and utilization, ICAR–IGFRI. Technical Publication Number- 12/2017-18

Abrol, D. P. (2012). Pollination Biology. Biodiversity Conservation and Agricultural Production. Springer, Dordrecht. doi:10.1007/978-94-007-1942-2

Allard, R. W. (1999). *Principles of Plant Breeding, 2nd Edition*. John Wiley and Sons. ISBN 978-0-471-02309-8.

Allard, R. W. (1999). *Principles of Plant Breeding, 2nd Edition*. John Wiley and Sons. p. 264. ISBN 978-0-471-02309-8.

Amoo, S. O., Okorogbona, A. O. M., Du Plooy, C. P. and Venter, S. L. (2017). *Sesamum indicum*. Medicinal Spices and Vegetables from Africa. pp. 549–579. https://doi.org/10.1016/b978-0-12-809286-6.00026-1

Ananda, G., Myrans, H., Norton, S. L., Gleadow, R., Furtado, A. and Henry, R. J. (2020). Wild Sorghum as a Promising Resource for Crop Improvement. *Frontiers in Plant Science*, **11**: 1108. https://doi.org/10.3389/fpls.2020.01108

Anderson, E. (1949). Introgressive Hybridization. John Wiley and Sons, New York

Ankineedu, G., Rao, N.G.P. (1973) Development of pistillate castor. *Indian Journal of Genetics and Plant Breeding*. **33**: 416–422.

Arora, A. (2012). Oat (*Avena sativa* L.). In: Bhardwaj D. N. (ed) Breeding of Field Crops. Agrobios (India), Jodhpur. ISBN 978-81-7754-474-g, pp. 785-816

Artschwager, E., Brandes, E. W. and Starrett R. C. (1929). Development of flower and seed of some varieties of sugar cane. *Journal of Agricultural Research.* **39**: 1-30. https://handle.nal.usda.gov/10113/IND43967678

Ashok K. A., Rakesh K. S., Govindraj, M., Sunita, G. (2019). Application of Plant Breeding and Genomics for improved Sorghum and Pearl Millets Grain Nutritional Quality. Book, *Sorghum and Millets,* 2nd Edition. Published by Elsevier Inc.Chep III. pp. 51-68. https://doi.org/10.1016/B978-0-12-811527-5.00003-4.

Ashok K. A., Rakesh, K. S., Govindraj, M., Sunita, G. (2019). Application of Plant Breeding and Genomics for improved Sorghum and Pearl Millets Grain Nutritional Quality. Book, *Sorghum and Millets,* 2nd Edition. Published by Elsevier Inc.Chep III. pp 51-68. https://doi.org/10.1016/B978-0-12-811527-5.00003-4.

Athwal, D. S. (1966). Current plant breeding research with special reference to *Pennisetum*. In *Indian Journal of Genetics and Plant Breeding.* pp. 73-85.

Auckland A. K., Van der Maesen, L. J. G. (1980) Chickpea. In: Fehr, W. R., Hadely, H. H. (eds). Hybridisation of crop plants. *American Society of Agronomy*, Madison, WI. pp. 249–259.

Auld, D. L., Zanotto, M. D., Mckeon and T. A. (2009). Castor. In: Vollman, J.; Rajcan, I. (eds) Oil crops breeding. Handbook of plant breeding. *Springer*. New York. pp. 316–332.

Barber, C. A. (1922) The classification of Indian canes. *International Sugar Journal.* **24**:18–21.

Barcaccia, G. and Albertini, E. (2013). Apomixis in plant reproduction: a novel perspective on an old dilemma. *Plant Reproduction.* **26**(3): 159–179. https://doi.org/10.1007/s00497-013-0222-y

Bateman, A. J. (1952). Self-incompatibility systems in angiosperms. *Heredity*. **6**: 285–310. https://doi.org/10.1038/hdy.1952.40

Bhardwaj, D. N. (2012). Pearl millet (*Pennisetum americanum* Leeke R. Br.). In: Bhardwaj D. N. (ed) Breeding of Field Crops. Agrobios (India), Jodhpur. ISBN 978-81-7754-474-g, pp.123-134.

Bhargava, A. and Srivastava, S. (2019). Plant Breeding. In: Participatory Plant Breeding: Concept and Applications. Springer, Singapore. https://doi.org/10.1007/978-981-13-7119-6 2

Bhat, S. R. and Pankaj, kumar, P. S. (2008). An improved cytoplasmic male sterile (*Diplotaxis berthautii*) *Brassica juncea*: Identification of restorer and molecular characterization. *Euphytica.* **159**: 145–152.

Bhatt, R. K., Juklani, A. K. and Roy, M. M. (2016). Clusterbean (*Cyamopsis tetragonoloba* [L.] Taub.), an important industrial arid legume: A review. *Legume Research.* **40**(2): 207–214.

Bicknell, R. A. (2004). *Understanding Apomixis: Recent Advances and Remaining Conundrums. The Plant Cell Online.* **16**(1): S228–S245. doi:10.1105/tpc.017921

Boavida, L. C., Vieira, A. M., Becker, J. D., and Feijo, J. A. (2005). Gametophyte interaction and sexual reproduction: how plants make a zygote. *The International Journal of Developmental Biology.* **49**(5-6): 615–632. doi:10.1387/ijdb.052023lb

Bohra, A., Jha, U. C., Adhimoolam, P., Bisht, D. and Singh, N. P. (2016). Cytoplasmic male sterility (CMS) in hybrid breeding in field crops. *Plant Cell Reports.* **35**(5): 967–993. doi:10.1007/s00299-016-1949-3

Bohra, A., Saxena, K. B., Varshney, R. K. and Saxena, R. K. (2020). Genomics-assisted breeding for pigeonpea improvement. *Theoretical and Applied Genetics.* **133**(5): 1721-1737. https://doi.org/10.1007/s00122-020-03563-7

Bouis Howarth, E. and Amy Saltzman. (2017). Improving nutrition through biofortification: A review of evidence from HarvestPlus, 2003 through 2016. *Global Food Security* **12**: 49–58. dx.doi.org/10.1016/j.gfs.2017.01.009.

Boukar, O., Fatokun, C. A., Huynh, B. L., Roberts, P. A. and Close, T. J. (2016). Genomic tools in cowpea breeding programs: status and perspectives. *Frontiers in plant science.* **7**: 757. https://doi.org/10.3389/fpls.2016.00757.

Boukar, O., Togola, A., Chamarthi, S., Belko, N., Ishikawa, H., Suzuki, K. and Fatokun, C. (2019). Cowpea [*Vigna unguiculata* (L.) Walp.] Breeding. In: Al-Khayri, J., Jain, S., Johnson, D. (eds) Advances in Plant Breeding Strategies: Legumes. *Springer.* Cham. https://doi.org/10.1007/978-3-030-23400-3_6.

Boukar, O., Fatokun, C. A., Huynh, B. L., Roberts, P. A. and Close, T. J. (2015). Cowpea. In: De Ron, A. (eds). Grain Legumes. Handbook of Plant Breeding, **10**. *Springer.* New York, NY. https://doi.org/10.1007/978-1-4939-2797-5_7

Bourland, F. and Myers G. (2015). Chapter 9. Conventional Cotton Breeding. In: Fang, D., Percy, R. (eds.). Cotton, 2nd Edition. American Society of Agronomy. Madison, Wis. Agronomy Monograph, Cotton. **57**: 205-228. https://doi.org/10.2134/agronmonogr57.2013.0025

Bradbury, L. M., Gillies, S. A., Brushett, D. J., Waters, D. L. and Henry, R. J. (2008). Inactivation of an aminoaldehyde dehydrogenase is responsible for fragrance in rice. *Plant molecular biology.* **68**(4): 439-449. https://doi.org/10.1007/s11103-008-9381-x

Bradshaw, J. E. 2017. Plant breeding: past, present and future. *Euphytica.* **213:** 60. https://doi.org/10.1007/s10681-016-1815-y

Burton, G. W. (1965) Pearl millet Tift 23A released. *Crops Soils.* pp. 17-19.

C. Sandhu, J. S., Gupta, S. K. and Kaur, L. (2007). Wide hybridisation in chickpea and pigeonpea. *Pulses at a glance.* Punjab Agricultural University, Ludhiana. pp. 32-37.

Chand, S., Roy, A. K., Singh, T., Agrawal, R. K., Yadav, V. K., Kumar, S., Malaviya, D. R., Chandra, A and Yadava, D. K. (2023) Twenty-four years lucerne (*Medicago sativa* L.) breeder seed production in India: a retrospective study. *Frontier Plant Science* 14:1259967. https://doi.org/10.3389/fpls.2023.1259967

Chandra, A. K. Chandora, R., Sood, S. and Malhotra, N. (2021). Global production, demand and supply. In *Millets and Pseudo Cereals.* pp. 7-18. https://doi.org/10.1016/B978-0-12-820089-6.00002-1

Chauhan, B. B., Gami, R. A., Prajapati, K. P., Patel, J. R. and Patel R. N. (2019). Study of *Per Se* Performance and Heterosis for Seed Yield and Component Traits in Sesame (*Sesamum indicum* L.). *Current Agriculture Research Journal.* **7**(3): 408-416.

Chavan, U. D., Patil, J. V. and Shinde, M. S. (2009). Nutritional and roti quality of sorghum genotypes. *Indonesian Journal of Agricultural Science.* **10**(2): 80-87. https://doi.org/10.21082/ijas.v10n2.2009.p80-87

Che, P., Zhao, Z. Y., Glassman, K., Dolde, D., Hu, T. X., Jones, T. J. and Albertsen, M. C. (2016). Elevated vitamin E content improves all-trans β-carotene accumulation and stability in biofortified sorghum. *Proceedings of the National Academy of Sciences.* **113**(39): 11040-11045.

Chen, Letian. and Liu, Yao-Guang. (2014). Male Sterility and Fertility Restoration in Crops. *Annual Review of Plant Biology.* **65**(1): 579–606. doi:10.1146/annurev-arplant-050213-040119

Chen, S. (2024). Advances in Molecular Breeding of Forage Crops: Technologies, Applications and Prospects. *Agriculture* 2024, **14:** 279. https://doi.org/10.3390/agriculture14020279

Cheng, S. H., Cao, L. Y., Zhuang, J. Y., Chen, S. G., Zhan, X. D., Fan, Y. Y. and Min, S. K. (2007). Super hybrid rice breeding in China: achievements and prospects. *Journal of Integrative Plant Biology.* **49**(6): 805-810. https://doi.org/10.1111/j.1744-7909.2007.00514.x

Chhabra, R., Muthusamy, V., Gain, N., Katral, A., Prakash, N. R., Zunjare, R. U. and Hossain, F. (2021). Allelic variation in sugary 1 gene affecting kernel sweetness among diverse-mutant and-wild-type maize inbreds. *Molecular Genetics and Genomics.* **296**(5): 1085-1102.

Chilcoat, D., Liu, Z. B. and Sander, J. (2017). Use of CRISPR/Cas9 for crop improvement in maize and soybean. *Progress in molecular biology and translational science.* **149**: 27-46.

Chowdhury, J. B., Mehta, R. K. and Joshi, A. B. (1966). Pollination in berseem. *Indian Journal of Genetics and Plant Breeding.* **26 (1)**: 118-120.

Christian, De Guzman and James, Oard. (2019). Genetics and Breeding System for Cytoplasmic and Genetic Male Sterility in Rice. doi: http://dx.doi.org/10.5772/intechopen.85191

Constable, G., Llewellyn, D., Walford, S. A. and Clement, J. D. (2015). Cotton Breeding for Fibre Quality Improvement. In: Cruz, V.M.V., Dierig, D.A. (eds) Industrial Crops. Handbook of Plant Breeding. *Springer*. New York. **9**. https://doi.org/10.1007/978-1-4939-1447-0_10

Darwin, C. (1876). The Effects of Cross and Self Fertilization in the Vegetable Kingdom. John Murray, London

Das, G., Patra, J. K. and Baek, K. H. (2017). Insight into MAS: A Molecular Tool for Development of Stress Resistant and Quality of Rice through Gene Stacking. *Front Plant Science*. **13**: 8-985. https://doi.org/10.3389/fpls.2017.00985.

Dass, S. (2012). Maize (*Zea mays* L.). In: Bhardwaj D. N. (ed) Breeding of Field Crops. Agrobios (India), Jodhpur. ISBN 978-81-7754-474-g, pp.75-98.

De Nettancourt, D. (2001). *The Basic Features of Self-Incompatibility. Incompatibility and Incongruity in Wild and Cultivated Plants.* Springer, Berlin, Heidelberg.1–24. doi:10.1007/978-3-662-04502-2_1

De Wet, J. M. J. and Harlan, J. R. (1971). The origin and domestication of *Sorghum bicolor. Economic Botany.* **25**: 128–135. doi: 10.1007/BF02860074

De Wet, J. M. J. and Harlan, J. R. (1971). The origin and domestication of Sorghum bicolor. *Economic Botany*. ***25***(2): 128-135. https://doi.org/10.1007/BF02860074

Dhillon, B. S. and Prasanna, B. M. (2001). Maize. In Breeding Field Crops, Theory and Practice edited by V. L. Chopra (ed). Oxford and IBH, New Delhi. pp. 147-189.

Dhillon, S. S., Kumar, P. R. and Gupta, N. (1993). Breeding objectives and methodologies. In: Labana, K. S., Banga, S. S., Banga, S. K. (eds) Breeding Oilseed Brassicas. *Monographs on Theoretical and Applied Genetics.* **19.** https://doi.org/10.1007/978-3-662-06166-4_2

Dillon, S. L., Lawrence, P. K. and Henry, R. J. (2001). The use of ribosomal ITS to determine phylogenetic relationships within Sorghum. *Plant Systematic and Evolution.* **230**: 97–110. doi: 10.1007/s006060170007

Dimitrijevic, A. and Horn, R. (2018). Sunflower Hybrid Breeding: From Markers to Genomic Selection. *Frontiers Plant Science*. **8**: 2238. https://doi.org/10.3389/fpls.2017.02238

Diptibala Rout, Debarchana Jena, Vineeta Singh, *et al.* (2020) Hybrid Rice Research: Current Status and Prospects. doi: 10.5772/intechopen.93668

Dixit, G. P., Srivastava, A. K. and Singh, N. P. (2019). Marching towards self-sufficiency in chickpea. *Current Science*. **116**(2): 239-242. https://doi.org/10.18520/cs/v116/i2/239-242

Dixit, G. P., Srivastava, A. K., Jayalakshmi, V., Bindra, S. and Singh, S. (2022). Chickpea Breeding. In: Yadava, D. K., Dikshit, H. K., Mishra, G. P., Tripathi, S. (eds) Fundamentals of Field Crop Breeding. *Springer*. Singapore. https://doi.org/10.1007/978-981-16-9257-4_20

Doebley, J. (2004). The genetics of maize evolution. *Annual review of genetics*. **38**(1): 37-59.

Doebley, J. F., Gaut, B. S. and Smith, B. D. (2006). The molecular genetics of crop domestication. *Cell*. **127**(7):1309-1321. https://doi.org/10.1016/j.cell.2006.12.006

Dong, Y., Yang, X. Liu and J. (2014). Pod shattering resistance associated with domestication is mediated by a NAC gene in soybean. *Nature Communications*. **5:** 3352 https://doi.org/10.1038/ncomms4352

Dong, Z., Alexander, M. and Chuck, G. (2019). Understanding grass domestication through maize mutants. *Trends in Genetics*. **35**(2): 118-128.

Dresselhaus, T., Sprunck, S. and Wessel, G. M. (2016). Fertilization Mechanisms in Flowering Plants. *Current Biology:* **26**(3): R125–R139. https://doi.org/10.1016/j.cub.2015.12.032

Driscoll, C. J. and Qualset, C. O. (1986). Nuclear male sterility systems in seed production of hybrid varieties. *Critical Reviews in Plant Sciences*. **3**(3): 227–256. doi:10.1080/07352688609382211

Dua, R. P., Chaturvedi, S. K. and Shiv, S. (2001). Reference varieties of chickpea for IPR regime. Indian Institute of Pulses Research, Kanpur duration genotypes in chickpea (*Cicer arietinum* L.). *Legume Research*. **21**(2): 121-124.

East, E. M. (1940). The distribution of self-sterility in flowering plants. *Proceedings of the American Philosophical Society*. **82**: 449.

East, E. M. and Mangelsdorf, A. J. (1925). A new interpretation of the hereditary behaviour of self-sterile plants. *Proceedings of the National Academy of Sciences of the United States of America*. **11**(2): 166–171. doi.org/10.1073/pnas.11.2.166

Elleuch, M., Besbes, S. and Roiseux, O. (2007). Quality characteristics of sesame seeds and by-products. *Food Chemistry*. **10**: 641–650.

Ellis, R.H., Hong, T.D., Roberts, E.H. (1990). An intermediate category of seed storage behavior? 1. Coffee. Journal of Experimental Botany **41**:1167–1174

Fang, Y., Wang, L., Sapey, E., Fu, S., Wu, T., Zeng, H., Sun, X., Qian, S., Khan, M. A. A., Yuan, S., Wu, C., Hou, W., Sun, S. and Han, T. (2021). Speed- Breeding System in Soybean: Integrating Off-Site Generation Advancement, Fresh Seeding, and Marker-Assisted Selection. *Frontiers Plant Sci*ence. **12**: 717077. https://doi.org/10.3389/fpls.2021.717077

Felicity, V. (2016). Changes in sunflower breeding over the last fifty years. *Oil seeds, fat crops and lipids.* **23**(2): 202. https://doi.org/10.1051/ocl/2016006

Frankel, O. H. and E. Bennett (Eds.). (1970). Genetic resources in plants - Their exploration and conservation. Blackwell, Oxford, U.K.

Frankel, R. and Galun, E. (1977). Allogamy. In: Pollination Mechanisms, Reproduction and Plant Breeding. *Monographs on Theoretical and Applied Genetics*, vol 2. Springer, Berlin, Heidelberg. https://doi.org/10.1007/978-3-642-81059-6_3

Frankel, R. and Galun, E. (1977). Autogamy. In: Pollination Mechanisms, Reproduction and Plant Breeding. *Monographs on Theoretical and Applied Genetics*, vol 2. Springer, Berlin, Heidelberg. https://doi.org/10.1007/978-3-642-81059-6_2

Frankel, R. and Galun, E. (1977). Introduction. In: Pollination Mechanisms, Reproduction and Plant Breeding. *Monographs on Theoretical and Applied Genetics*, vol 2. Springer, Berlin, Heidelberg. https://doi.org/10.1007/978-3-642-81059-6_1

Frankel, R. and Galun, E. (1977). Pollination Mechanisms, Reproduction and Plant Breeding. *Monographs on Theoretical and Applied Genetics*. Vol. 2. doi:10.1007/978-3-642-81059-6

Franklin-Tong, V. E. (2002). The difficult question of sex: the mating game. *Current Opinion in Plant Biology.* **5**(1): 14-18. doi: 10.1016/s1369-5266(01)00217-5

Franklin-Tong, V. E. and Franklin, F. C. (2003). The different mechanisms of gametophytic self-incompatibility. *Philosophical transactions of the Royal Society of London. Series B. Biological Sciences.* **358**(1434): 1025–1032. https://doi.org/10.1098/rstb.2003.1287

Gami, R. A., Chauhan, B. B. and Patel, R. N. (2020). Hayman's diallel analysis for yield and attributing traits in sesame (*Sesamum indicum* L.). *Electronic Journal of Plant Breeding.* **11**(2): 359-366. https://doi.org/10.37992/2020.1102.064

Ganapathy, K. N. (2017) Genetic improvement in little millet. In: Patil J. V. (ed). Millets and sorghum: biology and genetic improvement, 1st edn. John Wiley and Sons Ltd., Chichester. https://doi.org/10.1002/9781119130765.ch6

Ganapathy, K. N. (2017) Improvement in finger millet: status and future prospects. In: Patil J. V. (ed). Millets and sorghum: biology and genetic improvement, 1st edn. John Wiley and Sons Ltd., Chichester. https://doi.org/10.1002/9781119130765.ch3

Ganapathy, K. N., Hariprasanna K, Tonapi V.A. (2021) Breeding for enhanced productivity in millets. In: Singh, M. and Sood, S. (eds). Millets and pseudo cereals. Woodhead Publishing, Duxford. pp 39–63.

Gantait, S., Panigrahi, J., Patel, I. C., Labrooy, C., Rathnakumar, A. L. and Yasin, J. K. (2019). Peanut (*Arachis hypogaea* L.) Breeding. In: Al-Khayri, J., Jain, S.,

Johnson, D. (eds) Advances in Plant Breeding Strategies: Nut and Beverage Crops. *Springer.* Cham. https://doi.org/10.1007/978-3-030-23112-5_8

Gaur, P. M. (2012). Chickpea (*Cicer arientinum* L.). In: Bhardwaj D. N. (ed) Breeding of Field Crops. Agrobios (India), Jodhpur. ISBN 978-81-7754-474-g, pp.165-194.

Gemmell, D. J., Bradshaw, J. E., Hodgkin, T. *et al.* (1989). Self-incompatibility in *Brassica napus*: seed set on crossing 19 self-incompatible lines. *Euphytica.* **42**: 71–77. https://doi.org/10.1007/BF00042617

Girija shankar, V., Sharma, H. C., Sharma, K. K., Swathisree, V., *et al.* (2005). Development of transgenic sorghum for insect resistance against the spotted stem borer (*Chilo partellus*). *Plant Cell Reports.* **24**(9): 513-22. doi: 10.1007/s00299-005-0947-7.

Glaszmann, J. C. (1987). Isozymes and classification of Asian rice varieties. *Theoretical and Applied genetics.* **74**(1): 21-30. https://doi.org/10.1007/BF00290078

Gomashe, S. S. (2017) Barnyard Millet: Present Status and Future Thrust Areas. In: Patil JV (ed) Millets and sorghum: biology and genetic improvement, 1st edn. John Wiley and Sons Ltd., Chichester, pp 184-198.

Gomashe, S. S. (2017) Proso Millet, *Panicum miliaceum* (L.): Genetic Improvement and Research Needs. In: Patil JV (ed) Millets and sorghum: biology and genetic improvement, 1st edn. John Wiley and Sons Ltd., Chichester, pp 150–169

Gopala, K. S., Vinod, K. K., Bhowmick, P. K., Bollinedi, H., Ellur, R. K., Seth, R. and Singh, A. K. (2022). Rice Breeding. In *Fundamentals of Field Crop Breeding.* pp. 113-220. https://doi.org/10.1007/978-981-16-9257-4_3

Goulet, B. E., Roda, F. and Hopkins, R. (2017). Hybridization in Plants: Old Ideas, New Techniques. *Plant Physiology.* **173**(1), 65–78. https://doi.org/10.1104/pp.16.01340

Govindaraj, M., Rai, K. N., Cherian, B., Pfeiffer, W. H., Kanatti, A. and Shivade, H. (2019). Breeding biofortified pearl millet varieties and hybrids to enhance millet markets for human nutrition. *Agriculture.* **9**(5): 106. https://doi.org/10.3390/agriculture9050106.

Govindaraj, M., Rai, K. N., Shanmuga, S., P., Dwivedi, S. L., Sahrawat, K. L., Muthaiah, A. R. and Rao, A. S. (2013). Combining ability and heterosis for grain iron and zinc densities in pearl millet. *Crop Science.* **53**: 507–517.

Gowda, M. V. C., Motagi, B. N. and Naidu, G. K. (2002). GPBD4: A Spanish bunch groundnut genotype resistant to rust and late leaf spot. *International Arachis Newsletter.* **22**: 29–32.

Gupta Hari S., Babu Raman, Pawan K. Agrawal, Vinay Mahajan, Firoz Hossain and Nepolean Thirunavukkarasu. (2013). Accelerated development of quality protein maize hybrid through marker-assisted introgression of opaque-2 allele. *Plant Breeding.* **132:** 77–82. doi:10.1111/pbr.12009

Gupta, A., Sood, S., Agrawal, P. K. and Bhatt, J. C. (2011). Floral biology and pollination system in small millets. *The European Journal of Plant Science Biotechnology*.**6**: 81-86.

Gupta, D. S., Kumar, J., Parihar, A. K. and Gupta, S. (2020). Breeding for high-yielding and disease-resistant urdbean cultivars. In: Gosal S. S, Wani S. H. (eds). Accelerated plant breeding, **3**. *Springer*. Cham. https://doi.org/10.1007/978-3-030-47306-8_6

Gupta, D. S., Kumar, J., Parihar, A. K., Chandra, A., Sujayanand, G. K. and Gupta, S. (2022). Urdbean Breeding. In: Yadava, D. K., Dikshit, H. K., Mishra, G. P., Tripathi, S. (eds) Fundamentals of Field Crop Breeding. *Springer*. Singapore. https://doi.org/10.1007/978-981-16-9257-4_23

Gupta, H. S., Raman, B., Agrawal, P. K., Mahajan, V., Hossain, F. and Thirunavukkarasu, N. (2013). Accelerated development of quality protein maize hybrid through marker–assisted introgression of opaque–2 allele. *Plant Breeding*. **132**(1): 77-82. https://doi.org/10.1111/pbr.12009

Gupta, M., Bindra, S., Sood, A., Singh, I., Singh, G., Gaur, P. M. and Singh, S. (2017). Identifying new sources of tolerance to post emergence herbicides in chickpea (*Cicer arietinum* L.). *Journal of Food Legumes*. **31**(1): 5-9.

Gupta, S. (2012). Urdbean (*Vigna mungo* (L.) Hepper). In: Bhardwaj D. N. (ed) Breeding of Field Crops. Agrobios (India), Jodhpur. ISBN 978-81-7754-474-g, pp.237-254.

Gupta, S. and Kumar, S. (2006). Urdbean breeding. In: Ali M, Kumar S (eds) Advances in mungbean and urdbean. IIPR, Kanpur. pp. 149–168.

Gururaj, S., Kenchanagoudar, P. V. and Naragund, V. B. (2002). Identification of sources for field resistance to peanut bud necrosis disease in groundnut. *Karnataka Journal of Agricultural Sciences*. **15**: 646–648.

Gustafson, P., Raskina, O., Ma, X. and Nevo, E. (2009). Wheat evolution, domestication and improvement. *Wheat: science and trade*. Wiley, Danvers. pp. 5-30.

Gustafsson, A. (1946). Apomixis in higher plants. Part I. The mechanism of apomixis. Lunds Universitets Arsskrift, N.F. Avd. 2, **42**: 1-66.

Ha, J. and Lee, S. H. (2019). Mungbean (*Vigna radiata* (L.) R. Wilczek) Breeding. In: Al-Khayri, J., Jain, S., Johnson, D. (eds) Advances in Plant Breeding Strategies: Legumes. pp. 371-407. *Springer*. Cham. https://doi.org/10.1007/978-3-030-23400-3_10

Hamamura, Y., Nagahara, S. and Higashiyama, T. (2012). Double fertilization on the move. *Current Opinion in Plant Biology*. **15**(1):70-77. doi: 10.1016/j.pbi.2011.11.001

Hanna, W. W. (1991). Apomixis in Crop Plants–Cytogenetic Basis and Role in Plant Breeding. *Developments in Plant Genetics and Breeding*. pp 229–242. doi:10.1016/b978-0-444-88259-2.50015-1

Haring, V., Gray, J., McClure, B., Anderson, M. and Clarke, A. (1990). *Self-incompatibility: a self-recognition system in plants. Science.* **250**(4983): 937–941. doi:10.1126/science.2237440

Hariprasanna, K. (2017) Foxtail millet, *Setaria italica* (L.) P. Beauv. In: Patil J.V. (ed). Millets and sorghum: biology and genetic improvement, 1[st]edn. John Wiley and Sons Ltd., Chichester. pp. 112–149.

Hariprasanna, K. (2017) Kodo Millet, *Paspalum scrobiculatum* L. In: Patil J.V. (ed). Millets and sorghum: biology and genetic improvement, 1[st]edn. John Wiley and Sons Ltd., Chichester. pp. 199–225.

Hariprasanna, K. and Patil, J. V. (2015). Sorghum: Origin, Classification, Biology and Improvement. In: Madhusudhana R., Rajendrakumar P., Patil J. (eds) Sorghum Molecular Breeding. Springer, New Delhi. https://doi.org/10.1007/978-81-322-2422-8_1

Hariprasanna, K. and Patil, J. V. (2015). Sorghum: Origin, Classification, Biology and Improvement. In: Madhusudhana, R., Rajendrakumar, P., Patil, J. (eds) Sorghum Molecular Breeding. Springer, New Delhi. https://doi.org/10.1007/978-81-322-2422-8_1

Harlan, J. R. (1971). Agricultural Origins: Centers and Non centers. *Science.* **174** (4008): 468–474. doi:10.1126/science.174.4008.468

Harlan, J. R. and de Wet J. M. J. (1972) A simplified classification of cultivated sorghum. *Crop Science.* **12**(2):172–176. https://doi.org/10.2135/cropsci1972.0011183X001200020005x.

Harlan, J. R. and De Wet, J. M. J. (1971). Toward a rational classification of cultivated sorghums. *Crop Science.* **12**: 172–176. 10.2307/1218252

Hash, C. T., Sharma, A., Kolesnikova-Allen, M. A., SD, S., Thakur, R. P., A. G., B. R. and Witcombe, J. R. (2006). Teamwork delivers biotechnology products to Indian small-holder crop-livestock producers: Pearl millet hybrid "HHB 67 Improved" enters seed delivery pipeline. *Journal of SAT Agricultural Research.* **2**(1).

Haturvedi, S. K., Gaur, P. M., Sandhu, J. S. and Mishra, N. (2010). Efforts for improving seed size and yield of kabuli chickpea. *Progressive Research International Journal.* **5**(2): 5-8.

Hawkes, J. G. (1983). The diversity of crop plants. Harvard University Press, Cambridge U.K.

Hegde, D. M. and Lavanya, C. (2012). Castor (*Ricinus communis* L.). In: Bhardwaj D. N. (ed) Breeding of Field Crops. Agrobios (India), Jodhpur. ISBN 978-81-7754-474-g, pp.471-512.

Hegde, D. M. and Prabakaran, A. J. (2012). Sunflower (*Helianthus annuus* L.). In: Bhardwaj D. N. (ed) Breeding of Field Crops. Agrobios (India), Jodhpur. ISBN 978-81-7754-474-g. pp.385-424.

Hiscock, S. J. and Tabah, D. A. (2003). The different mechanisms of sporophytic self-incompatibility. *Philosophical transactions of the Royal Society of London. Series B. Biological Sciences.* **358**(1434): 1037–1045. https://doi.org/10.1098/rstb.2003.1297

Hong Yu and Jiayang Li. (2021). Short- and long-term challenges in crop breeding. *National Science Review.* **8:** (2) nwab002. https://doi.org/10.1093/nsr/nwab002

Hoskin, C. J. and Higgie, M. (2013). Hybridization: its varied forms and consequences. *The Journal of Evolutionary Biology.* **26**(2): 276-8. doi: 10.1111/jeb.12035.

Hossain, F., Muthusamy, V., Bhat, J. S., Zunjare, R. U., Kumar, S., Prakash, N. R. and Mehta, B. K. (2022). Maize Breeding. In: Yadava, D. K., Dikshit, H. K., Mishra, G. P., Tripathi, S. (eds) Fundamentals of Field Crop Breeding. *Springer.* https://doi.org/10.1007/978-981-16-9257-4_4

Hossain, F., Rakshit, S., Kumar, B. (2021). Molecular breeding for increasing nutrition quality in maize: recent progress. In: Henry, R., Alam, M., Seneweera, S., Rakshit, S. and Hossain A (eds). Molecular breeding in wheat, maize and sorghum: strategies for improving abiotic stress tolerance and yield. CABI. UK. pp. 360.

https://www.nordgen.org/vart-arbete/svalbard-global-seed-vault/

Huang, C., Sun, H. and Xu, D. (2018) *ZmCCT9* enhances maize adaptation to higher latitudes. *Proc National Academy Science*, USA. **115**: 334–341. https://doi.org/10.1073/pnas.1718058115

Hughes, M. B. and Babcock, E. B. (1950). Self-incompatibility in *Crepis foetida* (L.) subsp. rhoeadifolia (bieb.) Schinz et Keller. *Genetics.* **35**(5-1): 570–588. PMID: 14773776.

Husain, S. M. and Srivastava, S. K. (2012). Soybean (*Glycine max* (L.) Merr). In: Bhardwaj D. N. (ed) Breeding of Field Crops. Agrobios (India), Jodhpur. ISBN 978-81-7754-474-g, pp.321-338.

IARI, (2020). Changing the world with seeds: The breeding history of IR 8. *Rice Today.*

IIMR (2018) Annual report 2017-18. ICAR–Indian Institute of Millets Research, Hyderabad.

Islam, F., Gill, R. A., Ali, B., Farooq, M. A., Xu, L., Najeeb, U. and Zhou, W. (2016). Sesame. Breeding Oilseed Crops for Sustainable Production. pp. 135–147. https://doi.org/10.1016/b978-0-12-801309-0.00006-9

Janila, P., Nigam, S. N., Pandey and M. K. (2013). Groundnut improvement: use of genetic and genomic tools. *Frontiers in Plant Science.* **4**: 23. https://doi.org/10.3389/fpls.2013.00023

Janoria, M. P. (1989). Basic plant ideotype for rice. *International Rice Research Newsletter.* **14**(3): 12-13.

Jeffrey Ross-Ibarra, Peter, L. Morrell and Brandon, S. Gaut. 2007. Plant domestication, a unique opportunity to identify the genetic basis of adaptation. *Proceedings of the National Academy of Sciences of the United States of America.* **104** (1): 8641-8648; https://doi.org/10.1073/pnas.0700643104

Jegadeesan, S. and Yu, K. (2020). Food Grade Soybean Breeding, Current Status and Future Directions. In (Ed.), Legume Crops - Prospects, Production and Uses. *IntechOpen*. https://doi.org/10.5772/intechopen.92069

Jiang, S. Y., Ma, A., Xie, L. *et al.* (2016). Improving protein content and quality by over-expressing artificially synthetic fusion proteins with high lysine and threonine constituent in rice plants. *Scientific Reports.* **6:** 34427. https://doi.org/10.1038/srep34427.

John, W. Dudley. 1977. Quantitative Genetics and Plant Breeding. Advances in Agronomy. 59: 1-23. https://doi.org/10.1016/S0065-2113(08)60051-6.

Jorasch, P. (2020). Potential, Challenges, and Threats for the Application of New Breeding Techniques by the Private Plant Breeding Sector in the EU. *Frontiers in Plant science.* **11:** 582011. https://doi.org/10.3389/fpls.2020.582011

Jukanti, A. K., Pathak, R. and Mushyam, C. (2019). Cluster Bean [*Cyamopsis tetragonoloba* (L.) Taub] Breeding. In: Al-Khayri, J., Jain, S., Johnson, D. (eds) Advances in Plant Breeding Strategies: Legumes. *Springer.* Cham. https://doi.org/10.1007/978-3-030-23400-3_4.

Kalve, S. and Tadege, M. (2017). A comprehensive technique for artificial hybridisation in Chickpea (*Cicer arietinum*). *Plant Methods.* **13**(1): 1-9. https://doi.org/10.1186/s13007-017-0202-6

Kapadia, V. N. (2019). Breeding for high forage yield in lucerne. *The Indian Journal of Genetics and Plant Breeding.* 1-33.

Kaur, S., Rakshit, S., Choudhary, M., Das, A. K. and Kumar, R. R. (2021). Meta-analysis of QTLs associated with popping traits in maize (*Zea mays* L.). **19**: 6(8). https://doi.org/10.1371/journal.pone.0256389

Kawakatsu, T. and Takaiwa, F. (2010). Differences in transcriptional regulatory mechanisms functioning for free lysine content and seed storage protein accumulation in rice grain. *Plant and cell physiology.* **51**(12): 1964-1974. https://doi.org/10.1093/pcp/pcq164

Kellogg, E. A. (2015). Flowering Plants, Monocots, Poaceae. *The Families and Genera of Vascular Plants. Springer International Publishing, Switzerland.*pp. 416. https://doi.org/10.1007/978-3-319-15332-2_1

Khan, T. N. (1973). A new approach to the breeding of pigeonpea (*Cajanus Cajan* Millsp.): formation of composites. *Euphytica.* **22**(2): 373-377. https://doi.org/10.1007/BF00022649

Khoury, C. K., Castaneda-Alvarez, N. P., Achicanoy, H. A., Sosa, C. C., Bernau, V., Kassa, M. T. and Struik, P. C. (2015). Crop wild relatives of pigeonpea [*Cajanus cajan*

(L.) Millsp.]: Distributions, ex situ conservation status, and potential genetic resources for abiotic stress tolerance. *Biological Conservation. 184*: 259-270. https://doi.org/10.1016/j.biocon.2015.01.032.

Khush, G. S. (1997). Origin, dispersal, cultivation and variation of rice. *Plant molecular biology*. **35**(1): 25-34.

Khush, G. S. and Virk, P. S. (2000). Rice breeding: achievements and future strategies. *Crop improvement*. **27**(2): 115-144.

Kihara, H. (1983). Origin and history of Daruma-a parental of Norin 10. In *Proceedings of 6th International Wheat Genetics. Symposium.* pp. 13-19.

Kim, Y. J. and Zhang, D. (2018). Molecular Control of Male Fertility for Crop Hybrid Breeding. *Trends in Plant Science*. **23**(1): 53–65. doi:10.1016/j.tplants.2017.10.001

Knox, R, B., Zee, S. Y., Blomstedt, C. and Singh, M. B. (1993). Male gametes and fertilisation in angiosperms. *New Phytologist*. **125**(4): 679-694. doi:10.1111/j.1469-8137.1993.tb03917.x

Kölreuter, J. G. (1766). Vorläufige nachricht von einigen das geschlecht der pflanzen betreffenden versuchen und beobachtungen, nebst fortsetzungen 1, 2 und 3 (1761-1766). Wilhelm Engelmann, Leipzig, Germany

Koltunow, A. M. and Grossniklaus, U. (2003). Apomixis: a developmental perspective. *Annual Review of Plant Biology*. **54**: 547-74. doi: 10.1146/annurev.arplant.54.110901.160842

Kona, P., Mahatma, M. K. and Gangadhara, K. (2020). Variation for physical and nutritional quality traits in advanced breeding lines of groundnut. National seminar on technological innovations in oilseed crops for enhancing productivity, profitability and nutritional security. *Journal of Oilseeds Research*. **37**: 3.

Kranthi, K. R. and Manickam, S. (2012). Cotton (*Gossypium hirsutum* L.). In: Bhardwaj D. N. (ed) Breeding of Field Crops. Agrobios (India), Jodhpur. ISBN 978-81-7754-474-g, pp.591-628.

Krishnappa, G. (2022). Wheat Breeding. In: Yadava, D. K., Dikshit, H. K., Mishra, G. P. and Tripathi, S. (eds) Fundamentals of Field Crop Breeding. Springer, Singapore. https://doi.org/10.1007/978-981-16-9257-42

Kuhn, Maxmllian. (1867). Einige Bemerkungen uber Vandellia und den Bluiten-polymorphismus. *Botanische Zeitung*. **25**: 65-67.

Kumar, A. A. (2016). Botany, taxonomy and breeding. In *The Sorghum Genome.* pp. 27-45. https://doi.org/10.1007/978-3-319-47789-3_2

Kumar, A. A. (2016). Botany, Taxonomy and Breeding. In: Rakshit, S., Wang, YH. (eds) The Sorghum Genome. Compendium of Plant Genomes. *Springer*. Cham. https://doi.org/10.1007/978-3-319-47789-3_2

Kumar, A. A., Reddy, B. V., Ramaiah, B., Sahrawat, K. L. and Pfeiffer, W. H. (2013). Gene effects and heterosis for grain iron and zinc concentration in sorghum [*Sorghum bicolor* (L.) Moench]. *Field Crops Research.* **146:** 86-95.

Kumar, H., Akhatar, J. and Wani, S. H. (2020). Recent advances in cytoplasmic male sterility (CMS) in crop *Brassica* in *Brassica juncea.* In: Wani SH *et al.* (eds) Brassica Improvement. *Springer Nature. Switzerland AG, Berlin.* https://doi.org/10.1007/978-3-030-34694-2_6

Kumar, P. R. (2001). Rapeseed-mustard. In: Chopra VL (ed) breeding field crops. *Oxford-IBH Publication, New Delhi.* pp. 394–432.

Kumar, R., Pandey, M. K., Roychoudhry, S., Nayyar, H., Kepinski, S. and Varshney, R. K. (2019). Peg biology: deciphering the molecular regulations involved during peanut peg development. *Frontiers in Plant Science.* **10**: 1289. https://doi.org/10.3389/fpls.2019.01289

Kumar, S. and Singh, J. (2012). Sugarcane (*Saccharum* spp.). In: Bhardwaj D. N. (ed) Breeding of Field Crops. Agrobios (India), Jodhpur. ISBN 978-81-7754-474-g, pp.709-732.

Kumar, V., Rani, A. and Anshu, A. K. (2021). Marker-assisted stacking of null Kunitz trypsin inhibitor and off-flavour generating lipoxygenase-2 in soybean. *Journal of Agricultural Science.* **159**: 272–280.

Kumar, V., Rani, A. and Rawal, R. (2015). Marker assisted accelerated introgression of null allele of Kunitz trypsin inhibitor in soybean. *Breeding Science.* **65**: 447–452

Kushwah, A., Bindra, S., Singh, I., Dixit, G. P., Sharma, P., Srinivasan, S. and Singh, S. (2020). Advances in Chickpea Breeding and Genomics for Varietal Development and Trait Improvement in India. In: Gosal, S. S., Wani, S. H. (eds) Accelerated Plant Breeding *Springer.* **3**. https://doi.org/10.1007/978-3-030-47306-8_2

Ladizinsky, G. (1998). Plant Evolution under Domestication. The Netherlands: Kluwer Academic Publishers.

Lasa, J. M. and Bosemark, N. O. (1993). Male sterility. In: Hayward, M. D., Bosemark, N. O., Romagosa, I. and Cerezo, M. (eds) Plant Breeding. Plant Breeding Series. Springer, Dordrecht. https://doi.org/10.1007/978-94-011-1524-7_15

Lavanya, C. and Gopinath, V. (2008). Study on inheritance of morphological characters and sex expression in pistillate lines of castor. *Indian Journal of Genetics and Plant Breeding.* **68**: 275–282.

León-Martínez, G., and Vielle-Calzada, J. P. (2019). Apomixis in flowering plants: Developmental and evolutionary considerations. *Plant Development and Evolution.* pp 565–604. doi:10.1016/bs.ctdb.2018.11.014

Lertrat, K. and Pulam, T. (2007). Breeding for increased sweetness in sweet corn. *International Journal of Plant Breeding.* **1**(1): 27-30.

Lewis, D. (1954). Comparative incompatibility in angiosperms and fungi. *Advances in Genetics.* **6**: 235. 10.1016/s0065-2660(08)60131-5

Li L, Jiang, H. and Campbell, M. (2008). Characterization of maize amylose-extender (*ae*) mutant starches: part I. Relationship between resistant starch contents and molecular structures Carbohydrates Polymerase. **74**: 396 - 404.

Liu, K. (1997). Soybean Improvements through Plant Breeding and Genetic Engineering. In: Soybeans. *Springer, Boston, MA.* https://doi.org/10.1007/978-1-4615-1763-4_11

Loh, C. S., Shu, W. and Koh, W. L. (2004). Regeneration of Plants via Somatic Embryogenesis. In: Pua EC., Douglas C.J. (eds) Brassica. *Biotechnology in Agriculture and Forestry.* Vol 54. Springer, Berlin, Heidelberg. https://doi.org/10.1007/978-3-662-06164-0_7

Lopes da Silva, F., Borém, A., Sediyama, T. and Ludke, W. H. (eds.) (2017). *Soybean Breeding.* https://doi.org/10.1007/978-3-319-57433-2.

Lord, E. M. and Russell, S. D. (2002). The mechanisms of pollination and fertilization in plants. *Annual Review of Cell and Developmental Biology.* 18: 81-105. doi: 10.1146/annurev.cellbio.18.012502.0834t38

Lumpkin, T. A. (2015). How a Gene from Japan Revolutionized the World of Wheat: CIMMYT's Quest for Combining Genes to Mitigate Threats to Global Food Security. In: Ogihara Y., Takumi S., Handa H. (eds) Advances in Wheat Genetics: From Genome to Field. Springer, Tokyo. https://doi.org/10.1007/978-4-431-55675-6_2

Ma, H. and Sundaresan V. (2010). Development of flowering plant gametophytes. *Current Topics in Developmental Biology.* **91**:379-412. doi: 10.1016/S0070-2153(10)91013-2

Mackenzie, S. (2012). Male sterility and hybrid seed production. *Plant Biotechnology and Agriculture.* pp 185–194. doi:10.1016/b978-0-12-381466-1.00012-2

Mahadevaiah, C., Appunu, C., Aitken, K., Suresha, G. S., Vignesh, P., Mahadeva Swamy, H. K., Valarmathi, R., Hemaprabha, G., Alagarasan, G. and Ram, B. (2021). Genomic Selection in Sugarcane: Current Status and Future Prospects. *Frontiers in Plant Science.* **12.** https://doi.org/10.3389/fpls.2021.708233

Mahajan, V. and Gupta, A. J. (2016). Development of photo and thermo- insensitive cultivars of short-day onion in India. *Acta Horticlturae.* **1143**: 61-68. DOI: 10.17660/ActaHortic.2016.1143.10

Maheshwari, P. (1950). An Introduction to the Embryology of Angiosperms, McGraw-Hill, New York,

Mariani, C., Beuckeleer, M. D., Truettner, J., Leemans, J. and Goldberg, R. B. (1990). Induction of male sterility in plants by a chimaeric ribonuclease gene. *Nature.* **347**(6295): 737–741. doi:10.1038/347737a0

Mariani, C., Gossele, V., Beuckeleer, M. D., Block, M. D., Goldberg, R. B., Greef, W. D. and Leemans, J. (1992). A chimaeric ribonuclease-inhibitor gene restores fertility to male sterile plants. *Nature.* **357**(6377), 384–387. doi:10.1038/357384a0

Meena, H. P. and Sujatha, M. (2022). Sunflower Breeding. In: Yadava, D. K., Dikshit, H. K., Mishra, G. P. and Tripathi, S. (eds) Fundamentals of Field Crop Breeding. *Springer.* Singapore. https://doi.org/10.1007/978-981-16-9257-4_19

Meena, H. P., Sujatha, M. and Varaprasad, K. S. (2013). Achievements and bottlenecks of heterosis breeding of sunflower (*Helianthus annuus* L.) in India. *Indian Journal of Genetics and Plant Breeding.* **73**(2): 123–130.

Mehboob-Ur-Rahman, M. and Zafar, Y. (2018). Past, Present and Future Trends in Cotton Breeding. *IntechOpen.* https://doi.org/10.5772/intechopen.69672

Mehta, A. K. and Faruqui, S. A. (2012). Berseem (*Trifolium alexandrinum* L.). In: Bhardwaj D. N. (ed) Breeding of Field Crops. Agrobios (India), Jodhpur. ISBN 978-81-7754-474-g, pp. 773-784

Mehta, T. R. (1986). Early history of plant breeding in India. *The Indian Journal of Genetics and Plant Breeding.* **46**(1): 1-11.

Mekonnen, T. W., Gerrano, A. S., Mbuma, N. W., Labuschagne, M. T. Breeding of Vegetable Cowpea for Nutrition and Climate Resilience in Sub-Saharan Africa: Progress, Opportunities, and Challenges. *Plants* (2022), **11**, 1583. https://doi.org/10.3390/plants11121583

Melanie, L. Hand and Anna M. G. Koltunow. (2014). The Genetic Control of Apomixis: Asexual Seed Formation. *Genetics.* **197**(2): 441-450. *https://doi.org/10.1534/genetics.114.163105*

Mertz, E. T., Bates, L. S. and Nelson, O. E. (1964). Mutant gene that changes protein composition and increases lysine content of maize endosperm. *Science.* **145** (3629), 279-280.

Michael, T.P. and Jackson, S. (2013). The First 50 Plant Genomes. The Plant Genome, 6: plant genome 2013.03.0001in. https://doi.org/10.3835/plantgenome2013.03.0001in

Milani, M. and de Medeiros Nóbrega, M. B. (2013). Castor Breeding. In (Ed.), Plant Breeding from Laboratories to Fields. IntechOpen. https://doi.org/10.5772/56216

Millan, T., Clarke, H. J., Siddique, K. H., Buhariwalla, H. K., Gaur, P. M., Kumar, J. and Winter, P. (2006). Chickpea molecular breeding: new tools and concepts. *Euphytica.* **147**(1): 81-103. https://doi.org/10.1007/s10681-006-4261-4

Mirajkar, S. J., Devarumath, R. M., Nikam, A. A., Sushir, K. V., Babu, H. and Suprasanna, P. (2019). Sugarcane (*Saccharum* spp.): Breeding and Genomics. In: Al-Khayri, J., Jain, S., Johnson, D. (eds) Advances in Plant Breeding Strategies: Industrial and Food Crops. *Springer. Cham.* pp. 363-406. https://doi.org/10.1007/978-3-030-23265-8_11

Mishra, G. P., Dikshit, H. K., Sv, R., Tripathi, K., Kumar, R. R., Aski, M. and Nair, R. M. (2020). Yellow mosaic disease (YMD) of Mungbean (*Vigna radiata* (L.) Wilczek): current status and management opportunities. *Front Plant Science* **11**:918. https://doi.org/10.3389/fpls.2020.00918

Mishra, G. P., Dikshit, H. K., Tripathi, K., Aski, M. S., Pratap, A., Dasgupta, U. and Gupta, S. (2022). Mungbean Breeding. In: Yadava, D. K., Dikshit, H. K., Mishra, G. P., Tripathi, S. (eds) Fundamentals of Field Crop Breeding. *Springer*. Singapore. https://doi.org/10.1007/978-981-16-9257-4_22

Mittermeier, R. A., Myers, N., Thomsen, J. B., da Fonseca, G. A. B. and Olivieri, S. (1998). Biodiversity Hotspots and Major Tropical Wilderness Areas: Approaches to Setting Conservation Priorities. *Conservation Biology.* **12**(3): 516–520. doi:10.1046/j.1523-1739.1998.012003516.x

Mizukami, Y., Kato, M. and Takamizo, T. (2006). Interspecific hybrids between *Medicago sativa* L. and annual *Medicago* containing Alfafa weevil resistance. *Plant Cell Tissue Organ Cult*ure. **84**: 80-89. https://doi.org/10.1007/s11240-005-9008-8

Mogali, S. C. and Hegde, G. M. (2020). Recent Advances in Mungbean Breeding: A Perspective. In: Gosal, S. S., Wani, S. H. (eds) Accelerated Plant Breeding, **3**. *Springer*. Cham. https://doi.org/10.1007/978-3-030-47306-8_9

Moss, J. P., Ramanatha and Rao, V. (1995). The peanut-reproductive development to plant maturity. In: Pattee, H. E. and Stalker, H. T. (eds) Advances in peanut science. *American Peanut Research and Education Society. Stillwater*. pp. 1–13.

Murray, B. G. (2003). Crop Improvement/Hybridization and Plant Breeding. *Encyclopedia of Applied Plant Sciences.* pp 119–125. doi:10.1016/b0-12-227050-9/00244-1

Muthusamy Vignesh, Firoz Hossain, Nepolean Thirunavukkarasu, *et al.* (2015). Allelic variations for *lycopene-ε-cyclase* and β-*carotene hydroxylase* genes in maize inbreds and their utilization in β-carotene enrichment programme. *Cogent Food and Agriculture.* **1**: 1033141 http://dx.doi.org/10.1080/23311932.2015.1033141

Muthusamy, V., Hossain, F., Thirunavukkarasu, N., Choudhary, M., Saha, S., Bhat, J. S. and Gupta, H. S. (2014). Development of β-carotene rich maize hybrids through marker-assisted introgression of β-carotene hydroxylase allele. *PLoS One*. **9**(12), e113583. https://doi.org/10.1371/journal.pone.0113583

Mwangangi, I., Kiilu Muli, J. and Neondo, J. (2019). Plant Hybridization as an Alternative Technique in Plant Breeding Improvement. *Asian Journal of Research in Crop Science.* **4**(1): 1-11. https://doi.org/10.9734/ajrcs/2019/v4i130059

Myers, N. 1988. Threatened biotas: 'hot-spots' in tropical forests. *The Environmentalist.* **8**: 187–208.

Nabours, R. K. (1930). Plant Hybridization before Mendel. *Science.* **72**(1863): 274–275. doi:10.1126/science.72.1863.274

Nagaharu, U. (1935). Genomic analysis in *Brassica* with special reference to the experimental formation of *B. napus* and peculiar mode of fertilisation. *Pakistan Journal of Botany*. **7**: 389–452.

Nandini, C. and Bhat, S. (2019). Modified crossing (SMUASB) method for artificial hybridization in Proso millet (*Panicum miliaceum* L.) and Little millet (*Panicum sumatrense*). *Electronic Journal of Plant Breeding*. **10**(3): 1161-1170.

Narain, A. (1961) Inheritance of bloom in castor. *Indian Journal of Genetics and Plant Breeding*. **21**: 136–141.

Narayanapur, Vijayakumar, B., Suma, B. and Minimol, J. S. (2018). Self-incompatibility: a pollination control mechanism in plants. *International Journal of Plant Sciences*. **13**(1): 201-212. doi: 10.15740/HAS/IJPS/13.1/201-212.

Nasrallah, J. B. (2019). Self-incompatibility in the *Brassicaceae*: Regulation and mechanism of self-recognition. *Plant Development and Evolution*. pp 435–452. doi:10.1016/bs.ctdb.2018.10.002

Negm, M. (2020). Handbook of Natural Fibres. Cotton breeding, Volume 1: Types, Properties and Factors Affecting Breeding and Cultivation. pp. 579–603. https://doi.org/10.1016/B978-0-12-818398-4.00018-9

Nelson, O. E. (1969). Genetic modification of protein quality in plants. *Advances in Agronomy*. **21**: 171-194.

Nigam, S. N., Dwivedi, S. L. and Gibbons, R.W. (1991a). Groundnut breeding: constraints, achievements and future possibilities. *Plant Breeding Abstract*. **10** (61): 1128–1136.

Nigam, S. N., Dwivedi, S. L. and Reddy and L. J. (1989). An update on groundnut breeding activities at ICRISAT centre with particular reference to breeding and selection for improved quality. *Lilongwe. Malwi*. pp. 115–125.

Nigam, S. N., Waliyar, F. and Aruna and R. (2009). Breeding peanut for resistance to aflatoxin contamination at ICRISAT. *Peanut Science*. **36**(94): 42–49

Paroda, R. S. and Arora, R. K. (1988) Plant Genetic Resources: General Perspective. https://www.bioversityinternational.org/

Pattanashetti, S. K, Upadhyaya, H. D. and Dwivedi, S. L. (2016) Pearl millet. In: Singh M, and Upadhyaya, H. D. (eds) Genetic and genomic resources for grain cereals improvement. Academic Press, San Diego, CA, pp. 253–289.

Pereira, J. F., Zhang, J., Jessup, R. and Jones, C. S. (2022) Editorial: Forage crop improvement for improved livestock production and nutrition. *Frontier Plant Science*. 13:1082873. https://doi.org/10.3389/fpls.2022.1082873

Peter, K.V. (2003). Encyclopedia of Applied Plant Sciences, production systems and agronomy. *Plantation Crops and Plantations.* pp 956–960. doi:10.1016/b0-12-227050-9/00242-8

Poehlman, J. M. (1987). Breeding Cotton. In: Breeding Field Crops. *Springer*, Dordrecht. https://doi.org/10.1007/978-94-015-7271-2_20

Prabhakar, Aruna, C. and Bhat, V. (2015) Genetic improvement in Sorghum. In: Tonapi, V. A. and Patil, J. V. (eds) Millets-ensuing climatic resilience and nutritional security. ICAR–IIMR, Hyderabad. pp. 109–196.

Prabhakar, Madhusudhana, R. and Aruna, C. (2022). Sorghum Breeding. In: Yadava, D. K., Dikshit, H. K., Mishra, G. P., Tripathi, S. (eds) Fundamentals of Field Crop Breeding. Springer, Singapore. https://doi.org/10.1007/978-981-16-9257-4_7

Prakash, N. R., Zunjare, R. U., Muthusamy, V., Rai, M., Kumar, A., Guleria, S. K. and Hossain, F. (2021). A novel quantitative trait loci governs prolificacy in 'Sikkim Primitive'–A unique maize (*Zea mays* L.) landrace of North–Eastern Himalaya. *Plant Breeding*. **140**(3): 400-408. https://doi.org/10.1111/pbr.12924

Prakash, S., Ahuja, I., Upreti, H. C., Dinesh, K. V., Bhat, S. R., Kirti, P. B. and Chopra, V. L. (2001). Expression of male sterility in alloplasmic *Brassica juncea* with *Erucastrum canariense* cytoplasm and the development of a fertility restoration system. *Plant Breeding.* **120**: 479-482.

Prakash, S., Bhat, S. R. and Kirti, P. B. (2004). Rapeseed-mustard crops in India: history and improvement. *Brassica*. **6**(3-4): 1-54.

Prasanna, B. M., Cairns, J. E., Zaidi, P. H., Beyene, Y., Makumbi, D., Gowda, M. and Zhang, X. (2021). Beat the stress: breeding for climate resilience in maize for the tropical rainfed environments. *Theoretical and Applied Genetics.* **134**(6): 1729-1752. https://doi.org/10.1007/s00122-021-03773-7

Prasanna, B. M., Palacios-Rojas, N. l., Hossain, F., Muthusamy, V., Menkir, A., Dhliwayo, T. and Fan, X. (2020). Molecular breeding for nutritionally enriched maize: status and prospects. *Frontiers in Genetics.* **10**: 1392. https://doi.org/10.3389/fgene.2019.01392

Prasanna, B. M., Vasal, S. K., Kassahun, B. and Singh, N. N. (2001). Quality protein maize. *Current science*. 1308-1319.

Pratap, A. (2012). Mungbean (*Vigna radiata* (L.) Wilczek). In: Bhardwaj D. N. (ed) Breeding of Field Crops. Agrobios (India), Jodhpur. ISBN 978-81-7754-474-g, pp.217-236.

Priyadarshan P. M. (2019). *Introduction to Plant Breeding*. In: Plant Breeding: Classical to Modern. Springer, Singapore. https://doi.org/10.1007/978-981-13-7095-3_1.

Priyadarshan, P. M. (2019) Germplasm Conservation. In: Plant Breeding: Classical to Modern. Springer, Singapore. https://doi.org/10.1007/978-981-13-7095-3_3

Priyadarshan, P. M. (2019). Hybridization. In: Plant Breeding: Classical to Modern. Springer, Singapore. https://doi.org/10.1007/978-981-13-7095-3_9

Priyadarshan, P. M. (2019). Modes of Reproduction and Apomixis. In: Plant Breeding: Classical to Modern. Springer, Singapore. https://doi.org/10.1007/978-981-13-7095-3_4

Priyadarshan, P. M. (2019). Objectives, Activities and Centres of Origin. In: Plant Breeding: Classical to Modern. Springer, Singapore. https://doi.org/10.1007/978-981-13-7095-3_2

Qiu, F., Zheng, Y., Zhang, Z. and Xu, S. (2007). Mapping of QTL associated with waterlogging tolerance during the seedling stage in maize. *Annals of botany*. **99**(6): 1067-1081. https://doi.org/10.1093/aob/mcm055

Radhakrishnan, T., Kona, P., Ajay, B. C. and Kumar, N. (2022). Groundnut breeding. In: Yadava, D. K., Dikshit, H. K., Mishra, G. P., Tripathi, S. (eds) Fundamentals of Field Crop Breeding. *Springer*, Singapore. https://doi.org/10.1007/978-981-16-9257-4_16

Raghavendra, N. Kulkarni and Kuppusamy Baskaran (2013). From Herkogamy to Cleistogamy - Development of Cleistogamy in Periwinkle. *Journal of Heredity*. **104**(1): 140–148, https://doi.org/10.1093/jhered/ess077

Rai, K. N., Gupta, S. K., Sharma, R., Govindaraj, M., Rao, A. S., Shivade, H. and Bonamigo, L. A. (2014). Pearl millet breeding lines developed at ICRISAT: A reservoir of variability and useful source of non-target traits. *The Journal of Agricultural Science*. **1**: 1–13.

Rai, K. N., Gupta, S. K., Sharma, R., Govindaraj, M., Rao, A. S., Shivade, H. and Bonamigo, L. A. (2014). Pearl millet breeding lines developed at ICRISAT: a reservoir of variability and useful source of non-target traits. *SAT eJournal*, **1**(1): 1-13.

Raina, A., Khan, S., Wani, M. R., Laskar, R. A. and Mushtaq, W. (2019). Chickpea (*Cicer arietinum* L.) Cytogenetics, Genetic Diversity and Breeding. In: Al-Khayri, J., Jain, S., Johnson, D. (eds) Advances in Plant Breeding Strategies: Legumes. *Springer*. Cham. https://doi.org/10.1007/978-3-030-23400-3_3

Ram Bhajan. (2012). Rapeseed-Mustard (*Brassica* sp.). In: Bhardwaj D. N. (ed) Breeding of Field Crops. Agrobios (India), Jodhpur. ISBN 978-81-7754-474-g, pp.285-320.

Ram, B., Hemaprabha, G. and Singh, B. D. (2022). History and Current Status of Sugarcane Breeding, Germplasm Development and Molecular Biology in India. *Sugar Tech*. **24**: 4-29. https://doi.org/10.1007/s12355-021-01015-0

Ram, B., Karuppaiyan, R. and Hemaprabha, G. (2022). Sugarcane Breeding. In: Yadava, D. K., Dikshit, H. K., Mishra, G. P., Tripathi, S. (eds). Fundamentals of Field Crop Breeding. *Springer*, Singapore. https://doi.org/10.1007/978-981-16-9257-4_9

Ram, H. H. (2011) Sesame. In: Ram, H. H. (ed) Crop breeding and biotechnology. Kalyani Publishers, New Delhi. pp. 514–528.

Ramahatan, T., Durairaj, M. S. and Prabakaran, A. J. (1992). Induced male sterile in *Sesamum indicum* L. *Sesame and Safflower Newsletter.* **7**: 34–36.

Ramulu, K. S., Sharma, V. K., Naumova, T. N., Dijkhuis, P. and van Lookeren Campagne, M. M. (1999). Apomixis for crop improvement. *Protoplasma.* **208**(1-4): 196–205. doi:10.1007/bf01279090

Rani, A. and Kumar, V. (2022). Soybean Breeding. In: Yadava, D. K., Dikshit, H. K., Mishra, G. P., Tripathi, S. (eds) Fundamentals of Field Crop Breeding. *Springer.* Singapore. https://doi.org/10.1007/978-981-16-9257-4_17.

Rao, N. G. P. Tripathi, D. P. and Rana, B. S. (1984). Genetic analysis of cytoplasmic systems in sorghum. *Indian Journal of Genetics and Plant Breeding* (*India*). **44**: 480–496.

Rathnakumar, A. L. (2012). Groundnut (*Arachis hypogaea* L.). In: Bhardwaj D. N. (ed) Breeding of Field Crops. Agrobios (India), Jodhpur. ISBN 978-81-7754-474-g, pp.339-384.

Rauf, S. (2019). Breeding Strategies for Sunflower (*Helianthus annuus* L.) Genetic Improvement. In: Al-Khayri, J., Jain, S. and Johnson, D. (eds) Advances in Plant Breeding Strategies: Industrial and Food Crops. *Springer*. Cham. https://doi.org/10.1007/978-3-030-23265-8_16

Rauf, S., Shehzad, M., Al-Khayri, J.M., Imran, H.M. and Noorka, I.R. (2019). Cotton (*Gossypium hirsutum* L.) Breeding Strategies. In: Al-Khayri, J., Jain, S., Johnson, D. (eds) Advances in Plant Breeding Strategies: Industrial and Food Crops. *Springer*, Cham. https://doi.org/10.1007/978-3-030-23265-8_2

Rawal, R., Kumar, V. and Rani, A. (2020). Genetic elimination of off-flavour generating lipoxygenase-2 gene of soybean through marker assisted backcrossing and its effect on seed longevity. *Plant Breeding Biotechnology.* **8**(2): 162–173.

Rea, A. C. and Nasrallah, J. B. (2008). Self-incompatibility systems: barriers to self-fertilization in flowering plants. *The International Journal of Developmental Biology*. **52**(5-6): 627–636. doi:10.1387/ijdb.072537ar

Reitz, L. P. and Salmon, S. C. (1968). Origin, History, and Use of Norin 10 Wheat. *Crop Science.* **8**(6): 686-689.doi.org/10.2135/cropsci1968.0011183X000800060014x

Rhoades, M. M. (1950). Gene induced mutation of a heritable cytoplasmic factor producing male sterility in maize. *Proceedings of the National Academy of Sciences of the United States of America*. **36**: 634–635.

Rhodes, D. H., Hoffmann, L., Rooney, W. L., Herald, T. J., Bean, S., Boyles, R. and Kresovich, S. (2017). Genetic architecture of kernel composition in global sorghum germplasm. *BMC genomics*. **18**(1): 1-8.

Rieseberg, L. H, Archer, M. A. and Wayne, R. K. (1999). Transgressive segregation, adaptation and speciation. *Heredity*. **83**(4): 363-72. doi: 10.1038/sj.hdy.6886170

Roach, B. T. (1972). Nobilisation of sugarcane. Proceedings of the International Society of Sugarcane Technologists. **14**: 206–216.

Robert, E.H. (1973). Predicting the storage life of seeds. *Seed Science and Technology*. **1**: 499-514.

Robin, S., and Manonmani, S. (2012). Rice (*Oryza sativa*). In: Bhardwaj D. N. (ed) Breeding of Field Crops. Agrobios (India), Jodhpur. ISBN 978-81-7754-474-g, pp.23-74.

Rolf, H. J. Schlegel. (2018). History of Plant Breeding. 1st Edition. *CRC Press*. Boca Raton.

Roy, A. K., Agrawal, R. K., Chand, S., Ahmad, S., Kumar, R. V. and Mall, A. K. (2020). Database of forage crop varieties: 2020, in AICRP on Forage Crops and Utilization, vol. 362. (Jhansi, Uttar Pradesh, India: ICAR–IGFRI).

Sameer Kumar C. V., Satheesh Naik, S. J., Mohan, N., Saxena, R. K. and Varshney, R. K. (2019). Pigeonpea Improvement: An Amalgam of Breeding and Genomic Research. *Plant breeding*.**138** (4): 445-454. https://doi.org/10.1111/pbr.12656

Sameer Kumar, C. V., Satheesh Naik, S. J., Mohan, N., Saxena, R. K. and Varshney, R. K. (2017). Botanical Description of Pigeonpea [*Cajanus Cajan* (L.) Millsp.]. In: Varshney, R., Saxena, R., Jackson, S. (eds) The Pigeonpea Genome. Compendium of Plant Genomes. *Springer*. Cham. https://doi.org/10.1007/978-3-319-63797-6_3

Sane, A. P., Nath, P. and Sane, P. V. (1996). Cytoplasmic male sterility in sorghum: organization and expression of mitochondrial genes in Indian CMS cytoplasms. *Journal of Genetics*. **75**(2): 151-159. https://doi.org/10.1007/BF02931758

Sanjana Reddy, P., Satyavathi, C. T., Khandelwal, V., Patil, H. T., Gupta, P. C., Sharma, L. D. and Tonapi, V. A. (2021). Performance and stability of pearl millet varieties for grain yield and micronutrients in arid and semi-arid regions of India. *Frontiers in Plant Science*. pp. 985. https://doi.org/10.3389/fpls.2021.670201

Sanjeeva, R. D., Neeraja, C. N., Madhu Babu, P., Nirmala, B., Suman, K., Rao, L. S. and Voleti, S. R. (2020). Zinc biofortified rice varieties: challenges, possibilities, and progress in India. *Frontiers in Nutrition*. **7**: 26. https://doi.org/10.3389/fnut.2020.00026

Sankara Rao, K., Raja K. Swamy, Kumar, D., Singh A. R. and K. Gopalakrishna Bhat (2019). Flora of Peninsular India.

Satheesh Naik, S. J., Bohra, A., Singh, I. P. and Tiwari, A. (2022). Pigeonpea Breeding. In: Yadava, D. K., Dikshit, H. K., Mishra, G. P., Tripathi, S. (eds) Fundamentals of Field Crop Breeding. *Springer*. Singapore. https://doi.org/10.1007/978-981-16-9257-4_21

Satish, K. and Singh, S. S. (2012). Wheat (*Triticum aestivum* L.). In: Bhardwaj D. N. (ed) Breeding of Field Crops. Agrobios (India), Jodhpur. ISBN 978-81-7754-474-g, pp.1-22.

Satyavathi, C. T., Ambawat, S. and Khandelwal, V. (2021) Micronutrient rich pearl millet for nutritionally secure India. ICAR–All India Coordinated Research AICRP-PM (2017). In: *Proceedings of the 52nd annual group meeting of ICAR–All India Coordinated Research Project on pearl millet*, Ludhiana, India, 28–30 April 2017.

Satyavathi, C. T., Mukesh Sankar, S., Singh, S. P., Kapoor, C., Soumya, S. L. and Singhal, T. (2022). Pearl Millet Breeding. In: Yadava, D. K., Dikshit, H. K., Mishra, G. P., Tripathi, S. (eds) Fundamentals of Field Crop Breeding. *Springer*. https://doi.org/10.1007/978-981-16-9257-4_6

Saxena, K. B. (2014). Temperature sensitive male-sterility system in pigeonpea. *Current Science*. 107: 277–281.

Saxena, K. B., Choudhary, A. K., Saxena, R. K. and Varshney, R. K. (2018). Breeding pigeonpea cultivars for intercropping: synthesis and strategies. *Breeding science*. **68**(2): 159-167.

Saxena, K. B., Ravishankar, K., Vijaya Kumar, R., Sreejith, K. P. and Srivastava R. K. (2010). Vegetable Pigeonpea – a High Protein Food for all Ages. Information Bulletin No. 83. Patancheru 502 324, Andhra Pradesh, India: International Crops Research Institute for the Semi-Arid Tropics. pp. 124.

Saxena, K., Bohra, A., Choudhary, A. K., Sultana, R., Sharma, M., Pazhamala, L. T. and Saxena, R. K. (2021). The alternative breeding approaches for improving yield gains and stress response in pigeonpea (*Cajanus cajan*). *Plant Breeding*. **140**(1): 74-86. https://doi.org/10.1111/pbr.12863.

Saxena, R. K., Saxena, K. B. and Varshney, R. K. (2019). Pigeonpea (*Cajanus cajan* L. Millsp.): An Ideal Crop for Sustainable Agriculture. In: Al-Khayri, J., Jain, S., Johnson, D. (eds) Advances in Plant Breeding Strategies: Legumes. *Springer*, Cham. https://doi.org/10.1007/978-3-030-23400-3_11

Saxena, R. K., Saxena, K. B., Pazhamala, L. T., Patel, K., Parupalli, S., Sameerkumar, C. V. and Varshney, R. K. (2015). Genomics for greater efficiency in pigeonpea hybrid breeding. *Frontiers in Plant Science*. **6**: 793. https://doi.org/10.3389/fpls.2015.00793

Schertz K. F. and Ritchey J. M. (1978) Cytoplasmic-genic male sterility systems in sorghum. *Crop Science*. **18**: 890–893.

Sears, E. R. (1976). Genetic control of chromosome pairing in wheat. *Annual review of genetics*. **10**(1): 31-51. https://doi.org/10.1146/annurev.ge.10.120176.000335

Seetharam, A. (1982) Finger millet improvement. *Indian Farm.* **32**(3): 3-6.

Selvi, V. M., Nirmalakumari, A. and Subramanian, A. (2014). Genetics and interrelationships of yield traits for enhancing productivity of little millet. *Electronic Journal of Plant Breeding.* **5**(1): 82-86.

Senthilvel, S., Manjunatha, T. and Lavanya, C. (2022). Castor Breeding. In: Yadava, D. K., Dikshit, H. K., Mishra, G. P., Tripathi, S. (eds) Fundamentals of Field Crop Breeding. *Springer,* Singapore. https://doi.org/10.1007/978-981-16-9257-4_18

Shanthala, J., Savithramma, D. L., Gazala, P., Jambagi, B. K. and Desai, S. K. P. (2020). Genomics-Assisted Breeding Green Gram (*Vigna radiata* (L.) Wilczek) for Accelerating Genetic Gain. In: Gosal, S. S., Wani, S. H. (eds) Accelerated Plant Breeding, **3**. *Springer.* Cham. https://doi.org/10.1007/978-3-030-47306-8_5.

Sharma, P., Mehta, G., Muthusamy, S. K., Singh, S. K. and Singh, G. P. (2021). Development and validation of heat-responsive candidate gene and miRNA gene based SSR markers to analysis genetic diversity in wheat for heat tolerance breeding. *Molecular Biology Reports.* **48**(1): 381-393. https://doi.org/10.1007/s11033-020-06059-1

Sharma, P., Singh, I., Bohra, A., Singh, I. P., Tiwari, A., Sethi, M. and Singh, S. (2020). Updates of Pigeonpea Breeding and Genomics for Yield Improvement in India. In: Gosal, S. S., Wani, S.H. (eds) Accelerated Plant Breeding. *Springer.* **3**., Cham. https://doi.org/10.1007/978-3-030-47306-8_4

Sharp, W. R., Sondahl, M. R., Caldas, L. S. and Maraffa, S. B. (1980). The physiology of *in vitro* asexual embryogenesis. *Horticultural Reviews.* **2**: 268–310.

Shifriss, O. (1956). Sex instability in Ricinus. *Genetics.* **41**(2):265.

Shifriss, O. (1960). Conventional and unconventional system controlling sex variation in *Ricinus. Journal of Genetics.* **57** :361-388.

Shivanna, K. R. and Mohan Ram, H. Y. (2005). Contributions of Panchanan Maheshwari's school to angiosperm embryology through an integrative approach. *Current Science.* **89**(11): 1820-1834.

Shukla, P., Singh, N. K., Gautam, R., Ahmed, I., Yadav, D., Sharma, A. and Kirti, P. B. (2017). Molecular Approaches for Manipulating Male Sterility and Strategies for Fertility Restoration in Plants. *Molecular Biotechnology.* **59**(9-10): 445–457. doi:10.1007/s12033-017-0027-6

Shull, G. H. (1908). The composition of a field of maize. *Journal of Heredity.* **4**: 296–301

Shull, G. H. (1911). The genotypes of maize. *The American Naturalist.* **45**: 234–252

Sica, P. (2021). Sugarcane Breeding for Enhanced Fibre and Its Impacts on Industrial Processes. *IntechOpen.* https://doi.org/10.5772/intechopen.95884.

Siddiq, E. A., Saxena, S. and Malik, S. S. (2006). Rice. In Dhillon, B. S., Saxena, S., Agrawal A. and Tyagi, R. K. (Eds). Plant Genetic Resources: Foodgrain Crops. ISPGR, NBPGR, Narosa Publishing House. New Delhi. pp. 27-57

Simpson, M. G. (2010). Plant Reproductive Biology. *Plant Systematics.* pp 573–584. doi:10.1016/b978-0-12-374380-0.50013-0

Simpson, M. G. (2010). *Plant Reproductive Biology. Plant Systematics.* pp 573–584. doi:10.1016/b978-0-12-374380-0.50013-0

Singh, A. K. and Gopala, Krishnan, S. (2016). Genetic improvement of basmati rice-the journey from conventional to molecular breeding. In *Molecular Breeding for Sustainable Crop Improvement.* pp. 213-230. https://doi.org/10.1007/978-3-319-27090-6_10

Singh, A. K. and Krishnan, S. G. (2016). Genetic improvement of basmati rice - The journey from conventional to molecular breeding. In: Rajpal, V., Rao, S. and Raina, S. (Eds) Molecular breeding for sustainable crop improvement. Sustainable development and biodiversity. *Springer.* **11.** https://doi.org/10.1007/978-3-319-27090-6_10

Singh, B. P. and Srivastava Umesh. 2013. Plant Genetic resources in Indian Perspective Theory and Practices. pp 1-119, ICAR, New Delhi.

Singh, H.C. (2012). Sorghum (*Sorghum bicolor* Moench). In: Bhardwaj D. N. (ed) Breeding of Field Crops. Agrobios (India), Jodhpur. ISBN 978-81-7754-474-g, pp.113-122.

Singh, I. P. and Singh Farindra. (2012). Pigeonpea (*Cajanus cajan* (L.) Millsp). In: Bhardwaj D. N. (ed) Breeding of Field Crops. Agrobios (India), Jodhpur. ISBN 978-81-7754-474-g, pp.135-164.

Singh, P., Kairon, M. S. and Suman B. S. Breeding Hybrid Cotton. CICR technical bulletin no: 14.

Singh, S. K. (2006). Wheat. In Dhillon, B. S., Saxena, S., Agrawal A. and Tyagi, R. K. (Eds). Plant Genetic Resources: Foodgrain Crops, ISPGR, NBPGR, Narosa Publishing House. New Delhi. pp. 58-89.

Singh, S. K. (2012). Cowpea (*Vigna unguiculata* L.). In: Bhardwaj D. N. (ed) Breeding of Field Crops. Agrobios (India), Jodhpur. ISBN 978-81-7754-474-g, pp.275-284.

Skoric, D. (1992). Achievements and future directions of sunflower breeding. *Field Crop Research.* **30**: 231–270.

Slafer, G. A., Miralles, D. J. and Araus, J. L. (2003). *Flowering and reproduction Reproductive Strategies. Encyclopedia of Applied Plant Sciences.* pp 349–357. doi:10.1016/b0-12-227050-9/00018-1

Smartt, J. (1981). Gene pools in Phaseolus and Vigna cultigens. *Euphytica.* **30**(2): 445-449.

Smith, D. C. (1966). Plant breeding-Development and success. In "Plant Breeding" (K. J. Frey, ed.), pp. 3-54. Iowa State University Press, Ames. Iowa, United States.

Soltis, P. S. (2013). *Hybridization in Plants. Encyclopedia of Biodiversity.* pp 166–176. doi:10.1016/b978-0-12-384719-5.00202-1

Soltis, P. S. and Soltis, D. E. (2009). The Role of Hybridization in Plant Speciation. *Annual Review of Plant Biology.* **60**(1): 561–588. doi:10.1146/annurev.arplant.043008.092039

Sood, S., Khulbe, R. K., Gupta, A. K., Agrawal, P. K., Upadhyaya, H. D. and Bhatt, J. C. (2015). Barnyard millet–a potential food and feed crop of future. *Plant Breeding.* **134**(2): 135-147.

Southworth, D. (1996). Gametes and fertilization in flowering plants. *Current Topics in Developmental Biology.* **34**: 259-79. doi: 10.1016/s0070-2153(08)60713-9

Sudha, M., Karthikeyan, A., Anusuya, P., Ganesh, N. M., Pandiyan, M., Senthil, N. and Angappan, K. (2013). Inheritance of resistance to mungbean yellow mosaic virus (MYMV) in inter and intra specific crosses of Mungbean (*Vigna radiata*). *American Journal of Plant Sciences.* **4**(10):1924–1927. https://doi.org/10.4236/ajps.2013.410236

Sujatha, M., Meena, H. P. and Mukta, N. (2019). Varieties and hybrids of sunflower. ICAR–Indian Institute of Oilseeds Research, Hyderabad. pp. 60.

Swaminathan, M. S. (2017). 50 Years of Green Revolution. The Quest for a World without Hunger: Volume 1. https://doi.org/10.1142/10279

Tadesse, Y., Sagi, L., Swennen, R. and Jacobs, M. (2003). Optimization of transformation conditions and production of transgenic sorghum (Sorghum bicolor) via microparticle bombardment. *Plant Cell, Tissue and Organ Culture.* **75**(1): 1-18.

Tejveer Singh, A. Radhakrishna, D. Seva Nayak and Malaviya, D. R. (2019). Genetic Improvement of Berseem (*Trifolium alexandrinum*) in India: Current Status and Prospects. *International Journal of Current Microbiology and Applied Sciences.* **8(01)**: 3028-3036. https://doi.org/10.20546/ijcmas.2019.801.322

Thomas, R. Hargrove and Victoria, L. Cabanilla. (1979). The Impact of Semidwarf Varieties on Asian Rice-Breeding Programs. *BioScience.* **29**(12): 731-735. https://doi.org/10.2307/1307667

Tonapi, V. A., Ganapathy, K. N., Hariprasanna, K., Bhat, B. V., Amasiddha, B., Avinash, S., Deepika, C. (2022). Small Millets Breeding. In: Yadava, D. K., Dikshit, H. K., Mishra, G. P., Tripathi, S. (eds) Fundamentals of Field Crop Breeding. *Springer.* https://doi.org/10.1007/978-981-16-9257-4_8

Tripathi, A. (2012). Sesame (*Sesamum indicum* L.). In: Bhardwaj D. N. (ed) Breeding of Field Crops. Agrobios (India), Jodhpur. ISBN 978-81-7754-474-g, pp.535-566.

Tripathy, S. K., Kar, J. and Sahu, D. (2019). Advances in Sesame (*Sesamum indicum* L.) Breeding. In: Al-Khayri, J., Jain, S., Johnson, D. (eds) Advances in Plant Breeding Strategies: Industrial and Food Crops. *Springer.* Cham. https://doi.org/10.1007/978-3-030-23265-8_15

Tucker, M. R. and Koltunow, A. M. G. (2009). Sexual and asexual (apomictic) seed development in flowering plants: molecular, morphological and evolutionary relationships. *Functional Plant Biology*. **36**(6): 490-504. doi: 10.1071/FP09078

Van Dijk, P. J. and Underwood, C. J. (2018). Sexual Reproduction vs Unisexual; Asexual Reproduction. *Plants. Encyclopedia of Reproduction.* 10–17. doi:10.1016/b978-0-12-809633-8.20530-x

Van Slageren, M. W. (1994). Wild wheats: a monograph of *Aegilops L.* and Amblyopyrum (Jaub. and Spach) Eig (Poaceae). Wageningen: Agricultural University.pp. 1-512

Varshney, R. K. (2016). Exciting journey of 10 years from genomes to fields and markets: some success stories of genomics-assisted breeding in chickpea, pigeonpea and groundnut. *Plant Science*. **242**: 98–107. https://doi.org/10.1016/j.plantsci.2015.09.009

Varshney, R. K., Pandey, M. K. and Janila, P. (2014). Marker-assisted introgression of a QTL region to improve rust resistance in three elite and popular varieties of peanut (*Arachis hypogaea* L.). *Theoretical and Applied Genetics*. **127**(8): 1771–1781. https://doi.org/10.1007/s00122-014-2338-3

Vavilov (1951) by R. W. Schery, 1972. Plants for Man, Prentice Hall, Englewood Cliffs, N.J.

Vavilov, N. I. (1992). Origin and geography of cultivated plants. Cambridge University Press, Cambridge, UK, Transl. by D. Love.

Velu, G., Singh, R. P. and Balasubramaniam, A. (2015). Reaching out to farmers with high zinc wheat varieties through public-private partnerships-an experience from eastern-gangetic plains of India. *Advances Food Technology Nutrition Science*. **1**: 73–75.

Vilà, M., Weber, E. and Antonio, C. M. (2000). Conservation Implications of Invasion by Plant Hybridization. *Biological Invasions.* (**2**): 207–217. doi.org/10.1023/A:1010003603310

Voss-Fels, K. P., Stahl, A. and Hickey, L. T. (2019). Q and A: modern crop breeding for future food security. *BMC Biology*. **17:** 18. https://doi.org/10.1186/s12915-019-0638-4

Waghmare, V.N. (2022). Cotton Breeding. In: Yadava, D. K., Dikshit, H. K., Mishra, G. P., Tripathi, S. (eds) Fundamentals of Field Crop Breeding. *Springer*. Singapore. https://doi.org/10.1007/978-981-16-9257-4_11

Wang, J., Lin, Z., Zhang, X., Liu, H., Zhou, L., Zhong, S. and Lin, Z. (2019). krn1, a major quantitative trait locus for kernel row number in maize. *New Phytologist.* **223**(3): 1634-1646. https://doi.org/10.1111/nph.15890

Wang, Z. and S'akiroglu, M. (2021). "The origin, evolution, and genetic diversity of Alfalfa," in The Alfalfa Genome. Eds. L. X. Yu and C. Kole (Cham: Springer). https://doi.org/10.1007/978-3-030-74466-3_3

Watson, L. and Dallwitz, M. J. (1992). The families of flowering plants: descriptions, illustrations, identification, and information retrieval. Version: 25th November 2022.

Wei, X., Liu, K., Zhang, Y., Feng, Q., Wang, L., Zhao, Y., Li, D., Zhao, Q., Zhu, X., Zhu, X., Li, W., Fan, D., Gao, Y., Lu, Y., Zhang, X., Tang, X., Zhou, C., Zhu, C., Liu, L., Zhong, R., Tian, Q., Wen, Z., Weng, Q., Han, B., Huang, X. and Zhang, X. (2015). Genetic discovery for oil production and quality in sesame. *Nature Communications.* **6**: 8609. https://doi.org/10.1038/ncomms9609.

Weiss, E. A. (1971) Castor, sesame and safflower. Leonard Hill Books, London. pp. 901.

Weiss, E. A. (2000) Castor. In: Oilseed crops (world agriculture series), 2nd edition. Blackwell Science Ltd., Oxford. pp. 13–51.

Wilhelmi, L. K. and Preuss, D. (1999). The mating game: pollination and fertilization in flowering plants. *Current Opinion in Plant Biology.* **2**(1):18-22. doi: 10.1016/s1369-5266(99)80004-1

Yadav, A. K., Narwal, M. S. and Arya, R. K. (2013). Evaluation of pearl millet (*Pennisetum glaucum*) genotypes and validation of screening methods for supra-optimal temperature tolerance at seedling stage. *Indian Journal of Agricultural Sciences.* **83**(3): 260-271.

Yadav, O. P. and Rai, K. N. (2013). Genetic improvement of pearl millet in India. *Agricultural Research.* **2**(4): 275-292.

Yadav, O. P., Gupta, S. K., Govindaraj, M., Sharma, R., Varshney, R. K., Srivastava, R. K. and Mahala, R. S. (2021). Genetic gains in pearl millet in India: insights into historic breeding strategies and future perspective. *Frontiers in Plant Science.* pp. 396. https://doi.org/10.3389/fpls.2021.645038.

Yadav, R., Kalia, S., Rangan, P., Pradheep, K., Rao, G. P., Kaur, V., Pandey, R., Rai, V., Vasimalla, C. C., Langyan, S., Sharma, S., Thangavel, B., Rana, V. S., Vishwakarma, H., Shah, A., Saxena, A., Kumar, A., Singh, K. and Siddique, K. H. M. (2022) Current Research Trends and Prospects for Yield and Quality Improvement in Sesame, an Important Oilseed Crop. *Frontiers Plant Science.* **13**: 863521. https://doi.org/10.3389/fpls.2022.863521

Yadav, S. S. and Chen, W. (Eds.). (2007). *Chickpea breeding and management.* CABI, Wallingford.

Yadava, D. K., Yashpal, Saini, N., Nanjundan, J. and Vasudev, S. (2022). *Brassica* Breeding. In: Yadava, D. K., Dikshit, H. K., Mishra, G. P., Tripathi, S. (eds) Fundamentals of Field Crop Breeding. *Springer*. Singapore. https://doi.org/10.1007/978-981-16-9257-4_15

Yadava, D. K., Yashpal, Sujata Vasudev, Naveen Singh, Navinder Saini, *et al.* (2019). Indian mustard: Variety Pusa Double Zero Mustard-31 (PDZM-31). *Indian Journal of Genetics and Plant Breeding*. **79**(3): 636-637.

Yadava. D. K., Singh, N. and Sujata, V. (2014). Oilseed Brassica. In: Chopra VL (ed) *Breeding field crops II advances. Stadium Press LLC, Houston*. pp. 201–231.

Yamagishi, H. and Bhat, S. R. (2014). Cytoplasmic male sterility in Brassicaceae crops. *Breeding Science*. **64**(1): 38–47. https://doi.org/10.1270/jsbbs.64.38

Yuan, L. (2001). Breeding of super hybrid rice. In *Rice research for food security and poverty alleviation.* International Rice Research Institute (IRRI), Los Banos, Philippines. pp. 143-149.

Yuan, L. P. (1977). The execution and theory of developing hybrid rice. *Agriculture Science.* ***1***: 27-31.

Zeven, A. C. and de Wet, J. M. J. (1975). Dictionary of cultivated plants and their regions of diversity. Pudoc, Wageningen, the Netherlands

Zeven, A. C. and Zhukovsky, P. M. (1982). Dictionary of cultivated plants and their centres of diversity. Pudoc, Wageningen, the Netherlands.

Zhang, Y., Xu, W., Wang, H., Dong, H., Qi, X., Zhao, M. and Hu, L. (2016). Progress in genetic improvement of grain yield and related physiological traits of Chinese wheat in Henan Province. *Field Crops Research*. **199**: 117-128. https://doi.org/10.1016/j.fcr.2016.09.022

Zhukovsky, P. M. (1965). Main gene centres of cultivated plants and their wild relatives within the territory of the U.S.S.R. *Euphytica*. **14**: 177–188. https://doi.org/10.1007/BF00038985

Zimmerman L. H., Smith J. D. (1966). Production of F_1 seed in castor bean by use of sex genes sensitive to environment. Crop Science, **6**:406 -409. https://doi.org/10.2135/cropsci1966.0011183X000600050005x

Index

A

Abiotic stress resistance 42
Alcohol emasculation 17
Allogamy 6
Alloplasmic CMS 11
Amphidiploid 239
Anthesis 36, 48, 127
Apomictic lines 103
Artificial crossing 127, 141, 149
Artificial method 18
Asian rice 49
Autogamy 5
Autoplasmic CMS 10

B

Backcross method 26, 102
Barnyard millet 119
Berseem 278
Biofortification 41, 54, 68, 86
Biotechnological approaches 43, 60, 74, 133, 187
Biotic stress resistance 42, 70
Bisexual flower 7
Bisexuality 6
Black gram 155
Breeding approaches 43, 101, 131
Bulk selection methods 25

C

Castor 223
Catastrophic sexual transformation theory 64
Certified seed production programme 88
Chasmogamy 6
Chickpea 125
Cleistogamy 6
Cold/hot water treatment 16
Common (hexaploid) wheat 37

Conventional hybrids 28
Cotton 237
Cowpea 161
Crop breeding 3, 13, 19
Crop duration 4
Cross incompatibility 15
Crossing 14, 36
Crossing techniques 259
Cross-pollination 6, 8
Cytoplasmic genetic male sterility 11
Cytoplasmic male sterility 10

D

Dhanashakti 86
Dichogamy 7
Dicliny 8
Dioecism 8
Diploid hybrids 247
Direct methods 23
Distant hybridization 14, 187
Dominant gene 26
Double-cross 73
Double haploid techniques 74
Drought resistance 70

E

Emasculation techniques 16
Embryo rescue 197
Environmental sensitive genic (nuclear) male sterility 10

F

Fibre 173
Finger millet 117
Flowering pattern 137
Fodder quality 100
Forages 267
Foxtail millet 117

G

Gametocides 16
Gametophytic self-incompatibility 9
Gene editing technology 74
Gene pool 47, 126
Genetic emasculation 16
Genetic origin 177
Genetic variability 116
Genic (nuclear) male sterility 10
Genomic assisted breeding 44, 74, 198
Genomic selection 44
Germplasm collection 262
Grain quality 70
Green gram 147
Green revolution 38
Groundnut 175
Guar 167

H

Hand emasculation 16
Haploid production 14, 15
Health benefit 116
Herkogamy 8
Heterosis and hybrid wheat 43
Heterosis breeding 27, 102, 144, 247
Heterostyly 8
Hybrid cultivars 27
Hybrid inviability 15
Hybridization 13, 56, 263
Hybridization and selection 153
Hybridization technique 67, 258
Hybrid seed production 88

I

Inbred development 74
Indirect methods 23
Inter-generic hybridization 15
Intergenic amphidiploid hybrid 191
Intermediate races 93
Interspecific hybridization 15, 85, 101, 247
Inter-varietal hybridization 14, 186, 203
Intraspecific hybrids 247
Intra-varietal hybridization 14
Introgressive hybridization 16

K

Kodo millet 120

L

Lantern technique 259
Little millet 120
Lucerne 272

M

Maize 63
Male sterility 10
Marker-assisted selection 60
Mass selection 24, 210
Maturity duration 137
Monoecism 8
Monophyletic 49
Mustard 189
Mutation breeding 43, 57, 103, 144, 159, 263

N

Natural method 17
Next-generation agriculture 115
Non-conventional hybrids 28
Normal hybridization 14
Novel hybrids 15
Nutritional benefits 116
Nutritional significance 94

O

Oat 280
Oilseeds 173
One-line method 60
Open-pollinated cultivars 212

P

Parbhani Shakti 100
Pearl millet 77
Pedigree method 103
Pedigree selection 25
Photoperiod-sensitive genetic male sterility 10, 59
Pigeonpea 135
Pistillate condition 228
Plant selection 22
Pollination 127, 141, 251
Pollination nature 5
Polyphyletic 49
Population improvement 103
Population improvement approaches 20
Primary introduction 21
Primary triticale 39
Progeny selection 23
Proso millet 118
Protandry 7
Protogyny 7
Protoplast fusion 198
Pulses 123
Pure line selection 56
Pureline selection 23

Q

Quality protein 68

R

Recessive gene 27
Recurrent selection 24
Rice 45
Rooted stalk (marcotting) technique 260

S

Secondary introduction 22
Secondary triticale 40
Seed production 212
Self-incompatibility 9
Selfing 17, 36
Selfing techniques 13, 17
Self-pollination 5
Semi-dwarf rice 51
Semi-dwarf wheat 38
Semi-spreading 177
Sesame 215
Sex promoting environments 227
Sex revertant 226
Single or simple cross 73
Single-seed descent (SSD) method 26
Small millets 107, 117
Somaclonal variation 198
Sorghum 91
Soybean 199
Space isolation 17
Special breeding 20
Sporophytic self-incompatibility 9
Suction pressure method 16
Sugarcane 253
Sugar crops 173
Sulfurous acid technique 260
Sunflower 205
Synthetic varieties 28

T

Targeted mutagenesis 44
Teosinte theory 64
Tetraploid hybrids 247
Thermo-sensitive genetic male sterility 10, 59
Three-line hybrid breeding (CGMS) system 145
Three-line method 57
Three-way cross 73
Time isolation 17
Tissue culture techniques 197
Transgenic breeding 74, 198, 249
Transgenic technology 60
Tripartite theory 64
Tripsacum (*Zea diploperennis*) theory 64
Triticale 39

U

Unisexual flower 8

W

Waxy coat 230
Wheat 33